T0251830

SPICES

Agrotechniques for Quality Produce

Innovations in Horticultural Science

SPICES
Agrotechniques for Quality Produce

Amit Baran Sharangi, PhD
Suchand Datta, PhD
Prahlad Deb, PhD

AAP | APPLE
ACADEMIC
PRESS

Apple Academic Press Inc.
3333 Mistwell Crescent
Oakville, ON L6L 0A2 Canada

Apple Academic Press Inc.
9 Spinnaker Way
Waretown, NJ 08758 USA

© 2018 by Apple Academic Press, Inc.
First issued in paperback 2021
Exclusive worldwide distribution by CRC Press, a member of Taylor & Francis Group
No claim to original U.S. Government works
ISBN-13: 978-1-77463-075-4 (pbk)
ISBN-13: 978-1-77188-635-2 (hbk)

All rights reserved. No part of this work may be reprinted or reproduced or utilized in any form or by any electric, mechanical or other means, now known or hereafter invented, including photocopying and record-ing, or in any information storage or retrieval system, without permission in writing from the publisher or its distributor, except in the case of brief excerpts or quotations for use in reviews or critical articles.

This book contains information obtained from authentic and highly regarded sources. Reprinted material is quoted with permission and sources are indicated. Copyright for individual articles remains with the authors as indicated. A wide variety of references are listed. Reasonable efforts have been made to publish reliable data and information, but the authors, editors, and the publisher cannot assume responsibility for the validity of all materials or the consequences of their use. The authors, editors, and the publisher have attempted to trace the copyright holders of all material reproduced in this publication and apologize to copyright holders if permission to publish in this form has not been obtained. If any copyright material has not been acknowl-edged, please write and let us know so we may rectify in any future reprint.

Trademark Notice: Registered trademark of products or corporate names are used only for explanation and identification without intent to infringe.

Library and Archives Canada Cataloguing in Publication

Sharangi, A. B. (Amit Baran), author
Spices : agrotechniques for quality produce / Amit Baran Sharangi, PhD, Suchand Datta, PhD, Prahlad Deb, PhD.
(Innovations in horticultural science)
Includes bibliographical references and index.
Issued in print and electronic formats.
ISBN 978-1-77188-635-2 (hardcover).--ISBN 978-1-315-10125-5 (PDF)
1. Spice plants--India. 2. Spices--India. I. Datta, Suchand, author II. Deb, Prahlad, author III. Title.
IV. Series: Innovations in horticultural science

| SB306.I4S53 2018 | 633.8'30954 | C2018-900172-0 | C2018-900173-9 |

Library of Congress Cataloging-in-Publication Data

Names: Sharangi, A. B. (Amit Baran), author. | Datta, Suchand, author. | Deb, Prahlad, author.
Title: Spices : agrotechniques for quality produce / authors: Amit Baran Sharangi, Suchand Datta, Prahlad Deb.
Description: Waretown, NJ : Apple Academic Press, 2018. | Includes bibliographical references and index.
Identifiers: LCCN 2018000040 (print) | LCCN 2018000831 (ebook) | ISBN 9781315101255 (eb-ook) | ISBN 9781771886352 (hardcover : alk. paper)
Subjects: LCSH: Spice plants. | Spices.
Classification: LCC SB305 (ebook) | LCC SB305 .S47 2018 (print) | DDC 633.8/3--dc23
LC record available at https://lccn.loc.gov/2018000040

Apple Academic Press also publishes its books in a variety of electronic formats. Some content that appears in print may not be available in electronic format. For information about Apple Academic Press products, visit our website at **www.appleacademicpress.com** and the CRC Press website at **www.crcpress.com**

CONTENTS

ABOUT THE AUTHORS

Amit Baran Sharangi, PhD
Professor of Horticultural Science,
Head of the Department of Plantation, Spices,
Medicinal and Aromatic Crops,
Faculty of Horticulture,
Bidhan Chandra Krishi Viswavidyalaya
(Agricultural University), India

Amit Baran Sharangi, PhD, is a Professor of Horticultural Science and is presently the Head of the Department of Plantation, Spices, Medicinal and Aromatic Crops in the Faculty of Horticulture at Bidhan Chandra Krishi Viswavidyalaya (Agricultural University), India. He has been teaching for about 20 years and was instrumental in the process of coconut improvement, leading to the release of a variety Kalpamitra from the Central Plantation Crops Research Institute in India. He has spent time in Prof. Cousen's lab in Melbourne, Australia; Prof. Picha's lab in the USA; Dr. Dobson's lab in the UK; where he continued his research on herbs and spices.

Dr. Sharangi has published about 70 research papers in peer-reviewed journal, 50 conference papers, 18 books, and many book chapters from international publishers, and scores of popular scientific articles. Presently he is associated with 40 international and national journals in many roles, including editor in chief, regional editor, technical editor, editorial board member, and reviewer.

Professor Sharangi has visited abroad extensively on academic missions and has received several international awards, including the Endeavour Post-Doctoral Award (Australia), a INSA-RSE Visiting Scientist Fellowship (UK), Fulbright Visiting Faculty Fellowship (USA), Achiever's Award from the Society for the Advancement of Human and Nature (SADHNA), Man of the Year-2015 (Cambridge, UK), Outstanding Scientist (Venus International Research Award), Excellence Award (ARCC), Higher Education Leadership Award-2018 etc. He has delivered invited lectures in the UK, USA, Australia, Thailand, Israel, and Bangladesh on several aspects of herbs and spices.

Professor Sharangi is associated with a number of research projects as Principal and Co-Principal Investigator, that have academic and empirical implications. He is an elected Fellow of the West Bengal Academy of Science and Technology (FAScT). He is also a Fellow of the International Society for Research and Development (ISRD, UK), the Society of Applied Biotechnology (SAB), International Scientific Research Organization for Science Engineering and Technology (ISROSET), and Academy of Environment & Life Science (AELS) and Scientific Society for Advance Research and Social Change (SSARSC), etc. He is an active member of several science academies and societies, including the New York Academy of Science (NYAS), World Academy of Science, Engineering and Technology (WASET), The Association for Tropical Biology and Conservation (ATBC), African Forest Forum (AFF), to name a few.

Suchand Datta, PhD
Professor of Vegetable and Spice Crops,
Uttar Banga Krishi Viswavidyalaya, Pundibari, Cooch
Behar, West Bengal, India

Suchand Datta, PhD, is currently a Professor in vegetable and spice crops at Uttar Banga Krishi Viswavidyalaya, Pundibari, Cooch Behar, West Bengal, India. During his academic career of more than 15 years, he has guided a number of MSc and PhD students in his role as chairman and member of the advisory committee. He has published 60 research papers in national and international journals, several popular articles, and a large number of popular articles in local languages. He has participated and presented more than 30 research papers at national and international seminars, symposia, conferences, and world congresses. He acted as Co-Principal Investigator in the Central Scheme for the Development of Medicinal Plants, sponsored by the National Medicinal Plants Board, Department of Indian Systems of Medicine and Homoeopathy, Ministry of Health and Family Welfare, Government of India, and was a Scientist in the Integrated Programme for the Development of Spices (now renamed the Mission for Integrated Development of Horticulture).

Dr. Datta organized a national-level workshop, a National Consultative Meet on Large Cardamom and State-Level Seminars with special emphasis on spices. He also developed one ginger variety and registered one-leaf blotch and leaf spot resistant turmeric germplasm in the National Bureau of Plant Genetic Resources, New Delhi. At present he has taken the responsibility of In-Charge, All India Co-ordinate Research Project on Spices, and Principal Investigator and In-Charge of the Centrally Sponsored Scheme Mission for Integrated Development of Horticulture for spices in Uttar Banga Krishi Viswavidyalaya, Pundibari.

Dr. Datta has written several books and six book chapters and has received several awards for his work. He is a life member of five national and international professional societies. He has actively participated in different training programs for the benefit of the farmers.

Dr. Suchand Datta received his MSc degree in horticulture in spices and plantation crops from Bidhan Chandra Krishi Viswavidyalaya Haringhata, India, and his PhD in vegetable and spice crops from Uttar Banga Krishi Viswavidyalaya, Cooch Behar, India.

Prahlad Deb, PhD
Assistant Professor of Horticulture,
Institute of Agriculture (Palli Siksha Bhavana),
Visva-Bharati, Sriniketan, West Bengal, India

Prahlad Deb, PhD, is currently Assistant Professor of Horticulture at the Institute of Agriculture (Palli Siksha Bhavana), Visva-Bharati, Sriniketan, West Bengal, India. Dr. Deb has been teaching different undergraduate and postgraduate courses related to horticulture for several years. His major thrust of research is in minor horticultural crops, postharvest technology, and value-addition of horticultural crops. He is working on two major research projects funded by the Department of Atomic Energy (BRNS-BARC) and National Horticulture Mission, India. Dr. Deb has published more than thirty research papers in peer-reviewed national and international journals, eight book chapters, four technical bulletins, ten popular articles, and two newspaper articles. He has received several best presentation and best publication awards from recognized national societies. He is a life member of nine national and international professional societies, including the International Society for Horticultural Science (ISHS), Belgium. He has attended many international and national conferences and several scientific workshops. He has visited Bangladesh and Sri Lanka for deliberation at several international conferences.

Dr. Deb received his BSc in horticulture, MSc in horticulture (vegetable and spice crops) and PhD in pomology and postharvest technology from Uttar Banga Krishi Viswavidyalaya, Cooch Behar, India. He served as a Research Associate at the National Research Centre for Orchids for about two years.

INNOVATIONS IN HORTICULTURAL SCIENCE

About the Series
Editor-in-Chief:
Dr. Mohammed Wasim Siddiqui Assistant Professor-cum- Scientist
Bihar Agricultural University | www.bausabour.ac.in
Department of Food Science and Post-Harvest Technology
Sabour | Bhagalpur | Bihar | P. O. Box 813210 | INDIA
Contacts: (91) 9835502897
Email: wasim_serene@yahoo.com | wasim@appleacademicpress.com

The horticulture sector is considered as the most dynamic and sustainable segment of agriculture all over the world. It covers pre- and postharvest management of a wide spectrum of crops, including fruits and nuts, vegetables (including potatoes), flowering and aromatic plants, tuber crops, mushrooms, spices, plantation crops, edible bamboos etc. Shifting food pattern in wake of increasing income and health awareness of the populace has transformed horticulture into a vibrant commercial venture for the farming community all over the world.

It is a well-established fact that horticulture is one of the best options for improving the productivity of land, ensuring nutritional security for mankind and for sustaining the livelihood of the farming community worldwide. The world's populace is projected to be 9 billion by the year 2030, and the largest increase will be confined to the developing countries, where chronic food shortages and malnutrition already persist. This projected increase of population will certainly reduce the per capita availability of natural resources and may hinder the equilibrium and sustainability of agricultural systems due to overexploitation of natural resources, which will ultimately lead to more poverty, starvation, malnutrition, and higher food prices. The judicious utilization of natural resources is thus needed and must be addressed immediately.

Climate change is emerging as a major threat to the agriculture throughout the world as well. Surface temperatures of the earth have risen significantly over the past century, and the impact is most significant on agriculture. The rise in temperature enhances the rate of respiration, reduces cropping periods, advances ripening, and hastens crop maturity, which adversely affects crop productivity. Several climatic extremes such as droughts, floods, tropical cyclones, heavy precipitation events, hot extremes, and heat waves cause a negative impact on agriculture and are mainly caused and triggered by climate change.

In order to optimize the use of resources, hi-tech interventions like precision farming, which comprises temporal and spatial management of resources in horticulture, is essentially required. Infusion of technology for an efficient utilization of resources is intended for deriving higher crop productivity per unit of inputs. This would be possible only through deployment of modern hi-tech applications and precision farming methods. For improvement in crop production and returns to farmers, these technologies have to be widely spread and adopted. Considering the above-mentioned challenges of horticulturist and their expected role in ensuring food and nutritional security to mankind, a compilation of hi-tech cultivation techniques and postharvest management of horticultural crops is needed.

This new book series, Innovations In Horticultural Science, is designed to address the need for advance knowledge for horticulture researchers and students. Moreover, the major advancements and developments in this subject area to be covered in this series would be beneficial to mankind.

The probable coverage of the series would be as follows.
1. Importance of horticultural crops for livelihood
2. Dynamics in sustainable horticulture production
3. Precision horticulture for sustainability
4. Protected horticulture for sustainability
5. Classification of fruit, vegetables, flowers, and other horticultural crops
6. Nursery and orchard management
7. Propagation of horticultural crops
8. Rootstocks in fruit and vegetable production
9. Growth and development of horticultural crops
10. Horticultural plant physiology
11. Role of plant growth regulator in horticultural production
12. Nutrient and irrigation management
13. Fertigation in fruit and vegetables crops
14. High-density planting of fruit crops
15. Training and pruning of plants
16. Pollination management in horticultural crops
17. Organic crop production
18. Pest management dynamics for sustainable horticulture
19. Physiological disorders and their management
20. Biotic and abiotic stress management of fruit crops
21. Postharvest management of horticultural crops
22. Marketing strategies for horticultural crops
23. Climate change and sustainable horticulture
24. Molecular markers in horticultural science
25. Conventional and modern breeding approaches for quality improvement
26. Mushroom, bamboo, spices, medicinal, and plantation crop production

BOOKS IN THE SERIES

- Spices: Agrotechniques for Quality Produce
  Amit Baran Sharangi, PhD, S. Datta, PhD, and Prahlad Deb, PhD
- Sustainable Horticulture, Volume 1: Diversity, Production, and Crop
  Improvement
  Editors: DebashisMandal, PhD, Amritesh C. Shukla, PhD, and
  Mohammed Wasim Siddiqui, PhD
- Sustainable Horticulture, Volume 2: Food, Health, and Nutrition
  Editors: DebashisMandal, PhD, Amritesh C. Shukla, PhD, and
  Mohammed Wasim Siddiqui, PhD
- Underexploited Spice Crops: Present Status, Agrotechnology, and Future
  Research Directions
  Amit Baran Sharangi, PhD, Pemba H. Bhutia, Akkabathula Chandini Raj,
  and Majjiga Sreenivas

LIST OF CONTRIBUTORS

Suchand Datta
Department of Vegetable and Spices Crops (Research), Faculty of Horticulture, UBKV (Agricultural University), Pundibari, Cooch Behar–736165, West Bengal, India, Tel.: + 91-9434228494, Fax: + 91-3582-270632 (Office), E-mail: suchanddatta@rediffmail.com; suchanddatta@gmail.com

Prahlad Deb
Division of Horticulture, Institute of Agriculture (Palli Siksha Bhavana), Visva-Bharati, Sriniketan–731236, Birbhum, West Bengal, India, Tel.: + 91-9434484303, E-mail: debprld@yahoo.com

Amit Baran Sharangi
Department of Spices and Plantation Crops, Faculty of Horticulture, BCKV (Agricultural University), Mohanpur–741252, Nadia, West Bengal, India, Tel.: + 91-3473-222659, Fax: + 91-3473-222659, E-mail: dr_absharangi@yahoo.co.in, absharangi@gmail.com

LIST OF ABBREVIATIONS

ASTA	American Spice Trade Association
BAP	benzyl amino purine
BHA	butylated hydroxyanisole
BHT	butylated hydroxytoluene
CEF	curcuminoids enriched fraction
CFTRI	Central Food Technological Research Institute
CHR	Cardamom Hill Reserves
CM	chloroform: methanol
CMD	carrot motley dwarf
CMoV	carrot mottle virus
CP 7	compound panicle 7
CRLV	carrot red leaf virus
DAP	days after planting
DAS	days after sowing
DMH	1,2-dimethylhydrazine
DRIS	Diagnosis and Recommendation Integrated System
DW	dry weight
ESA	European Spice Association
FAO	Food and Agricultural Organization
FYM	farmyard manure
GAE	gallic acid equivalents
GBC	Grenada broken and clean
GUNS	Grenada unassorted nutmegs
HCA	hydroxyl citric acid
HW	hand weeding
IPC	International Pepper Community
ISG	International Spice Group
IU	international units
NAA	naphthalene acetic acid
NL	neutral lipid
NPOP	National Programme for Organic Production
NPV	nuclear polyhedrosis virus

NSC	National Steering Committee
NSKE	neem seed kernel extract
OC	organic carbon
PAR	photosynthetic active radiation
PCA	potato carrot agar
PCA	principal component analysis
PG	propyl gallate
PGRs	plant growth regulators
RDF	recommended dose of fertilizer
RH	relative humidity
RSM	response surface methodology
RT-PCR	reverse transcription-polymerase chain reaction
SB	sorghum-based
SCWE	subcritical water extraction
SUNS	sound unassorted nutmegs
TAG	triacylglycerol
TC	talc-based
TG	tellicherry garbled
TGEB	tellicherry garbled extra bold
TGSEB	tellicherry garbled black pepper special extra bold
THQ	thymohydroquinone
TKP	tamarind kernel powder
TQ	thymoquinone
TSS	total soluble solids

FOREWORD

Spices have a fascinating history. Early documentation suggests that hunters and gatherers wrapped meat in the leaves of bushes and accidentally discovered its magical power to enhance the taste of meat. Over the years, spices and herbs were used for medicinal purposes. As research is progressing, more evidence is supporting some of the anecdotal information supplied by our ancestors. Nowadays, spices are regarded as high-value and low-volume commodities of commerce in the global market. The fast-growing food industries around the world depend a lot on spices for imparting taste and flavor. Thus, spices emerge as one of the crucial building blocks in almost all food products in addition to their obvious use as functional foods, nutraceuticals, and sources of several high-value phytochemicals.

India has been a top and leading producer, consumer, and exporter of spices in the world, and almost all states in the country produce one or the other spices. After huge domestic consumption of around 90% of the spices produced, India still remains as the largest exporter of spices and spice value-added products. India contributes 48% of the total world trade in terms of quantity and 42% in value terms.

This book precisely highlights several important spices with their systematic position, national and international status, diverse uses, proximate composition, bioactive principles, agro-techniques for quality production, and value addition in a comprehensive and readable form. The efforts made by the authors are commendable, and I am sure this book will be helpful for students, researchers, and industrialists to gain and improve their knowledge in this direction and ultimately provide a new treat for the livelihood of millions of farmers reeling under social and economic stresses. I congratulate Prof. Sharangi of BCKV, Dr. Datta of UBKV, and Dr. Deb of Visva Bharati for their tremendous effort in contributing the book in its present format. The publisher, Apple Academic Press, needs special appreciation for shaping the book in its present format.

— **C. Chattopadhyay**
Vice Chancellor
UBKV, West Bengal

PREFACE

Indian spices are famous across the globe and so also is their legacy of being in use for a considerable time. The aromas and flavors of spices have attracted food lovers and continue to do so consistently. With the increasing awareness of health and nutraceuticals, people are now more conscious about spicy benefits. The past few years have witnessed several pioneering research works in this area with various spices. However, very few comprehensive exercises have been made so far to collect and collate the wisdom of the past and blend with the technological progress of today. Dissemination of the knowledge pool is as important as the research itself. We feel privileged to present this volume to the spice lovers of the world.

This volume contains six chapters. The first one is an introductory chapter that gives an overview of the important flavor compounds contained in different spices, present status of spices, and glimpses of the world scenario on the export and import of spices, major markets, etc. The second chapter deals with classification of spices, condiments, and herbs. The third chapter is the major chapter: It precisely describes agro-techniques or production technology of fifty individual spices, including of three major spices, three rhizomatous spices, six bulbous spices, eight tree spices (six aromatic and two acidulant), eleven seed spices, twelve leafy or herbal spices or aromatic herbs, four lesser known spices, and three other spices, with due consideration to quality and value addition issues. This chapter also gives a general discussion on the systematic position, composition, uses, export-import scenario, medicinal values, etc. The subsequent chapters deal with recent research approaches to spices around the world, promises of organic spices, and future thrust areas in this direction.

We hope that this compilation will be useful to all those who are interested in spices in one way or the other. We hope that this volume will be read and referred to by students, teachers, researchers, amateur readers, policymakers, and as well as those in farming communities.

— **Amit Baran Sharangi, PhD**
Suchand Datta, PhD
Prahlad Deb, PhD

CHAPTER 1

SPICE CROPS: SCENARIO AND SIGNIFICANCE

CONTENTS

1.1 INTRODUCTION

Spices are high value and low volume commodities of commerce in the world market. The fast growing food industries around the world depend a lot on spices for imparting taste and flavor. Moreover, the increased preference of natural colors and flavors of plant origin over the synthetic products by the health conscious consumers in developed countries is slowly bringing momentum in this direction. Thus, spices emerge as one of the crucial building blocks in almost all food products in addition to their obvious use as functional foods, nutraceuticals, and sources of several high value phytochemicals.

India is the largest producer of spices with an annual production of 6.1 million MT during 2014–15 from an area of 3.3 million hectares (Tamil Selvan, 2016). Black pepper, ginger, turmeric, cardamom, and tree spices such as nutmeg, cinnamon, garcinia, and tamarind are the tropical spices of importance in Indian context. Coriander, cumin, fennel, and fenugreek are important seed spices; and mint is a herbal spice of repute. Garcinia, black cumin, ajwain, saffron, mint, oregano, lavender, and star anise are considered promising among the other spices. Area, production and productivity of different spices in India has been presented in Table 1.1. India

TABLE 1.1   Present Status of Area, Production and Productivity of Major Spice Crops (2013–14)

Crop	Area (thousand ha)	Production (thousand tons)	Productivity (kg/ha)
Black pepper	123.8	50.87	410.9
Cardamom	92.8	21.28	229.3
Ginger	132.6	655.00	4939.7
Turmeric	232.6	1189.90	5115.6
Nutmeg	18.9	12.78	676.2
Clove	2.1	1.07	509.5
Coriander	447.1	313.6	701.4
Cumin	858.9	513.8	598.2
Fennel	54.2	70.12	1293.7
Fenugreek	70.1	65.94	940.7
Ajwain	89.6	26.67	297.7

has been a top and leading producer, consumer, and exporter of spices in the world and almost all states in the country produce one or the other spices. After a huge domestic consumption of around 90% of the spices produced, India still remains as the largest exporter of spices in all its forms; raw spices, ground spices, processed product of different spices, and as active ingredient isolates from the different spices. India contributes 48% of the total world trade in terms of quantity and 42% in terms of value. The spice industry in India and trade has shown stunning progress over the last 5 years with 120% increase in revenue, which is expected to touch $3 billion by 2017. East Asia is the major market for spices, followed by America and the European Union; and the world spice trade is expected to touch $17 billion by 2020. It is estimated that, we may have a population of about 1.69 billion during 2050 (with around 0.9 billion urban population, the second largest in the world) (Figure 1.1) and approximately the per capita consumption of black pepper, cardamom, turmeric, and ginger is expected to be about 148 g, 54 g, 1.6 kg, and 1.2 kg, respectively.

The main flavor compounds found in the major herbs and spices used by the food industry are summarized in Table 1.2.

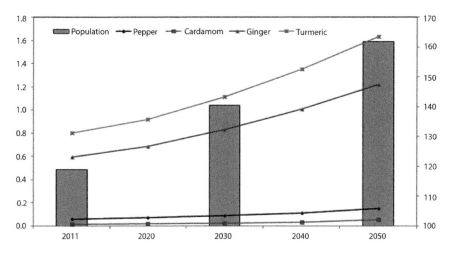

FIGURE 1.1   Estimated per capita demand for major spices in 2050 (After Krishnamurthy et al., 2015).

TABLE 1.2   Important Flavor Compounds in Spices and Herbs

Spice and Herbal spices	Important Flavor compounds	Chemical structure
A. Spices		
Allspice	Eugenol, caryophyllene	
Anise	(E)-anethole, methyl chavicol	
Black pepper	Piperine, S-3-Carene, β-caryophyllene	
Caraway	d-carvone, carone derivatives	

TABLE 1.2 (Continued)

Spice and Herbal spices	Important Flavor compounds	Chemical structure
Cardamom	α-terpinyl acetate, 1–8-cineole, linalool	
Cinnamon, cassia	Cinnamaldehyde, eugenol	
Chili	Capsaicin, dihydro capsaicin	
Clove	Eugenol, eugeneyl acetate	
Coriander	d-linalool, C10-C14–2-alkenals	
Cumin	Cuminaldehyde	
Dill	d-carvone	
Fennel	(E)-anethole, fenchone	

TABLE 1.2 (Continued)

Spice and Herbal spices	Important Flavor compounds	Chemical structure
Ginger	Gingerol, Shogaol, neral, geranial	
Mace	α-pinene, sabinene, 1-terpenin-4-ol.	
Mustard	Allyl isothiocynate	
Nutmeg	Sabinine, α-pinene, myristicin	
Parsley	Apiol	
Saffron	Safranal	

TABLE 1.2   (Continued)

Spice and Herbal spices	Important Flavor compounds	Chemical structure
Turmeric	Turmerone, Zingeberene, 1,8-cineole	Ar-turmerone α-turmerone 1,8-cineole zingiberene
Vanilla	Vanillin	Vanillin

B. Herbal Spices

Basil, Sweet	Methylchavicol, linalool, methyl eugenol	Linalool Methyl Eugenol
Bay laurel	1, 8-cineole	1,8-cineole
Marjoram	e- and t-sabinene hydrates, terpinen-4-ol	(+)-terpinen-4-ol

TABLE 1.2 (Continued)

Spice and Herbal spices	Important Flavor compounds	Chemical structure
Oregano	Carvacrol, thymol	
Origanum	Thymol, carvacrol	
Rosemary	Verbenone, 1–8-cineole, camphor, linanool	
Sage, Clary	Salvial-4 (14)-en-1-one, linalool	
Sage, Dalmation	Thujone, 1, 8-cineole, camphor	
Sage, Spanish	1, 8-cineole, camphor	

TABLE 1.2   (Continued)

Spice and Herbal spices	Important Flavor compounds	Chemical structure
Savory	Carvacrol	\n\ncarvacrol
Tarragon	Methyl chavicol, anethole	
Thyme	Thymol, carvacrol	\n\nThymol carvacrol
Peppermint	1-menthol, menthone, menthfuran	\n\n(-)-Menthol l-Menthone (+)-Menthofuran
Spear mint	1-carvone, carvone derivatives	\n\nR-(-)-carvone spearmint

1.2   INDIAN SPICE MARKET

Total spice exports from India crossed $2 billion ($2.4 billion, to be precise) heading on its way to meet the target set for the year 2017, which is $3 billion. Presently, India enjoys 48 and 43% share in terms of volume and value in the global spice market, respectively; and commands a formidable position in the world spice trade. India exports only 12% of the spices produced and the remaining 88% is for domestic consumption. The USA is the biggest importer of spice products, followed by Germany and Japan. The European Union has the largest imports of spices in value terms. Other major importing regions are the Middle East and North Africa, whilst there are growing markets in other countries (Table 1.3). During 2014–15, a total of 8,93,920 tonnes of spices and spice products valued Rs. 14,899.67 crores (US$2432.85 million) has been exported from India as against 8,17,250 tonnes valued at Rs. 1,37,353.92 crores (US$ 2267.67 million) in 2013–14 (Table 1.4).

The import of spices during 2014–15 was 1,38,715 tonnes valued at 3843.82 crores (US$ 629.36 million). The import of spices is for domestic consumption, value addition and re-exports. Spices such as clove, poppy

TABLE 1.3   Major Market Items for Indian Spices

Country	Major items
USA	Mint items, spices oils and oleoresins, pepper, chili, turmeric
China	Mint products, chili, spices oils and oleoresins
Malaysia	Chili, turmeric, coriander, cumin, fennel
UAE	Turmeric, chili, nutmeg, curry powder, cumin
UK	Spices oils and oleoresins, mint products, chili
Bangladesh	Chili, turmeric, garlic, ginger, cumin
Germany	Spices oils and oleoresins, mint products, turmeric
Pakistan	Chili, large cardamom, cumin, coriander
Japan	Spices oils and oleoresins, mint products, turmeric
Sri Lanka	Chili, turmeric, coriander, cumin, fennel
Saudi Arabia	Small cardamom, curry powder, turmeric, ginger
Singapore	Spices oils and oleoresins, mint products, chili
South Africa	Spices oils and oleoresins, turmeric, chili, coriander
Netherlands	Spices oils and oleoresins, mint products, turmeric, pepper, chili
Mexico	Spices oils and oleoresins, mint products, cumin, chili
Brazil	Spices oils and oleoresins, mint products, cumin, chili

TABLE 1.4   Export of Spices from India, During 2013–14 and 2014–15

Spice	2013–14		2014–2015	
	Qty. in tonnes	Value in Rs. lakhs	Qty. in tonnes	Value in Rs. lakhs
Pepper	21,250.00	94,002.34	21,450.00	120,842.16
Cardamom (S)	3,600.00	28,380.88	3,795.00	32,346.75
Cardamom (L)	1,110.00	7,961.15	665.00	8,403.90
Chili	3,12,500.00	272,227.20	347,000.00	351,710.00
Ginger	23,300.00	25,614.27	40,400.00	33,133.00
Turmeric	77,500.00	66,675.85	86,000.00	74,435.00
Coriander	45,750.00	37,185.65	46,000.00	49,812.50
Cumin	121,500.00	160,006.00	155,500.00	183,820.00
Celery	5,600.00	3,661.48	5,650.00	4,302.10
Fennel	17,300.00	16,001.42	11,650.00	13,165.50
Fenugreek	35,575.00	13,378.37	23,100.00	13,947.63
Other seeds (1)	27,800.00	15,425.65	28,250.00	16,512.50
Garlic	25,650.00	8,387.05	21,610.00	8,183.00
Nutmeg and mace	4,450.00	26,285.62	4,475.00	26,797.50
Other spices (2)	34,700.00	41,846.80	36,500.00	44,915.00
Curry powder	23,750.00	40,132.03	24,650.00	47,626.00
Mint products (3)	24,500.00	343,042.20	25,750.00	268,925.00
Spices oils and oleoresins	11,415.00	173,324.85	11,475.00	191,090.00
Total	**817,250**	**1,373,539.26**	**893,920**	**1,489,967.53**
Value in million US$	**2267.67**		**2432.85**	

(1) Includes mustard, aniseed, Bishops weed (ajwain seed), dill seed, poppy seed, etc.

(2) Includes asafoetida, cinnamon, cassia, cambodge, saffron, spices, etc.

(3) Includes mint oils, menthol, and menthol crystals.

Source: DGCI&S, Kolkata/shipping bills/exporter's returns.

seed, cinnamon, ginger fresh, and cardamom (large) are from neighboring countries for domestic consumption and black pepper, crude spice extracts for value addition, and re-export. Dehydrated spices like onion, garlic, turmeric, and chili products have very large extent of applications and today's world of *Ready to Eat* and *Ready to Cook* food segment, which is growing like anything. As a savory tastemaker, they are essential food ingredient of any food products.

India is facing stiff challenges by some new entrants in the global economy like Vietnam, China, Indonesia, Guatemala, etc., so far as the leadership in this sector is concerned. It starts from soil fertility issues viz., high acidity, poor drainage, low nutrient status, etc. The national productivity of all major spices is significantly lower than the competing countries. For example, in case of black pepper productivity, Thailand and Vietnam is superior (2000 kg/ha) to India (500 kg/ha). Similarly productivity of ginger and garlic in India is only 5300 kg/ha and 5000 kg/ha, respectively compared to China (5300 kg/ha and 24,000 kg/ha, respectively). The other notable concerns are: shifting of interests of growers to more profitable and or less risky crops, cyclic market fluctuations at international and national levels, insufficient inflow of information among the different stakeholders in the industry, emergence, and epidemics of pests and diseases, pesticide residues and mycotoxin contaminants in the products and lack of MRL and ADI standards in some of the pesticides used in spices, new stricter legislations and regulations, severe adulteration and incidents of contaminants in spices (aflatoxin, pesticide, illegal dyes, microbials, etc.) and last but not the least, climate change resulting in drought or excess moisture, high or low temperature during critical periods, etc.

To keep ourselves afloat as a major player in the global scenario, a coordinated approach is required from the research, development, marketing and policy making agencies working in the spice sector. A cost-effective, eco-friendly, and production-oriented technologies starting from the development of healthy disease free drought resistant planting materials are to be developed, which will be workable at the farm level. More investment in the spice processing sector and more support to the concerned entrepreneurs are the need of the hour. The global spice trade is expected to increase with the growing consumer demand in importing countries for more exotic, ethnic tastes in food. In the UK, for example, spice imports have increased by 27% in the last five years, mainly through the growth in cinnamon, cloves, garlic, and seed spices.

KEYWORDS

- **area**
- **flavor chemicals**
- **production, export**

- significance
- spice crops

REFERENCES

1. Krishnamurthy, K. S., Biju, C. N., Jayashree, E., Prasath, D., Dinesh, R., Suresh, J., & Nirmal Babu, K. (Eds.), (2015). Souvenir and Abstracts, National Symposium on Spices and Aromatic Crops (SYMSAC VIII): Towards 2050 Strategies for Sustainable Spices Production, Indian Society for Spices, Kozhikode, Kerala, India. p. 263.
2. Tamil Selvan, M., (2016). Status and Strategies of Planting Material Production in Spices in India. Indian Journal of Arecanut, Spices & Medicinal Plants, 18(2), 3–12.

CHAPTER 2

SPICE CROPS: CLASSIFICATION

CONTENTS

2.1 INTRODUCTION

The word "spice" derived from the Latin word 'species,' which means a commodity of value and distinction. There are 109 spices grown in the world. Out of these, 63 spices are grown in India. These spices can be classified or grouped into the different groups according to the different system of classification. None of the classification is complete, since each classification has got some gap or overlapping. Spices can be classified differently according to plant parts used, climatic requirement, life cycle duration, economic importance, nature of flavor, botanical classification, etc.

2.2   CLASSIFICATION BASED ON PLANT PARTS USED

Spices can be classified depending on the parts of the plant that are to be used. Different plant parts like leaf, root, bulb, fruit, seed, etc. are used as spice. This method of classification helps in describing their method of cultivation. For example, bulbous spices require similar type of cultural operations for their successful growth and development.

1. Seed: Coriander, cumin, fenugreek, fennel, ajwain, ajmund, black cumin (Nigella), aniseed, caraway, celery dill.
2. Flower and floral parts: Clove, saffron, caper, savory.
3. Fruit: Small cardamom, large cardamom, chili, kokum.
4. Berry: Black pepper, juniper.
5. Bark: Cinnamon, cassia.
6. Bulb: Onion, garlic, leek, chive, shallot.
7. Root: Angelia, sweet flag, horseradish, lovage.
8. Rhizome: Turmeric, ginger, galangal.
9. Leaves: Mint, curry leaf, bay leaf, chive, rosemary, sweet basil, marjoram.
10. Pod: Vanilla, tamarind
11. Kernel: Nutmeg
12. Aril: Mace
13. Latex: Asafoetida

2.3   CLASSIFICATION BASED ON ECONOMIC IMPORTANCE

Spices can be classified according to the value of national and international trade in the following ways:

1. Major spices: Black pepper, small cardamom, chili, ginger, and turmeric.
2. Seed spices: Coriander, cumin, fenugreek, fennel, dill, aniseed, etc.
3. Tree spices: Cinnamon, cassia, clove, nutmeg, all spice, curry leaf, kokum, tejpat.
4. Herbal spices: Thyme, rosemary, sweet basil, marjoram, sage.
5. Other spices: Vanilla, saffron, asafoetida, etc.

2.4   CLASSIFICATION BASED ON THE CLIMATIC REQUIREMENT

In this method of classification, spices can be classified into three different groups (tropical, temperate, and sub temperate), according to their suitability in their growing condition. Depending on suitable climatic conditions like temperature, sunlight, humidity, and air of a particular climatic zone, spices are grouped. Such type of classification is helpful for identifying the spice crop suited for cultivation in a particular zone.

2.4.1   TROPICAL SPICES

Spices of this category need high temperature and abundant humidity and easily damaged by low temperature. Tropical spices are ginger, turmeric, black pepper, cinnamon, kokum, galangal, small cardamom and clove, etc.

2.4.2   SUBTROPICAL SPICES

Sub-tropical climate is found where three distinct seasons like winter, summer, and monsoon are found. Low temperature prevails in winter and high temperature during summer. Most of the spices under this category require relatively low temperature during their vegetative or early growth stage and high temperature in reproductive stage. The examples of sub-tropical spices grown in winter are cumin, fennel, coriander, fenugreek, onion, and garlic. Certain sub-tropical spices grown during summer are turmeric and ginger.

2.4.3   TEMPERATE SPICES

Spices of this type can withstand low temperature and frosty weather but are damaged easily in hot and warm weather. Examples of temperate spices are thymes, saffron, caraway, and asafoetida.

2.5   CLASSIFICATION BASED ON TYPE OF FLAVOR

Depending on the origin and flavor content of the spices, they can be classified as follows:

1. Aromatic spices: Cardamom, aniseed, clery, cumin, coriander, fenugreek, and cinnamon.
2. Colored spices: Turmeric, paprika, and saffron.
3. Pungent spices: Ginger, chili, black pepper, and mustard.
4. Phenolic spices: Clove and allspice.

2.6   CLASSIFICATION BASED ON THE LIFE SPAN OF THE SPICES

Depending upon the requirement of growing season of the spices, the spices can be grouped into the following ways:

1. Annual spices: Such types of spices complete their life cycle in one growing season. Example of this type of spices is coriander, cumin, fennel, fenugreek, ajwain, and black cumin.
2. Biennial spices: It needs two growing seasons to complete the life cycle. Examples of biennial spices are onion and parsley.
3. Perennial spices: Perennial spices are those that live for more than two years. Example of perennial spices is black pepper, clove, nutmeg, cinnamon, cassia, and saffron.

2.7   CLASSIFICATION BASED ON GROWTH HABIT

The classification can be made on the following ways depending upon the growth habit of the spices:

1. Herbs: Coriander, cumin, fennel, and parsley.
2. Shrub: Rosemary, perennial chilies, and curry leaf.
3. Trees: Cinnamon, clove, nutmeg, tamarind, and kokum.
4. Climbers: Black pepper and vanilla.
5. Perennial herb: Turmeric, ginger, and mango ginger.

2.8   BOTANICAL CLASSIFICATION

Based on the botanical position spices, we can classify the species in different ways. Angiosperms or flowering plants are the most dominant group of the vascular plant world. They are classified in two different parts according to the number of cotyledons present in them. The majority of the spices can be divided into two groups namely monocotyledonae and dicotyledonae spices. This system classification may not be useful with relation to cultivation technique. However, this method has botanical vale by which we can easily identify their botanical position.

2.8.1 MONOCOTYLEDONAE SPICES

Monocotyledonae type spices has single cotyledon in the embryo, flower parts mostly in three or its multiple, parallel leaf vein, single furrow or pore pollen, and adventitious root system. It includes the following families:

1.	Family	Araceae
	Sweet flag	*Acorus calmus*
2.	Family	Alliaceae
	Onion	*Allium cepa*
	Garlic	*Allium sativum*
	Leek	*Allium porum*
3.	Family	Irridaceae
	Saffron	*Crocus sativus*
4.	Family	Orchidaceae
	Vanilla	*Vanilla planifolia*
5.	Family	Zingiberaceae
	Turmeric	*Curcuma longa*
	Ginger	*Zingiber officinale*
	Mango ginger	*Curcuma amada*
	Small cardamom	*Elettaria cardmomum*
	Large cardamom	*Amomum subulatum*

2.8.2 DICOTYLEDONAE SPICES

Dicotyledonae type spices has two cotyledon in the embryo, flower parts mostly in four, five or their multiple, reticulated or netted leaf vein, three furrows pollen, and radicle root system. It includes the following families:

1.	Family	Apiaceae (Umbelliferae)
	Ajwain	*Trachyspermum ammi*
	Aniseed	*Pimpenella anisum*
	Asafoetida	*Ferula foetida*
	Caraway	*Carum carvi*
	Celery	*Apium graveolens*
	Coriander	*Coriandrum sativum*
	Cumin	*Cuminum cyminum*
	Fennel	*Foeniculum vulgare*
	European dill	*Anethum graveolens*
	Indian dill	*Anethum sowa*
	Parsley	*Petroselinum crispum*
2.	Family	Brassicaeae
	Indian mustard	*Brassica juncea*
	True mustard	*Brassica nigra*
3.	Family	Capparidaceae
	Caper	*Capparis spinosa*
4.	Family	Compositae
	Tarragon	*Artemesia dracunculus*
5.	Family	Guttiferaeae
	Kokum	*Garcinia indica*
6.	Family	Fabaceae
	Fenugreek	*Trigonella foenum graecum*
	Tamarind	*Tamarindus indica*
7.	Family	Laurceae
	Cinnamon	*Cinnamonum verum*
	Chinese cassia	*Cinnamonum aromaticum*
	Indian cassia	*Cinnamonum tamala*
	Bay leaf	*Laurus nobilis*
8.	Family	Lamiaceae
	Japanese mint	*Mentha arvensis*
	Pepper mint	*Menthe piperita*
	Lemon mint	*Menthe citrate*
	Sweet basil	*Ocimum bassilicum*
	Thyme	*Thymus vulgaris*
	Rosemary	*Rosamarinus officinalis*
	Sage	*Salvia officinalis*

	Marjoram	*Marjorana hortensis*
9.	Family	Myristicaeae
	Nutmeg	*Myristica fragrans*
10.	Family	Myrtaceae
	Clove	*Eugenia caryophyllus*
11.	Family	Piperaceae
	Black pepper	*Piper nigrum*
	Long pepper	*Piper longum*
12.	Family	Runnanculaceae
	Black cumin (Nigella)	*Nigella sativa*
13.	Family	Rutaceae
	Curry leaf	*Murraya koenigii*
14.	Family	Solanaceae
	Chili	*Capsicum annuum*
	Paprika	*Capsicum annuum*

KEYWORDS

- **botanical**
- **classification**
- **climatic**
- **economic**
- **parts**
- **seasonal**
- **spices**

CHAPTER 3

AGROTECHNIQUES OF DIFFERENT SPICES

CONTENTS

3.1 MAJOR SPICES

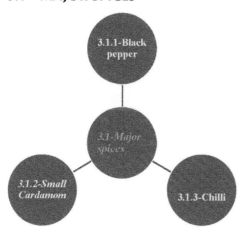

3.1.1 BLACK PEPPER (PIPER NIGRUM L.)

3.1.1.1 Systematic Position

Kingdom: Plantae
Division: Manoliophyta
Class: Magnoliopsida
Order: Piperales
Family: Piperaceae
Genus: *Piper*
Species: *nigrum*

3.1.1.2 About the Crop, National, and International Scenario, Uses, and Composition

Black pepper (*Piper nigrum* L.) is one of the oldest and world most important spices and is called' 'King of Spices.' The word "Peperi" in Greek, Pepper in English and Greek, and Piper in Latin were derived from the Sanskrit word "Pippali," which was the name of Long Pepper. In pepper cultivation, vines are trailed over living trees or on columns of 5–6 m tall for support. The climbing woody stems have swollen nodes with clinging roots, which helps in anchoring the vine to the support trees.

It is indigenous to the tropical forest of the Malabar Coast from where it is spread through out the tropics. India has the largest area under this crop (2.17 lakh ha) and it is the largest producer (65,000 MT). In India, black pepper is cultivated in Kerala, Karnataka, Tamil Nadu, Andhra Pradesh, Maharashtra, Orissa, West Bengal, Assam, and Andaman & Nicobar Islands. Pepper is cultivated in over 25 countries. India, Indonesia, Malaysia, and Brazil are the traditional producers. Countries like Madagascar, Sri Lanka, Vietnam, Thailand, China, and Mexico are also producing pepper on a commercial scale. Among the spices, pepper accounts for about 34% in international spice market. Indian pepper, mainly imported by USA, Russia, Canada, Germany, Italy, Netherlands, France, Japan, Morocco, Poland, UK, Canada, and Saudi Arabia. India, Brazil, Malaysia, Indonesia, Vietnam, and Thailand, are the leading suppliers to the world market. The quality of pepper is added by two components, piperine that contributes the pungency, which is the major alkaloid present in pepper and volatile oil that is responsible for the aroma and flavor to the black pepper. The global trade in black pepper is controlled by International Pepper Community (IPC). India, Brazil, Malaysia, and Indonesia are the member of IPC.

India is one of the major exporters of whole pepper and pepper products (mainly powder and oleoresin) across the world. On an average India exports around 22,000 tonnes of pepper and imports around that of 17,000 tonnes to the different parts of the world. Growth in Indian pepper exports is just 0.2% (CAGR) in the last 10 years, whereas, in value terms it is growing at a rate of 18%.

It contains lignans, alkaloids, flavonoids, aromatic compounds and amides. It also contains essential oil constitutes sabinene, pinene, phellandrene, linalool, and limonene. It also has piperine, which is a weak basic substance. Chavicine is an isomer of piperine. Piperine and chavicine are not responsible for the aroma of the black pepper but piperine imparts pungency to the black pepper. Piperine content in berry ranges from 1.2 to 9.5% but most common one is 4–6%. Oleoresin content ranges from 0.7 to 3.9% while the range of essential oil is 0.4 to 5.5%. Choline, Folic acid, Niacin, Pyridoxine, Riboflavin, Thiamin, Vitamin A, and Vitamin C, Vitamin E, and Vitamin K are the major vitamins found in the black pepper. Calcium, copper, iron, magnesium, manganese, phosphorus, and zinc are the main minerals found in the black pepper.

3.1.1.2.1 Use

i.   Black pepper constitutes an important component of culinary preparation and seasoning and as an essential ingredient of numerous commercial food stuffs.

ii. It acts as a preservative for meat and other perishable food preservation.

iii. It is an important constituent of whole, pickling spice, curry powder and spice formulae for seasoning sausages, poultry dressing, hamburger, etc.

iv. Pepper oil is used as a valuable food adjunct for flavoring agent in different foodstuff, beverages, and liquors. It is also used in perfumery and medicine industries.

v. Piperine is used to impart pungent test to brandy.

vi. Oleoresin also used as flavoring agent.

vii. Black pepper and its oil are used to cure dyspepsia, malaria, cold cough, etc. it also helps in improving the impotency and muscular pain.

3.1.1.3   Soil and Climatic Requirements

Crop grows well on soils ranging from heavy clay to light sandy clays rich in humus; and friable in nature, well drained but good water holding capacity. Soils with high organic matter and high base saturation with Ca and Mg enhances the productivity. Soil pH should be 4.5 to 6.0 and soils with pH above 7.5 inhibit growth. Loamy humus nourishes the crop well but the best crop could be obtained in virgin forest soil.

Black is a plant of humid tropics and requires a heavy and well-distributed rainfall and high temperature. It grows successfully between 20°N and 20°S of equator and upto 1500 m MSL, but thrives best at 500 m or below. Being a tropical crop, pepper requires 200–300 cm rainfall and high relative humidity with a little variation in day length throughout the year. A relative humidity of 60–95% is optimum at various stages of growth. Rainfall of 70 cm received in 20 days during May–June proves sufficient for triggering off flushing and flowering process in the plant. But once the process is set off, there should be continuous shower until fruit ripening. Any dry spell even 4 days within this critical period of 16 weeks may result in low yield. The crop may tolerate temperature between 10–40°C and ideal temperature is 23–32°C with an average of 28°C. The optimum temperature for root growth is 26–28°C.

3.1.1.4   Agrotechniques for Quality Production

3.1.1.4.1   Propagation

Generally black pepper is propagated through runner or vine cutting. In certain parts of the India it is propagated through rapid multiplication techniques. Recently, black pepper is being successfully propagated through serpentine layering method.

3.1.1.4.2   Fertilizer Requirement

Soil pH should be maintained around neutral to obtain yield responses from N fertilizer application. The amounts of macro and micronutrients removed through harvested produce were directly proportional to the yield, indicating the need for yield based fertilizer recommendations in black pepper. The magnitude of nutrient removal by harvested produce followed the decreasing order: N>K>Ca>Mg>P>S>Fe>Mn>Zn

a. General age of the recommendation: NPK at 100:40:140 g/mature vine/year.

b. Soil with poor fertility status: NPK at 140:55:270 g/mature vine/ year.

For 1-year-old plant $1/3^{rd}$ dose is applied and $2/3^{rd}$ dose for 2-year-old plant. The full dose is given from the 3 years and onwards. The fertilizer may be applied in split doses first in May–June with receipt few soaking rains and second in August–September. Apart from the inorganic fertilizers apply FYM/compost at the rate of 10 kg per vine. It is desirable to apply lime @ 500 g/vine in April–May in alternate years, fertilizer are to be applied (on northern side) at a distance of 30–45 cm away from the vine (in semicircular band) and cover with a thin layer of soil (Sadanandan, 1991).

3.1.1.4.3   Irrigation

There is a distinct soil moisture stress period of about five months (November–December to April–May) and consequent death of vines. To overcome the moisture period there should be sufficient moisture in soil. Moisture condition of soil enhances the entry of water in plant system and better translocation of nutrients for the metabolic activity of the plant. Irrigation studies at PRS, Panniyur showed that irrigation in Panniyur-1 and Karimunda about 100 L of water at IW/CPE ratio of 0.25 during December to March has increased the yield of pepper by 90% (Hort Net Kerala, 2016).

3.1.1.4.4   Cultural Practices

As the cutting grows, the shoots are tied to the standard as often as required. The young vine should be protected from direct hot sun during summer by providing them with artificial shade. Regulation of shade by lopping the branches of standards is necessary not only for providing optimum light to the vines but also for enabling the standards to grow straight. Adequate mulch with green leaf or organic matter should be applied towards the end of northeast monsoon. The base of the vines should not be disturbed as to avoid root damage. In the second year, practically the same cultural practices are repeated. However, lopping of standard should be done carefully from the 4th year onward, not only to regulate the standard but also to shade the pepper vines optimally. Excessive shading during flowering and fruiting encourage pest infestation. From the 4th year onwards usually two diggings are given during onset and end of southwest monsoon. Growing cover crops like *Calapogonium mucunoides,* and *Mimosa invisia* are also recommended under west coast conditions to provide an effective cover to prevent soil erosion during rainy season. Further, they dry during summer, leaving thick organic mulch.

3.1.4.4.5   Improved Varieties

Many varieties of black pepper are found in the native regions. Some good ones are popular and are in use evolved through natural as well as artificial selection. Many more varieties have also been developed through other breeding methods. Some of the most popular and improved varieties of black pepper have been discussed in tabular format (Table 3.1) highlighting their parentage, special features, average yield (kg/ha), yield/plant (kg, green), drying percentage (%), piperine (%), oleoresin (%), and oil (%).

3.1.1.5   Plant Protection

Phytopthora foot rot: The disease is caused by *Phytopthora capsici*. One or more black spots appear on the leaves, which have a characteristic fine fiber like projections at the advancing margins, which rapidly enlarge and cause defoliation. The tender leaves and succulent shoot tips of freshly emerging runner shoots trailing on the soil turn black when infected. If the main stem at the ground level or the collar is damaged, the entire vine wilts followed by shedding of leaves and spikes with or without black spots. The branches break up at nodes and the entire vine collapses within a month.

TABLE 3.1 Improved Varieties of Black Pepper

Variety	Parentage	Special features	Average Yield (kg/ha)	Yield/pl (kg, green)	Drying (%)	Piperine (%)	Oleoresin (%)	Oil (%)
Panniyur-1	Uthirankotta × Cheriyakaniyakkadan	Long spike, high yield potential under open condition	1242	5.0	35.3	5.30	11.8	3.5
Panniyur-2	Open pollinated seedling of Balankotta	Tolerates shade	2570	4.5	35.7	6.60	10.9	3.4
Panniyur-3	Uthirankotta × Cheriyakaniyakkadan	Long spike, bold berries prefers open condition	1953	4.4	27.8	5.20	12.7	3.17
Panniyur-4	Selection from Kuthiravally types	Tolerates adverse climatic conditions	1277	2.3	34.7	5.00	9.2	3.12
Panniyur-5	Open pollinated seedling of perumkodi	Tolerates adverse climatic conditions	2352	3.19	35.71	5.30	12.33	3.8
Panniyur-6	Selection from Karimunda types	Stable and regular yield, tolerates partial shade	2127	6.46	32.93	4.94	8.27	1.33

TABLE 3.1 (Continued)

Variety	Parentage	Special features	Average Yield (kg/ha)	Yield/pl (kg, green)	Drying (%)	Piperine (%)	Oleoresin (%)	Oil (%)
Panniyur-7	Open pollinated progenies of Kulluvally	Stable and regular yield, tolerates partial shade	1410	4.2	33.57	5.57	10.61	1.5
Sreekara	A selection from Karimundu (KS.14)	Tolerant to drought	2352	48	35	5.0	13	7
Subhakara	A selection from Karimundu (KS.27)	Tolerant to drought	2677	4.2	35	3.4	12	6
Panchami	A selection from Aimpiriyan coll.856	A high yielding clone already in cultivation	2828	5.2	34	4.7	12.5	3.4
Pournami	A selection from Ottaplackal type col. No.812	Tolerants to nematode	2333	4.7	31	4.1	13.87	3.4
PLD-2	Clonal selection from Kottanadan	Late maturity, suitable for plains and higher elevation	2475		-	3.0	15.45	4.8
IISR Shakti	OP seedling of Permbramundi	Tolerant to P. capsaici	5575 (potential)		43	3.3	10.2	3.7

Variety	Parentage	Special features	Average Yield (kg/ha)	Yield/pl (kg, green)	Drying (%)	Piperine (%)	Oleoresin (%)	Oil (%)
IISR Thevam	Clonal selection form Thevamundi	Field tolerant to *P. capsaicin*, stable yielder, suitable for higher elevation.	2437		32.5	1.6	8.15	3.1
IISR Malabar Excel	Cross between Cholamundi × Panniyur-1	Suitable for rainfed condition including, tea, and coffee plantation.	1453		32.3	2.4	11.7	2.8
IISR Girimunda	Cross between Cholamundi × Panniyur-1	Suitable for rainfed condition including, tea, and coffee plantation.	2880		32	2.2	9.65	3.4

To manage the disease avoid unnecessary tilling of soil, which can be conducive to spreading the pathogen; a cover crop of grass can help prevent water splash on the plants and thus the spread of the fungi; amending the soil with neem cake suppresses the *Phytophthora* and provides nutrients to the vines; systemic fungicides such as metalaxyl and fosetyl can give some measure of control; efforts are being made to establish resistant varieties.

Pollu disease/Anthracnose: It is caused by *Colletotrichum gloeosporioides*. The affected berries show brown sunken patches during early stages and their further development is affected. In later stages, the discoloration gradually increases and the berries show the characteristic cross splitting. Finally, the berries turn black and dry. The fungus also causes angular to irregular brownish lesions with a chlorotic hollow on the leaves.

As anthracnose is primarily a disease that occurs during the rainy season, systemic fungicides are required to prevent chemicals leaching from the plant; 1% Bordeaux mixture can be applied during monsoon season; metalaxyl and fosetyl are also effective.

Leaf rot and leaf blight: It is caused by *Rhizoctonia solani*. Greyish sunken spots and mycelial threads appear on the leaves and the infected leaves are attached to one another with the mycelial threads. On stems, the infection occurs as dark brown lesions, which spread both upwards and downwards. The new flushes subtending the points of infection gradually droop and dry up. The disease can be effectively managed by prophylactic spray with Bordeaux mixture 1%.

Stunt: It is caused by cucumber mosaic virus and pepper yellow mottle virus. The vines exhibit shortening of internodes to varying degrees. The leaves become small and narrow with varying degrees of deformation and appear leathery, puckered, and crinkled. Chlorotic spots and streaks also appear on the leaves occasionally. The yield of the affected vines decreases gradually.

Identification and use of virus free plants is an important aspect of stunt disease management of black pepper. Potential weeds and host cops, which act as reservoir for virus are to be removed and burnt. Field and nursery areas are to be monitored regularly for vectors like aphid and mealy bug and dimethoate 0.05% is sprayed, if necessary.

Slow decline: It is caused by attack of the nematodes *Radopholus similis and Meloidogyne incognita*. Foliar yellowing, defoliation, and die-back are the aerial symptoms of this disease. The affected vines exhibit varying degrees of root degeneration due to infestation by plant parasitic nematodes. The root system of diseased vines show varying degrees of necrosis and presence of root galls due to infestation by plant parasitic nematodes such as *Radopholus similis* and *Meloidogyne incognita* leading to rotting of feeder roots.

In areas severely infested with root knot nematodes, cuttings of the resistant variety 'Pournami' may be planted. Biocontrol agents like *Pochonia chlamydosporia* or *Trichoderma harzianum* can be applied @ 50 g/vine twice a year (during April–May and September–October) for management of nematodes.

Phyllody disease: It is caused by *Phytoplasma* belonging to aster yellows group. The affected vines exhibit varying stages of malformation of spikes. Some of the floral buds are transformed into narrow leaf like structures. Such malformed spikes show leafy structures instead of floral buds, exhibiting Phyllody symptoms. In advanced stages, the leaves become small and chlorotic, and the internodes are also shortened. The affected fruiting laterals give a witches broom appearance. Severely affected vines become unproductive. The infected vine becomes unproductive within two to three years. The infected vines are to be destroyed to prevent further spread of the disease.

Pollu beetle: The adults feed on tender shoots, spikes, and berries. The infested shoots and spikes turn black and drop. The grub on emergence bore into the berries, feed on the internal contents and make them hollow. The infested berries turn yellow initially and then black and crumble when pressed. The pest population is more severe in shaded areas. Spraying of Quinolphos @ 0.05% twice a year in June–July and September–October helps to manage the pest.

Top shoot borer: The top shoot borer is found more in younger plantations. The caterpillars of the moth bore into tender shoots, which turn black and dry up. When successive new shoots are infected; the growth of the vine is affected. Management of the pest can be done by spraying 0.05% endosulphan.

Leaf gall thrips: The feeding of thrips on tender leaves causes the leaf margins to curl down and inwards resulting in the formation of marginal leaf galls. The infested leaves become thick, malformed, and crinkled. Life stages of the insect can be seen within the gall. In severe cases of infestation, the growth of young vines is affected. Spraying of Dimethoate @ 0.05% during emergence of new flushes in young vines in the field and cuttings in the nursery is found effective in managing the pest.

Scale insect: Scale insects appear as encrustations on stems, leaves, and berries. They feed on plant sap resulting in yellowing and drying of infested portions of the vines. Management of scale insect is done by spraying Dimethoate @ 0.05%

Stem borer: The grubs bore into nodal region of the climbing and flower-ing shoots resulting in wilting of infested shoots. The entire upper portion of the vine collapses and dies eventually. When the berries are damaged, they either fall prematurely or do not attain their normal size. The female bores into the tissues and makes a hole. Once eggs are laid, the tissues around the hole become darker to black immediately after few hours.

Shade regulation by pruning the live standards is important to reduce and suppress the insect population. Spraying Dimethoate @ 0.05% is found effective. *Beauvaria bassiana* is a potential biocontrol agent and it has been recommended for application during morning especially during wet period, which ensures positive multiplication of the insect pathogen.

3.1.1.6   Harvesting and Yield

Flowering in pepper starts during May–June. The crop takes about 6–8 months from flowering to harvest. The harvest season extends from November to January in plains and January to March in hills. During harvesting the whole spike is hand picked when one or two berries in the spike turn orange or purple. Harvesting in immature stage reduces bulk density. The berries are separated from spike and dried in sun for 3 to 5 days. The average yield of black pepper (dry) obtained from vines under ideal management conditions is 2 kg/year. However, yield varies with age, vigor of vine, cultivars, edaphic condition and management level. The dry recovery of pepper varies from 30–35% depending upon the variety.

To improve the color and appearance of dry berries and reduce the drying time, harvested green pepper is to be soaked in hot water and then dried in sun. It also minimizes microbial contamination and gives hygienic product, which can be preserved easily.

3.1.1.7   Bush Pepper

Black pepper can be grown as bush in pots known as bush pepper. In this case fruiting branches are used as planting materials. Bush pepper pro-duces flower and fruit in the same year. It also continues to flower around the year.

One-year-old healthy fruiting branches are selected with 3–5 nodes and all the leaves are removed. The cuttings are treated with 0.2% copper oxy chloride for 20–30 minutes. The lateral should be given a slanting cut 2 cm below the nodal region and dipped in 1000 ppm IBA for 45 seconds and planted in shaded area in the nursery either in trenches or in poly bags (45 × 30 cm) containing moist decomposed coir dust. After planting, the trenches are covered with polythene sheets and in case of poly bags, the mouth is tightly tied to avoid moisture loss.

After 30–50 days these rooted cuttings are planted in pot or in field after sufficient hardening treatment. Pepper grown in pot has longer spike but where as pepper grown in field condition has more number of spikes per bush and yield of green and dry pepper is also higher. This could be grown as an inter crop in the coconut and arecanut gardens and bund crops. A quantity of 2–5 kg FYM along with 10 g NPK @ 1:1:2 ratios may be given per bush at 3 months interval. Watering and plant protection may be adopted according to necessity.

3.1.1.8   Yield

On an average 500 g to 1 kg of green pepper can be harvested during 2nd year of planting. Yield increase with the advance of age.

3.1.1.9   Post Harvest Technology

3.1.1.9.1   Curing of Black Pepper

After harvesting the spikes are heaped for 1–2 days and the berries are sepa-rated from the spikes and dried in the sun for 4–5 days on mats or clean con-crete floor, until the outer skin become dark brown to black and shriveled. To improve the color and appearance of dried berries and to reduce drying time, harvested green pepper is to be soaked in boiling water for 1 minute and then dried in shade. During drying blackening occurs due to enzymatic oxidation of phenols, which present in epicarp and mesocarp.

3.1.1.9.2   Different Steps of Processing

i) Drying
    a) Sun drying
    b) Improve method of drying/CFTRI method

   c) Mechanical drying: 30–40 tonnes of pepper can be dried in 8 hours

ii) Cleaning: Separation of various fractions to get rid of extraneous matter such as dirt, stones, stalks, leaves, etc.

iii) Grading

iv) Physical cleaning washing and drying, if required

v) Packing

   a) Bulk: Poly-lined gunny bag

   b) Retail

      A. Whole pepper: HDPE pouches of 200 gauges.

      B. Ground pepper: Packed in laminated heat sealable aluminum pouches (polythene coated) or moisture proof cellulose film.

3.1.1.9.3 Grading

According to Agmark grading, following grades have been formulated:

i) Whole pepper

   a) Malabar garbled black pepper (MG Grade 1 and 2)

   b) Malabar ungarbled black pepper (MGU Grades 1 and 2)

   c) Tellicherry garbled black pepper special extra bold (TGSEB)

   d) Tellicherry garbled extra bold (TGEB)

   e) Tellicherry garbled (TG)

   f) GL special, GL grades 1 and 2

   g) UGL special, UGL grades 1 and 2

   h) PH grade special, PHG grade-1

   i) Black pepper (non-specified) (NS grade X)

ii) Ground black pepper: standard and general grade.

3.1.1.9.4 Value Addition

Dehydrated green pepper: To stabilize the green color of pepper, a process has been developed involving blanching and sulphiting combined with controlled drying and reduction of moisture (Thomas and Gopalakrishnan, 1992). For producing dehydrated green pepper, slightly immature green pepper (20 to 30 days prior to full maturity) is mostly preferred. The cleaned pepper berries were blanched in boiling water for 10–30 minutes till the enzymes responsible for blackening the pepper are inactivated and polyphenols washed out of the berries. The blanched pepper was immediately cooled in water and subjected to sulphiting in potassium metabisulphite solution to fix the green color. Potassium metabisulphite has a phenolase inhibiting property and an ability to deter non-enzymatic browning (Varghese 1991). The sulphited berries were then washed and dried in a cabinet dryer at 50–55°C for 12–15 hours to get uniform green colored berries. Good quality dehydrated green pepper should contain less moisture (<8%). Total heat inactivation of the enzyme was obtained after 10 minutes of boiling (Mathew 1994), though the boiling time depends on the maturity of the berries.

Green pepper in brine: Four- to five-months old immature spikes are harvest and cleaned thoroughly. Light pepper, pinhead or broken berries are considered as extraneous matter. 12–14% common salt and citric acid not exceeding 0.6% by mass of the packing media are used for the preparation. It is also one of the important value added spice items of export. Vasantkumar (2006) reported that the pepper in brine, dehydrated green pepper, freeze dried green pepper, and white pepper earn appreciable quantity of money through export of these products to the different countries.

Canned green pepper: Chlorinated water (containing 20 ppm residual chlorine) is used to separate the berry from spike and clean the berries by immersing them for about an hour followed by their immersion in 2% hot brine (containing 0.2% citric acid), exhausting at 80°C, sealing properly, and processing in boiling water for 20 minutes. Canned pepper is then cooled immediately in a stream of running cold water. Pepper harvested one month prior to maturity is reported to be ideal for the manufacture of canned green pepper (Narayanan et al., 2000).

Bottled green pepper: Fresh green pepper berries of uniform size and maturity are separated the from spike immediately after harvest followed by cleaning, washing, and steeping in 20% brine solution containing citric acid. This is allowed to cure for 3 to 4 weeks. After draining off the liquid, fresh brine (16%) is added together with 100 ppm sulphur dioxide and 0.2% citric acid. The resulting product is stored in containers protected from sunlight (Pruthi, 1997), which is pale green to green and should preferably possess the characteristic odor and flavor of green pepper.

Freeze dried green pepper: Freeze dried green pepper retains the natural form of the green pepper and is believed to be far superior to dehydrated green pepper for its better color, flavor, essential oil and piperine content.

On rehydration it retains the original green color and shape of green pepper. It finds a wide application in instant soups and dry meals for its special characters and subtle flavor. It is also used in cheese industry. It is produced by vacuum drying at sub freezing temperatures ranging from –12 to –40°C.

Frozen green pepper: Frozen green pepper is considered superior to green pepper in brine or dehydrated green pepper due to the following advantages:

a) It has much better flavor, color, and texture and natural appearance.
b) Packaging cost is less due to cheaper container price.
c) It is a see-through container (unlike cans) and customer can see as to what they are buying.

Freezing is done at 40°C by covering the green pepper of desirable maturity in sodium chloride (2%), citric acid (0.25%) and ascorbic acid (0.1%). Frozen green pepper is used mostly in fresh salad and frozen meals.

Green pepper pickle (in oil, vinegar or brine): Green pepper is prepared in 15–16% brine, acidified, and in vinegar as in other vegetable. Kerala, Karnataka, Tamil Nadu, Gujarat, and Maharashtra are some of the states where this kind of preparation is popular. The recipe is very much similar to that of mango pickles and is enjoyed by many people mainly as an appetizer.

White pepper: Many people prefers white pepper over black pepper in light-colored preparations viz., sauces, cream soups, etc., where dark colored particles are undesirable. It is actually the white inner corn obtained through removal of the pericarp of the berries. A few varieties like Balankotta and Panniyur 1, with large sized berries, are best suited for making white pepper where moisture level of the finished product is brought down below 11% through drying. It imparts pungency and a modified natural flavor to the foodstuff (Sudharshan, 2000). The recovery from fresh pepper to dry white pepper is 28% on an average. Major producers of white pepper are Malaysia, Indonesia, and Brazil, while west European countries, USA, and Japan are the chief consumers.

For preparing white pepper the following techniques (Pruthi, 1980) may be used:

1. Water steeping and rotting technique (retting)
   a. From fresh ripe berries: After steeping of freshly harvested ripe berries in water tanks or under running water for 7–10 days, the outer rinds are removed by rubbing them in hands or trampling. The deskinned berries are then washed, drained, and put into

galvanized iron vessel containing a solution of bleaching powder for 2 days followed by draining, drying in the sun and cleaning.

b. From dried berries: After steeping the dried berries in water for 10–15 days they are removed, rubbed, washed thoroughly, steeped again in bleaching solution for 2–3 days, drained, dried in sun and sold as white pepper.

2. Steaming or boiling technique: This process has been developed at CFTRI where fresh berries after steaming or boiling (for about 15 minutes) are passed through a fruit pulping machine for outer skin removal. The deskinned berries are then washed, bleached, and dried in sun to get white pepper.

3. Chemical technique: The whole dried black pepper is steeped in five times of its weight of water for 4 days, mixed with NaOH solution (4%) and boiled (Joshi, 1962).

4. Decortications technique: Here non pathogenic bacterium *Bacillus subtilis* are used to act on black pepper at pH 6.8–7.0 in minimal nutrient medium to deskin them within 2–4 days. The resulting white pepper should contain 3.2% volatile oil and 4.5% piperine.

5. Pit method: Compared to water steeping method, this method is less time requiring. In this case, gunny bags with fully mature ripe berries are placed in the pit covered with soil and watered for microbial reaction with the berries. The berries with rotten pericarp are washed, bleached, and dried properly.

Ground pepper: Ground pepper is produced by grinding pepper in mill without adding any foreign matter. TNAU has developed a low temperature device in, which dry ice used to cool the grinding zone of the mill (Anonymous, 2001). More volatile oils may be retained in natural composition of the powder through cryogenic grinding with liquid nitrogen (Pruthi, 1980).

Pepper essential oil: The characteristic aroma of black pepper is due to its volatile oil content, which is mainly obtained through steam distillation or water distillation techniques (Pruthi, 1977). The oil is slightly greenish in color with a mild non-pungent taste. The main component of the essential oil is α-phellandrene, sabinene, α-pinene and β-pinene.

Oleoresin: When ground pepper is extracted with solvents like hexane, ethanol, acetone, ethylene dichloride, ethyl acetate, etc. generally at 55–60°C temperature, a concentrated product is obtained, which is known as oleoresin. It is a dark green viscous liquid with a strong aroma and pungent taste.

Normally solid to solvent ratio varies from 1:3 (Narayanan et al., 2000). High yields of oleoresin (12–14%) containing 19–35% volatile oil and 40–60% piperine are obtained with good quality pepper (Lakshmanachar, 1993). A significant variation in quality characters of some black pepper varieties was found under Calicut, Kerala condition of India (Zachariah, 2005). Oleoresin content of black pepper varied from 9.50 to 17.0%.

Dehydrated salted green pepper: De-hydrated salted green pepper is a 100% substitute for green pepper in brine. It is easier for transportation and storing as it does not involve any brine solution. It is a product, which can be used instead of pepper in brine, as it contains both pepper and salt in the same proportion and also maintains the natural green color.

Piperine: It is the main principle component of black pepper and is mainly used in food, pharmaceutical, and nutraceuticals industries. The content of piperine varies from 3–6%. It is sparingly soluble in water but readily soluble in alcohol. Its concentrated form can be produced by centrifuging the black pepper oleoresin in basket centrifuge.

3.1.1.10   Other Novel Products

Other novel products viz., pepper concrete, pepper absolute, pepper essence, pepper paste and pepper emulsion, etc. have also been prepared and marketed around the world.

3.1.1.11   Pepper By-Products

For economic use of pepper as a condiment and to replace it in times of scarcity, many products having the characteristic taste and pungency of pepper have been prepared as by-products.

Pep-sal or pepper sal: pepper-sal is a flavoring substance prepared from waste black pepper and common salt. It has found acceptance as a flavoring agent for salads, drinks, and meat dishes.

Pepper hulls: During the preparation of white pepper, pepper hulls or shells are removed and separately sold as a light to dark brown powder with a very pungent odor and taste.

Light powder: Some pepper berries remain underdeveloped due to incomplete photosynthesis and lack in starch synthesis. They are very light, but rich in oil, piperine, and oleoresin content.

Pepper pinheads: Pepper pinhead (berries of the size of 'pin-heads') is fairly rich in oil and oleoresin but their quality is somewhat inferior.

Pepper in curry powder-spice blends: Many spice mixtures or blends like curry powder, garam masala, and various other masalas, which are flooding markets in India and abroad, have pepper as one of the important components. They also find use in soups, pickles, sauces, and chutneys, etc.

3.1.2 SMALL CARDAMOM (AMOMUM SUBULATUM L.)

3.1.2.1 Systematic Position

Kingdom: Plantae
Division: Tracheophyta
Class: Magnoliopsida
Order: Zingiberales
Family: Zingiberaceae
Genus: *Amomum*
Species: *subulatum*

3.1.2.2 About the Crop, National and International Scenario, Uses, and Composition

Small cardamom, *Elettaria cardamomum* Maton belonging to the family *Zingiberaceae*. Botanically it is a dried fruit of the tall perennial herbaceous plant and popularly known as the 'Queen of Spices.' It is native to the evergreen rainforests of Western Ghats of Southern India from where it spread to other tropical countries such as Sri Lanka, Tanzania, and a few Central American countries. Before 1980, India was the main producer and exporter of cardamom. This is one of the most important spices after black pepper and it is popularly known as "Queen of Spices." Of late Guatemala stands up as a powerful competitor to Indian cardamom in the international spice market. Apart from these, it is also cultivated in Tanzania, Sri Lanka, El Salvador, Vietnam, Laos, Cambodia, and Papua New Guinea. In India, cardamom is mainly cultivated in the southern states of Kerala, Karnataka, and Tamil Nadu. Kerala accounts for near about 60% of the cultivation and production followed by Karnataka 30% and Tamil Nadu 10%. In India it is cultivated with an area of 85,000 ha, with a production of 14,000 metric tonnes. The productivity of Guatemala is about 300 kg/ha, whereas in India the productivity is 156 kg/ha.

On an average, export of cardamom from India during 2004 to 2009 was 1.9 thousand tonnes. But, during the past five years there has been a phenomenal rise in cardamom exports amounting almost 3.1 thousand tonnes (2010–2015). Cardamom exports have shown positive growth rate around 24% (in term of volume) and 34% (in term of value) (2004–15). However, on an average, cardamom imports during 2004 to 2015 was 0.5 thousand tonnes. Saudi Arabia and UAE are the major cardamom importers from India.

The basic cardamom aroma produced by a combination of the major components, α-terpinyl acetate and 1,8-cineole, which is principal components in the cardamom volatile oil.

3.1.2.2.1 Use

Cardamom is used for flavoring various food preparations like cakes, curries, breads, etc., and confectionaries, beverages, perfumery, and liquors. In the Middle East countries (like Iran, Iraq, and Arab), it is used for making a cardamom-flavored coffee. In some countries, like Britain and USA, it is used as aromatic stimulant. It is used as a masticatory and may be included in betel leaf with arecanut. In Sri Lanka, cardamom is used for manufacturing liquors. It is also used as an important ingredient of curry powder and other mixed masala. Cardamom essential oil is used in flavoring processed food, liquors, pharmaceuticals, perfumery, and beverage industry. It is also used in various forms of ayurvedic.

3.1.2.3   Soil and Climatic Requirements

Humus rich loamy soil under the canopy of the evergreen forest trees is ideal for this crop provided that minimum disturbances, good mulch and adequate moisture are available. A pH range of 6–7 is ideal for the availability and effectiveness of most of the nutrients. Usually cardamom-growing soils are acidic within 5.0 to 5.5 pH range and therefore use of lime is recommended in such areas to increase pH.

The natural habitat of cardamom is the evergreen forest of Western Ghats. Cardamom is pseophyte, i.e., shade-loving plant in, which excess or low shade will have an adverse influence of its growth and performance

Cardamom is grown at an elevation from 600–1200 m from MSL under a well-distributed rainfall of 150–400 cm and a temperature of 10–35°C. It has been observed that distribution of rainfall is more important rather than high annual rainfall for a good crop. Its optimum growth and development is observed in warm and humid places under a canopy of evergreen forest trees with 40–60% shade.

3.1.2.4   Varieties and Types of Small Cardamom

There are two botanical varieties of *Eletaria cardamomum*. They are:

VAR *major*: This is a wild cardamom of Sri Lanka and cultivated occasionally. It is a robust and with a plant of 3 m height with broad leaves and erect panicles, fruit size larger, i.e., 2.5–5 cm. This is known as long wild native cardamom

VAR *minor*: This includes most of the cultivated races, plant height varies from 2.5–5 cm, panicle erect in nature and longer and having more number of flowers, fruits are 1–2 cm long with few, small more aromatic seed.

Types of small cardamom: Based on the nature of panicles cardamom can be divided into three different groups.

i.   Malabar type: It has spreading type panicles and can grow in low altitude 600–900 m from MSL.
ii. Mysore type: It is vigorous in nature having erect type panicles and can grow well at higher altitude, i.e., 900–1200 m.
iii. Vazhuka type: It is a natural hybrid of Malabar and Mysore type as the plant exhibit intermediate characters, prefers high altitude, and panicles semi-erect in nature. Vazhuka type varieties are: NCC-200 (Njallani), MCC-12, and MCC-16.

The improved varieties of cardamom with their parentage, average yield (kg/ha), dry recovery (%), oil (%), 1,8 cineole (%), α-terpenyl acetate (%) and salient features have been discussed in Table 3.2.

3.1.2.5   Propagation

Cardamom is propagated through seed, suckers, rhizomes, and tissue culture plantlets. Plants mature in about 20–22 months after planting. Economic yield starts from 3rd year upwards and continues upto 8–10 years. The total

TABLE 3.2 Improved Varieties of Small Cardamom

Sl. No.	Variety	Pedigree/ Parentage	*Average Yield (kg/ha)	Dry recovery %	Oil%	1, 8 Cineol %	α-terpenyl acetate %	Salient features
1	Mudigere 1	Clonal selection from Malabar type	300	20	8	36	42	Erect and compact panicle, suitable for high density planting, moderately tolerant to thrips, hairy caterpillar and white grubs, pubescent leaf. Short panicle, pale green, oval bold capsule
2	Mudigere 2	Clonal selection from open pollination of Malabar type	475	–	8	45	38	Early maturing variety, suitable for high-density planting, round/oval bold capsules.
3	PV 1	A selection from Walayar collection, Malabar type	260	19.9	6.8	33	46	An early maturing type, short panicle, elongated slightly ribbed light green capsules, Long, bold capsule

Sl. No.	Variety	Pedigree/ Parentage	*Average Yield (kg/ha)	Dry recovery %	Oil%	1, 8 Cineol %	α-terpenyl acetate %	Salient features
4	PV 2	Selection from OP Seedlings of PV-1, a Malabar type	982	23.8	10.45	–	–	Early maturing, unbranched lengthy panicle, Long bold capsules, high dry recovery percentage, field tolerant to stem borer and thrips, suitable from elevation range of 1000–1200 m above MSL.
5	ICRI 1	Selection from Chakkupalam collection, a Malabar type	325	22.9	8.7	29	38	An early maturing type globose, round, and extra bold dark green capsules; medium sized panicle with profusely flowering, early maturing type, round, and bold capsule.
6	ICRI 2	Clonal selection from germplasm collection, Mysore type	375	22.5	–	–	–	Performs well under high altitude and irrigated condition, medium long panicles, oblong bold and parrot green capsules, tolerant to azhukal disease

TABLE 3.2 (Continued)

Sl. No.	Variety	Pedigree/ Parentage	*Average Yield (kg/ha)	Dry recovery %	Oil%	1, 8 Cineol %	α-terpenyl acetate %	Salient features
7	ICRI 3	Selection from Malabar type	440	22	6.6	54	24	Early maturing long pubescent leaves, tolerant to rhizome rot disease, oblong, bold parrot green capsules. Suitable for hill zone of Karnataka
8	ICRI 4 TDK4	Clonal selection from Vadagaraparai area of lower pulleys, a Malabar type	455	-	6.4	—	—	Early maturity, medium sized panicles, Globose bold capsules. Suitable for low rainfall areas, relatively tolerant to rhizome rot and capsule borer
9	ICRI 5							
10	IISR KodaguSuvasini (CCS-1)	Selection from OP progeny of CL-37 from RRS Mudigere, Malabar type	745	22	8.7	42	37	Early maturing, suitable for high density planting, long panicle, tolerant to rhizome rot, thrips, shoot/panicle/capsule borer.

Sl. No.	Variety	Pedigree/ Parentage	*Average Yield (kg/ha)	Dry recovery %	Oil%	1, 8 Cineol %	α-terpenyl acetate %	Salient features
11	IISR Avinash (RR-1)	A selection from OP progeny of CCS-1, a Malabar type	847	20.8	6.7	30.4	35.5	Has extended flowering period, dark green capsules and retains its color even after processing. Tolerant to rhizome rot, shoot/ panicle/capsule borer. High suitable for planting in valleys and rhizome rot prone areas and intensive cultivation.
12	IISR Vijetha (NKE-12)	Clonal selection from field resistant plants for Katte, a Malabar type	643	22	7.9	45	23.4	Virus resistant selection with high percentage of bold capsules. Recommended to moderate rainfall areas with moderate high shaded and mosaic infected areas, field tolerant to thrips and borer as well as mosaics.

life span of cardamom plant is about 15–20 years. Different methods of propagation are given below.

i.   Sucker: Each sucker must have two old shoots and 1 or 2 new shoots. Planting material with two or more suckers has more survival possibilities. Use of sucker for cardamom propagation is cheap and they commence bearing about a year earlier than transplanted seedlings. It is advantageous in area where 'katte' disease is not a problem.

ii. Rhizome: Planting materials from rhizomes are collected by uprooting from 2 or 2 and half years old clumps. Such planting materials having uniform growth habit with early bearing in nature.

iii. Tissue culture: Cardamom being a cross pollinated crops, clones are ideal for generating true to type planting material from in vitro culture plantlets. High rate of multiplication coupled with additional advantage of a preferred method over conventional method. Now this method is commercially utilized for propagation of small cardamom.

iv. Seed: Propagation by seed prevents sprayed of 'katte' disease. This is a most common and widely prevalent method among planters. A large number of seedlings can be raised. The main disadvantage is that the progeny is highly variable with no uniformity in yield.

Nursery selection: Raised beds are prepared with a height of 3°C, width of 1 m of convenient length (5–6 m). A fine layer of humus rich forest soil is sprayed over the bed. Drenching of soil in the seedbed to a depth of 15 cm with formaldehyde solution (2%) is found to be effective in controlling damping off of seedlings. After this bed are covered with polythene sheets for 3 days for effective fumigation. The seeds are sown after 2 weeks

Seed selection and treatment: The seeds are extracted manually from capsule, which are fully mature, bold, disease free and high yielding mother plants of known source. The seeds should be sown immediately after harvest to obtain higher germination. Mother plants should preferably be more than 5 years old.

Seed is recalcitrant in nature. Seed treatment with GA, scarification, hot water and cow dung slurry recorded higher germination percentage. Seed coat of cardamom being hard and need to soften before sowing or seed can be washed and later soaked in water overnight. This seeds become ready for sowing next day. Seed treatment with 25% Acetic acid, 25% Nitric acid, 25% citric acid or 5% Hydrochloric acid for 10 minutes improve germination.

3.1.2.6   Sowing

Time of sowing varies according to areas. In Kerala sowing is taken up during Nov–Jan, Karnataka Aug–Sept and early part of October is found ideal. Sowing of seed is done in rows, spaced 15 m apart and 1–2 cm apart within row. The common seed rate is 2 gm/sq/m. 1–6 months aged old seedlings are transplanted in the secondary nursery bed, when they attain 4–6 leaf stage. Spacing 20–25 cm × 20–25 cm. 18–12 months old seedlings are transplanted in main field.

3.1.2.7   Shade Management

Cardamom is a pseophyte plant; light shade is more favorable at initial stage to enhance vegetative growth. Shade pattern normally influences the humidity temperature and evapotranspiration rate and than provides a congenial growth and yield. To manage the shade providing filtered sunlight to a level of 40–60% depending upon the slope, availability of irrigation facility and variety. Normally southwest slopes needs more shade than northeast slopes. Some of the suitable trees are *Acrocarpus fraxinifolius*, *Vernonia arborea*, *Cedrella toona*, *Albizia lebbek*, and *Erythrina lithospermum*. It is advisable to avoid momoculture of shade trees.

3.1.2.8   Nutrient Management

Balance application of nutrient based on soil test and plant requirement will ensure better yield. Application of organic manure like neem cake @ 2 kg/plant or FYM or compost @ 5 kg/plant once in a year during May–June is beneficial. Under rainfed condition application of NPK @ 75:75:150 kg/ha and incase of irrigated condition NPK @ 125:125:250 kg/ha is beneficial. 1/3rd dose from 3rd year and onwards.

   In case of rained fertilizer should be applied in 2 splits but 3 splits incase of irrigated condition. Application of micronutrients like Zn as zinc sulphate @ 250 gm/100 L of water as foliar spray or molybdenum as ammonium molybdate @ 0.5 kg/ha is beneficial for growth and development of the crop.

3.1.2.9   Water Management

Cardamom should be irrigated during dry season, i.e., from November–December to April. This is the period when development of young 'Tillers' and elongation of panicle take place. If the plants are exposed to water stress during this period the growth will be affected and as a result yield is reduce. During dry period application of water @ 10 L/plant/day increase the yield by 200–275%. Critical stages of irrigation are: Germination, panicle initiation, flowering, and fruiting stage.

3.1.2.10   After Care

The after care of cardamom consists of weeding, removal of old dried stem, mulching, earthing up, gap filling. The plantation requires regular weeding during 1st and 2nd year of planting and thereafter as per requirement. Under forest system mulching is the natural process. The leaf litter, twigs, removed weeds are regularly sprayed over the soil surface, which conserve and improve the fertility of soil. Earthing up is an important interculture operation incase of cardamom. The underground rhizome and roots of cardamom are earthened up by using humus rich topsoils.

3.1.2.11   Bee Keeping

Bee colonies, which are very important incase of cardamom production because they are beneficial agent for better pollination. Honeybees are the principle pollinator of cardamom to an extent of 80–90%. Among the three species *Aphis cerenea* is the major one. It has been noted that an increase of the productivity of 9% could be easily achieved through keeping beehives in the cardamom plantation.

3.1.2.12   Use of Growth Regulator

Ethrel @ 25 ppm has been found to increase sucker (tiller) production. NAA @ 400 ppm and 2,4-d @ 4 ppm increase plant height, enhance the production of panicles and flower, reduce capsule drop and increase yield. Application of 2,4-D @4 ppm and NAA@ 40 ppm is reported to be effective in enhancing essential oil content in capsule.

3.1.2.13   Plant Protection

Katte disease: it is important and devastating disease of small cardamom. It is caused by infection of cardamom mosaic virus. *Pentalonia nigronervosa* (Banana bunchy top aphid) is the vector of this viral disease. The first visible symptom appears on the youngest leaf of the affected tiller as spindle shaped slender chlorotic flecks measuring 2–5 mm in length. Later these flecks develop into pale green discontinuous stripes. The stripes run parallel to the vein from the midrib to leaf margin. All the subsequently emerging new leaves show characteristic mosaic symptoms with chlorotic and green stripes. As the leaf matures, the mosaic symptoms are more or less masked.

Disease affected clumps are to be eradicated and disease free plants to be replanted in a phased manner. Raising nurseries near disease-affected plantation is to be avoided. Collecting planting material from diseased gardens is also to be avoided.

Leaf spots: Phyllosticta elettariae are the causal organism this disease. Disease appears as small round or oval spots, which are dull, white in color. These spots later become necrotic and leave a hole (shot hole) in the center. The spots may be surrounded by water soaked area.

Application of copper fungicides is the most effective method of controlling this disease. As the disease is mostly prevalent during the rainy months, periodical sprays at fortnightly intervals during such periods would ensure considerable protection from the disease. Removal of severely infected seedlings in the nursery will help in arresting the further spread of the disease.

Capsule rot/azhukal disease: It is caused by *Phytophthora* spp. On the infected leaves, water soaked lesions appear first followed by rotting and shredding of leaves along the veins. The infected capsules become dull greenish brown and decay. This emits a foul smell and subsequently shed. Infection spreads to the panicles and tillers resulting in their decay.

Affected capsules/spikes/clumps are to be removed from the plantation before the onset of monsoon. Shade regulation is ensured by pruning branches of overhead shade trees. Foliage and spikes are to be sprayed with 1% Bordeaux mixture once during May–June and again during

August–September. The base of clump is drenched with 0.2% Copper oxychloride @ 3 L/clump.

Clump rot/ rhizome rot: It is caused by *Pythium vexans, Rhizoctonia solani,* and *Fusarium* sp. Decay of the tillers starting from the collar region and toppling of tillers. Early symptoms on leaves appear as pale yellow color, partial of leaf margins and withering. Rotting or decay starts at the collar region and it spreads to rhizomes and roots. In severe cases, the collar region breaks off and the seedling collapse is the common symptom. Nursery beds are mixed with biocontrol inoculum (*Trichoderma harzianum*) @ 50 g/m³ before sowing for management of this disease.

Cardamom vein clearing: This viral disease is caused by infection of cardamom vein clearing virus. Its characteristic symptom "hook-like tiller" it is locally called as "Kokke Kandu." The characteristic symptoms are continuous or discontinuous intraveinal clearing, stunting, resetting, loosening of leaf sheath, shredding of leaves and clear mottling on stem. Clear light green patches with three shallow grooves are seen on the immature capsules. Cracking of fruits and partial sterility of seeds are other associated symptoms.

For management of this disease use of healthy and virus-free rhizomes and seedlings for planting is the effective way. Periodic removal of senile old parts, which serve as breeding sites for aphids, is also helpful for effective management of vector of this viral disease. Application of neem products at 0.1% concentration helps in reducing aphid population. Essential oil of turmeric acts as a repellent against the aphid.

Leaf blight: It is caused by *Colletotrichum gloeosporioides*. The infection starts on the young middle aged leaves in the form of elongate or ovoid, large, brown colored patches, which soon become necrotic and dry. These necrotic dry patches are seen mostly on leaf margins and in severe cases the entire leaf area on one side of the midrib is found affected.

As a prophylactic measure, Bordeaux mixture (1%) @ 500 mL to one liter/plant should be sprayed during May–June before the onset of monsoon, which may be repeated during the months of August–September. Once leaf blight appears in the field, fungicide sprays with the combination product of Carbendazim and Mancozeb (Companion) 0.1% or Carbendazim (Bavistin) 0.2% @ 500–750 mL/plant may be adopted.

Cardamom thrips: Wilting of panicles, scabbing, and splitting of capsules. Adults and larvae lacerate and feed on plant sap from tender shoots, panicles, and capsules. For management of the pest shade is to be regulated in the plantation. Monocrotophos (0.025%) or Quinolphos (0.025%) or Phosalone (0.07%) or Fenthion (0.05%) during March, April, May, August, and September as spray may be found effective. All dry leaf sheaths are to be removed before first spray.

Shoot, panicle, capsule borer: Yellow and/or withered central leaf bore hole is found on the capsules and pseudostem with extrusion of frass. Larvae bore into panicles, capsules, and pseudostem and feed on internal tissues. Monocrotophos (0.075%) or Fenthion (0.075%) during January–February and September–October as spray can effectively manage this pest.

Root grubs: Stunted plants with yellow leaves. Larvae feed on leaves and rhizomes and retard plant growth. Collection and destruction of adult beetles during peak periods of emergence is advocated. Application of Phorate 10G @ 20–40 g/clump is found to be beneficial.

Rhizome weevil: Grubs tunnel and feed on the rhizome causing death of entire clumps of cardamom. Affected plants/seedlings are to be destructed. Drenching the base of the clump with Malathion 1.25 L or Carbaryl (50 WP) 1.25 kg in 1000 L of water/ha is also practiced.

3.1.2.14   Harvest and Post-Harvest Technology

The peak period for harvesting capsules is October–November. Just ripened or physiologically ripened capsules are generally harvested. Freshly harvest capsules are washed in water to remove the adhering soil and treated with 2% washing soda (alkali) for 10 minutes, which helps to retain the green color as well as prevents growth of mould. The average yield of cardamom is around 150 kg (dry)/ha. Capsules are cured at an optimum temperature of 50° to reduce moisture content from 80% to 12% so as to retain its green color to the maximum extent. Curing of cardamom is to be done by three ways.

a. *Natural (sun) drying*: This is cheapest method of curing, it requires 5–6 days for drying but green color is faded by this method.
b. *Electrical drier*: In this type, about fifty kg capsules can be dried within 10–12 hrs at 45–50°C temperature, which retaining the green color of the capsules.
c. *Flue pipe curing.* It is most convenient method of curing, which gives high-quality green small cardamom. The structure consists of walls made of bricks or stones and tiled roof with ceiling. An iron pipe or zinc sheet starting from the furnace passes through the chamber and

opens outside the roof. The heated air current generated in the furnace passes through the pipe, which increases the temperature of the room and fans located either side of the wall helps to spread the temperature evenly. Inside the room the cardamom to be dried is kept in wooden or aluminum trays arranged in racks. The fire in the furnace is adjusted to maintain the temperature between 45–50°C and takes about 18 to 22 hours for drying (Korikanthimath, 1993).

3.1.2.14.1 Value Addition

Essential oil: Crushed fruits of cardamom are subjected to steam distillation for commercial production of Cardamom oil. The oil content varied from 3.1–10% and the main components of the essential oil are 1,8 cineole (25–45%), α-terpinyl acetate (28–34%), linyl acetate (1–8%), sabinene (2%), limonene (2–12%), linalool (1–4%). In general, spice from recent harvest, which did not suffer excessive volatile oil loss, should be employed for in order to obtain good yield. The best oil yield is obtained with Alleppey Green cultivar. The distillation time required for extraction of essential oil is 4 hours.

Oleoresin: On solvent extraction small cardamom produce 8–9% oleoresin. It is dark green in color and pungent in nature and replacement strength is 1:20.

Decorticated seeds and seed powder: Decorticated cardamom seed generally commands a disproportionately lower price than whole cardamoms due to the fairly rapid loss of volatile oil during storage and transportation. Seed powder is also sold to some extent.

3.1.3 CHILI (CAPSICUM SP L.)

3.1.3.1 Systematic Position

Kingdom: Plantae
Division: Tracheophyta
Class: magnoliopsida
Order: solanales
Family: solanaceae

Genus: *Capsicum*
Species: *annuum*

3.1.3.2 About the Crop, National and International Scenario, Uses, and Composition

Five species are commonly recognized as domesticated within the genus *Capsicum viz., Capsicum annuum, C. baccatum, C. chinense, C. frutescens,* and *C. pubescens*, while approximately 20 wild species have been documented. The word 'chili' (so spelt in Asia, but usually spelt as 'chile' in Mexico, Central America and the Southwestern USA) represents any pungent variety of any *Capsicum* species, but primarily *C. annuum,* whereas 'pepper' is a generic term describing the fruits of any Capsicum species, both pungent and non-pungent. In Hungary, all *C. annuum* fruits are called 'paprika,' but paprika is defined in the world market as a ground, red powder derived from dried fruits with the desirable color and flavor qualities. Bird's eye chilies are small-fruited, highly pungent forms of *C. annuum* or *C. frutescens*.

In India, chilies are grown in almost all states of the country. On an average (of 10 years) the country exports around 222 thousand tonnes of chili worth of 309 million dollar. Among the spice exports, chili share is about 22%. Chili exports are growing of the rate 11% (in terms of volume) and 18.5% (in terms of value), respectively. Major importing countries are Malaysia, Sri Lanka, UAE, US, etc.

The alkaloid compounds called capsaicinoids, produced in the fruit, are the main source of pungency in peppers. Capsaicin ($C_{18}H_{27}NO_3$, trans-8-methyl-N-vanillyl-6-nonenamide) is the most abundant one, followed by dihydrocapsaicin, with minor amounts of nordihydrocapsaicin, homocapsaicin, homodihydrocapsaicin, and others. The red color of mature pepper fruits is due to several related carotenoid pigments, including capsanthin, capsorubin, cryptoxanthin, and zeaxanthin. Capsanthin and its isomer capsorubin, make up 30–60% and 6–18%, respectively, of the total carotenoids in the fruit. It is also a good source of riboflavin, niacin, magnesium, and potassium, and a very good source of dietary fiber, vitamin A, vitamin C, vitamin E (Alpha Tocopherol), vitamin K, vitamin B, iron, and manganese.

Chili oleoresin is obtained by the extraction of chilies (the fruit of red pepper, *Capsicum annuum* L. or *Capsicum frutescens* L.) with approved food grade solvent and subsequent careful removal of the solvent by distillation. Besides intense pungency due to capsaicin and small quantities of allied alkaloids, the chili oleoresin is dark red in color due to carotenoid pigments. Color and pungency are important from the point of view of quality of the product. Color value of chili oleoresin ranges from 4000 to 20,000 and pungency is 2,40,000 Schoville heat unit. Capsaicin content of chili oleoresin is generally 1.5%.

It is mainly used for its pungency and pleasant flavor. It is essentially used in vegetable, meat, fish, and egg preparation. It is also used for preparation of pickles, sauce, etc. It is one of the major ingredients of curry powder and different spice mixture. Capsaicin has antibacterial, anticarcinogenic, analgesic, and antidiabetic properties. It also found to reduce low density lipid, cholesterol levels in obese individuals. Antioxidant substances in chili helps to protect the body from injurious effects of free radicals generated during stress, diseases conditions.

3.1.3.3   Soil and Climatic Requirements

Chili can be grown in wide range of soil starting from lateritic loam to black soil. However, it can be grown best in sandy loam soil with moderate pH. Friable and loose soil with good drainage is ideal as oxygen demand in root is more. Water stagnation for short duration also results in yellowing and shedding of leaves. Blossom end rot is a common phenomenon in heavy soil. Deep, fertile, sandy loam to clayey loam soil is ideal for rainfed crop while light sandy loam, alluvial loam or red loamy soil is good for irrigated crop.

It is basically warm loving plant, although it is grown in wide range of climates. It can be grown upto 1,500 m from mean sea level in tropical and subtropical region. Warm areas experiencing well distributed 60 to 150 cm rainfall is suitable for chili cultivation. Erratic and heavy rainfall is very detrimental for its growth and fruiting. Plant can withstand high temperature during growing period but fruit setting is hampered due to high temperature and heat wave. 21°C temperature is ideal for vegetative growth of the plant and 26.7°C is best for good fruiting. Desiccation of flower and fruit drop is common during heat wave.

The improved varieties of chili with their maturity (days), average yield (q/ha) and salient features have been discussed in Table 3.3.

The improved varieties of chili developed by the private companies) with source, maturity (days) and salient features have been summarized below (Table 3.4):

TABLE 3.3 Improved Varieties of Chili

Sl. No.	Variety	Maturity	Average Yield (q/ha)	Silent features
1	Pusa Jwala	130–150	Fresh: 75–80, Dry: 7–8	Dwarf variety, suitable for both the seasons in plains and March–April in the hills
2	Pusa Sadabahar	150–170	Fresh: 110–125, Dry: 15–20	Highly pungent variety, suitable for both the seasons
3	Bhagyalakshmi (G-4)	150–170	Fresh: 110–130, Dry: 10–13	Suitable for sowing both the season and fairly tolerant to diseases and insects
4	Sindhur (CA-960)	130–180	Fresh: 120–140, Dry: 11–12	Mildly pungent variety suitable for growing both the seasons
5	Pant C-1	90–130	Fresh: 100–120, Dry: 9–11	Erect bearing habit, highly pungent, suitable for both the seasons and tolerant to leaf curl virus
6	Punjab Lal	120–180	Fresh: 100–120, Dry: 9–10	Dwarf variety, erect fruit bearing, rich in capsaicin content suitable for both the seasons
7	Kashi Anmol	50–55	250 (Green fruits)	Determinate variety, pungent, capsaicin (0.5%), suitable for long distance transport.
8	Kashi Vishwanath	220 (red ripe fruits)	200	Suitable for dry fruit production
9	Kashi Early	240	300	For both green and dry fruits

TABLE 3.3 (Continued)

Sl. No.	Variety	Maturity	Average Yield (q/ha)	Silent features
10	Arka Lohit	180	Green: 25 t/ha, dry: 37 t/ha	Highly pungent, 0.205% capsanthin, suitable for irrigated and rainfed cultivation
11	Arka Suphal (PMR 57/88k)	180	Green: 25 t/ha, Dry: 3 t/ha	Suitable for irrigated and rain fed cultivation, tolerant to powdery mildew
12	Arka Meghana (F1 Hybrid)	120–140	Green: 33.5 t/ha, Dry: 5 t/ha	Early and spreading type variety, tolerant to powdery mildew and viruses
13	Arka Harita (F1 hybrid)	150–160	Green: 38.2 t/ha, Dry: 5.86 t/ha	Tall, spreading variety, resistant to powdery mildew and viruses
14	Arka Abhir (Paprika)	160–180	Dry: 2 t/ha	It has low pungency (0.05%), suitable for oleoresin extraction
15	K1	215	19 (dry chili)	Shiny red fruits with capsaicin content of 0.35 mg/g
16	K2	210	15 (dry chili)	Bright red fruits with capsaicin content of 0.49 mg/g
17	Andhra Jyoti (G-5)		11–12 (rainfed crop), 30–32 (irrigated crop)	High yielding variety, bright red fruits with capsaicin content of 0.65 mg/g

Sl. No.	Variety	Maturity	Average Yield (q/ha)	Silent features
18	CO 1	210	21 (dry chili)	Bright red and shiny at ripening with capsaicin content of 0.72 mg/g
19	CO 2		33.5 (green chili), 8.7 (dry chili)	Suited for both green dry purpose, thick red with more seeded variety
20	NP 46 (G-3)		Dry chili, 10–12 (rainfed crop), 23–30 (irrigated crop)	Attractive red on ripening with capsaicin content of 0.53 mg/g
21	MDU-1	210	18 (dry chili)	Long shiny fruits with capsaicin content of 0.70 mg/g, tolerant to fruit rot and die-back disease and also thrips
22	Kiran (X200)		Dry chili, 12–14 (rainfed crop), 30–35 (irrigated crop)	Fairly tolerant to thrips, mites, and aphids
23	Aparna (CA 1068)		Dry chili, 12–14 (rainfed crop), 30–35 (irrigated crop)	Variety can withstand with drought as well as high moisture conditions and some extent to salinity
24	DH-76.6	180	20	Bushy growth habit, highly branched

TABLE 3.4 Improved Varieties of Chili (Developed by Private Companies)

Sl. No.	Variety	Source	Maturity (Days)	Salient features
1	Pragati	Namadhari Seeds Pvt. Ltd., Bangalore	70 (green fruits)	Tall, early variety, tolerant to virus complex, suitable for fresh chili
2	Goli	Namadhari Seeds Pvt. Ltd., Bangalore	80 (green fruits)	Medium pungent, suitable for dry chili
3	NS 1701	Namadhari Seeds Pvt. Ltd., Bangalore	75 (green fruits), 85 (Red fruits)	Very highly pungent, highly tolerant to thrips, virus complex and high temperature, suitable for both green and dry purpose
4	NS 1101	Namadhari Seeds Pvt. Ltd., Bangalore	70 (green fruits), 80 (Red fruits)	Dual-purpose variety, very high pungency, tolerant to high temperature, thrips, and virus complex.
5	NS 7510	Namadhari Seeds Pvt. Ltd., Bangalore	75 (green fruits), 85 (Red fruits)	Early, tall spreading hybrid, high pungency and suitable for red fresh produce
6	Akash	Namadhari Seeds Pvt. Ltd., Bangalore	75 (green fruits), 85 (Red fruits)	It is very highly pungent and tolerant to virus complex
7	Indam-5	Indo American Hybrid Seed, Bangalore	–	Suitable for dry and green chili, medium pungent fruits, tolerant to thrips and virus complex
8	Indam-6	Indo American Hybrid Seed, Bangalore	–	Early variety with highly pungent fruits
9	Agnirekha	Syngeneta Seeds, Pune	–	Suitable for green as well as dry chilies and starts bearing 50–55 days after transplanting
10	Delhi Hot	Seminis Seeds, Aurangabad	–	Strong pungency
11	Golden Hot	Seminis Seeds, Aurangabad	–	Mild pungency suitable for fresh market

Sl. No.	Variety	Source	Maturity (Days)	Salient features
12	Wonder Hot	Seminis Seeds, Aurangabad	-	Dual purpose variety having moderate pungency
13	Super Hot	Seminis Seeds, Aurangabad	-	Strong pungency
14	Tejaswani	Seminis Seeds, Aurangabad	-	Dual purpose variety with high pungency
15	Ujala	Nunhems Proagro Seeds, Bangalore	-	Light green fruits with high pungency, suitable for fresh and dry chili
15	Devanus Deluxe	Nunhems Proagro Seeds, Bangalore	-	Well suited for rainfed conditions, dual purpose variety having glossy green, wrinkled, and pungent fruits
17	NS 211	Namadhari Seeds Pvt. Ltd., Bangalore	-	Fruits are dark green turning to deep red, highly pungent, tolerant to virus, high yielding variety
18	Kohinoor 76	Royal Sluis Ltd.	-	Dark green, thick wall, blocky fruits with 3–4 lobes, having 200–220 g weight per fruit
19	NS 631	Namadhari Seeds Pvt. Ltd., Bangalore	-	Dark green, thick wall, blocky fruits with 3–4 lobes, having 220–250 g weight per fruit
20	NS 632	Namadhari Seeds Pvt. Ltd., Bangalore	-	Dark green, thick wall, blocky fruits with 3–4 lobes, having 220–250 g weight per fruit

TABLE 3.4 (Continued)

Sl. No.	Variety	Source	Maturity (Days)	Salient features
21	SARPAN F1. hyb. Chili SH-26	Sarpan Seeds Co. Pvt. Ltd., Dharwad	-	Extra early, high yielding, dual purpose variety, suitable for grown in all seasons, fruiting starts 45–55 days after transplanting, light parrot green color, glossy, pungent, thick skinned, smooth fruits, highly tolerant to leaf curl, excellent keeping quality, very good for long distance shipment
22	SARPAN F1. hyb. Chili SH-48	Sarpan Seeds Co. Pvt. Ltd., Dharwad	250–280	Thin long, glossy, parrot green, partially wrinkled, pungent, early fruiting (50–60 days), highly tolerant to leaf curl disease complex, good keeping quality and high yielder, best suited for early kharif, rabi, and early summer
23	SARPAN F1. hyb. Chili SH-60	Sarpan Seeds Co. Pvt. Ltd., Dharwad	250–280	Fruits thin long, glossy, medium pungent, very high yielder, dual purpose variety, highly tolerant to leaf curl disease, suitable for growing in early kharif, rabi, and early summer
24	SARPAN F1. hyb. Chili JH-20 (Jwala F1)	Sarpan Seeds Co. Pvt. Ltd., Dharwad	-	Thin long, wrinkled, glossy, light green color fruits, early fruiting, suitable for all season, high tolerance to leaf curl and high heat.
25	SARPAN F1. hyb. Paparika SPH-9	Sarpan Seeds Co. Pvt. Ltd., Dharwad	-	Mild paprika F1 hybrid with excellent color (250–300 + ASTA), acidic flavor with negligible pungency, suitable for oleoresin extraction

Sl. No.	Variety	Source	Maturity (Days)	Salient features
26	SARPAN F1. hyb. Chili SH-72	Sarpan Seeds Co. Pvt. Ltd., Dharwad	-	Mild paprika F1 hybrid, long parrot green, juicy, low in pungency, early fruiting, high yielder, tolerant to leaf curl disease. Best in early kharif, early rabi and early summer, suitable for fresh green and for pickling
27	SARPAN F1. hyb. Chili SH-12	Sarpan Seeds Co. Pvt. Ltd., Dharwad	-	Hot red dry spicy hybrid, medium pungent, wrinkled fruits with excellent flavor and spice quality and keeping quality, high yielding, highly tolerant to leaf curl complex
28	SARPAN F1. hyb. Chili SH-82 (Paprika)	Sarpan Seeds Co. Pvt. Ltd., Dharwad	-	Early bearing (50–60 days), excellent yielder with high quality, dry fruits, high oil with 250 ASTA color values mild pungency, highly tolerant to leaf curl disease
29	SARPAN F1. hyb. Chili SH-92 (Paprika)	Sarpan Seeds Co. Pvt. Ltd., Dharwad	-	Good for spice powder and color oleoresin with high ASTA color units 250–300+, also suitable for fresh market as well as red chili, highly tolerant to leaf curl
30	SARPAN F1. hyb. Chili NG-530	Sarpan Seeds Co. Pvt. Ltd., Dharwad	-	Best in all seasons, suitable for fresh green and dry red fruits, highly tolerant to leaf curl

3.1.3.4 Agrotechniques for Quality Production

Transplanting of seedlings is the better method for production of chili. Nursery raising of seedlings is done in raised beds situated in high land with light sandy loam soil. The area is ploughed for 3–4 times to get fine tilth and leveled after removal of weeds and clods. Beds of 1.2 m width and convenient length raised to 15 cm are prepared for sowing seeds. 2 kg seeds are sown in 250 square meter nursery bed to obtain seedlings for 1 ha area. For main crop seeds are sown during the month of May-June and transplanted during middle of June or July. Seed sowing is done during August for autumn- winter crop and end of October to November for spring summer crop. March–April is ideal time for seed sowing of chili in hills. Seeds are treated with fungicides (Agrason-GN or Copper oxychloride solution) before sowing. Watering is done just after sowing of seeds and seedbeds are often covered with straw or other mulches for better and quick germination, which is also removed just after completion of germination. Seedlings become ready within 30 to 40 days after sowing of seeds to transplant in the main field. Chili can also be grown by direct seeding method, which is popular as 'Chutki' method, which is generally done during March–April.

Seedlings are transplanted in the main field after 4–5 ploughing and harrowing. 20 tonnes of well rotten farmyard manure is applied during the field preparation. Seedlings are spaced at 45 to 60 cm between lines and 30 to 45 cm within lines. For better establishment of seedlings in the main field irrigation in nursery bed should be stopped 10–12 days before transplanting and only sought, straight seedlings should be transplanted during the cloudy evening. A light irrigation is necessary after transplanting. Bhagyalakshmi, Pusa Jwala, Pusa Sadabahar, Utkal abha, Pant C-1, K-1, K-2, Andhra Joyti Punjab Lal, Kiran, Aprna, MDU-1, Kashi Vishwanath, Kasi earlyArka Lohit, Arka Suphal, Arka Meghna, Arka Abir, and Arka Harita are the important improved chili Varieties in India.

Chili requires a high amount of manures and fertilizers since it is long growing season crop. Among the different major nutrient, chili responds well to nitrogen and potash rather than phosphorus. But respond depends upon the inherent fertility status of the soil. Apply higher amount of nutrient under irrigated condition than the rainfed condition. Response of chili crop to fertilizers depends upon the variety, type of soil, fertility status of the soil and soil moisture. Normally 15–20 tonnes of well decomposed farmyard manure or compost applied at the time of final land preparation. For irrigated transplanted crop apply 100 kg N, 50 kg P & 50 kg K should be applied per hectare. Half of the nitrogenous fertilizers are applied as basal during land preparation along with full phosphatic and potasic fertilizers. Remaining nitrogenous fertilizers are applied in two equal splits at 1 month and 2 months after transplanting.

Chili cannot tolerate moisture stress as well as excess moisture in the soil. Thus, judicious irrigation is necessary depending upon the condition of soil. Ten leaf stage, flowering, and after each periodical harvest are critical stage of irrigation for chili crop. Irrigation should be given in these stages if soil moisture is deficit. Generally fortnight irrigation during winter and weekly irrigation during summer is generally recommended. Additionally to keep the crop free from weed, two hands weeding and three hoeing is necessary for keep the field free from weed and loosening of the soil, respectively.

3.1.3.5   Plant Protection

Foot rot and die back: it is caused by *Colletotrichum capsici.* As the fungus causes necrosis of tender twigs from the tip backwards the disease is called die-back. Flowers drop and dry up. There is profuse shedding of flowers. This drying up spreads from the flower stalks to the stem and subsequently causes die-back of the branches and stem and the branches wither. On the surface of the soil the necrotic areas are found separated from the healthy area by a dark brown to black band.

This disease can be managed by seed treatment with Thiram or Captan 4 g/kg is found to be effective in eliminating the seed-borne inoculums. Good control of the disease has been reported by three sprayings with Ziram 0.25%, Captan 0.2%, or miltox 0.2%. Chemicals like wettable sulphur 0.2%, copper oxychloride 0.25% and Zineb 0.15% not only reduced the disease incidence but also increased the fruit yield.

Leaf curl: This viral disease is caused by Leaf curl virus. Leaves curl towards midrib and become deformed. Stunted plant growth due to short-ened internodes and leaves greatly reduced in size. Flower buds abscise before attaining full size and anthers do not contain pollen grains. The virus is generally transmitted by whitefly. So control measures of whitefly in this regard would be helpful.

Bacterial leaf spot: This bacterial disease is caused by *Xanthomonas campestris pv. Vesicatoria.* The leaves exhibit small circular or irregular dark brown or black greasy spots. As the spots enlarge in size, the center becomes lighter surrounded by a dark band of tissue. The spot coalesce to form irregular lesions. Severely affected leaves become chlorotic and fall off. Petioles and stems are also affected. Seed treatment with 0.1% mercuric chloride solution for 2 to 5 minutes is effective. Seedlings may be sprayed with Bordeaux mixture 1% or copper oxychloride 0.25% for management of this disease.

Fusarium wilt: This fungal disease is caused by *Fusarium oxysporum.* Fusarium wilt is characterized by wilting of the plant and upward and inward rolling of the leaves. The leaves turn yellow and die. By the time above-ground symptoms are evident; the vascular system of the plant is discolored, particularly in the lower stem and roots. Use of wilt resistant varieties is necessary for controlling this disease. Drenching with 1% Bordeaux mixture or Blue copper @ 0.25% may give protection. Seed treatment with 4 g *Trichoderma viride* formulation or Carbendazim 2 g/kg seed is effective.

Cercospora leaf spot: It is caused by *Cercospora capsici.* Leaf lesions are typically brown and circular with small to large light grey centers and dark brown margins. The lesions may enlarge to 1cm or more in diameter and sometimes coalesce. Stem, petiole, and pod lesions also have light grey centers with dark borders, but they are typically elliptical. Severely infected leaves drop off prematurely resulting in reduced yield. Spraying twice at 10–15 days interval with Mancozeb 0.25% or Chlorothalonil (Kavach) 0.1% is an effective management of this disease.

Damping off: It is very much common in nursery and caused by the fungi *Pythium aphanidermatum.* Seedlings are killed before emergence. Water soaking and shriveling of stem is also occurred. Soil drenching with Copper oxychloride 0.25% is effective for controlling this disease.

Thrips (Scirtothrips dorsalis): The infested leaves develop crinkles and curl upwards and petiole elongated. Buds become brittle and drop down. Early stage, infestation leads to stunted growth and flower production, fruit set are arrested. The pest can be effectively controlled by spraying with 0.05% Endosulfan, 0.03% Dimethoate, 0.02% Phosphamidon or 0.05% Monocrotophos. Seed treatment with Imidacloprid 70% WS @ 12 g/kg of seed is also effective.

Mites (Polyphagotarsonemus latus): Downward curling and crinkling of leaves, leaves with elongated petiole and stunted growth are the common symptom of mite infestation. The activity of predatory mite *Amblyseius ovalis* is encouraged for effective management of the mite. Spraying 0.2% Sulphur or 0.03% Dicofol and Sulphur dusting @ 20–25 kg/ha also gives satisfactory control of the pest.

Aphid (*Myzus persicae*): The infested plants turn pale with sticky appearance. The leaves are curled and crinkled. Honeydew excretes – development of sooty mould symptom develops and stunted growth of the plant. Seeds are treated with Imidacloprid 70% WS @12 g/kg of seed. Spraying with 0.05% Endosulfan, 0.02% Phosphamidon, 0.03% Dimethoate, Methyl demeton or Thiometon control the pest effectively.

Tobacco cut worm (*Spodoptera litura*): Newly hatched larvae scrap the green matter in the leaf. Affected leaf looks like a papery white structure. Later instars larvae feed by making small holes. In severe infestations they feed voraciously on the entire lamina and petiole. The soil is ploughed to expose and kill pupae. Planting castor as a trap crop also reduces infestation. Pheromone traps @15/ha is recommended to be set up. The egg masses, gregarious larvae and grown up caterpillars are to be collected and destroyed.

3.1.3.6   Harvest and Post Harvest Technology

Typically fruits are harvested when they are partially dry on the plant and kept in well-ventilated areas receiving direct sunlight for drying. Sun drying can result in bleached fruits, especially if rainfall is received during the drying period and the fruits may have extraneous matter adhering to them.

Ripe fruits are plucked and harvested indoor by heaping for 3–4 days for uniform ripening as well as color development. White patches are developed after drying if this technique is not followed. For sun drying, chilies are spread on ground preferably on concrete floor and dried till the moisture level come down from 70–80% to 10%. This is especially essential for good storage life and retain red color and luster. The duration of sundrying may vary from 3 to 15 days. Sometimes unfavorable conditions like cloudy weather bring about discoloration. During this process, there may be a seed loss of 10 kg/ton of dry chilies. CFTRI suggested a process of improved sundrying. This process uses a specific emulsion known as "DIPSOL," which reduces the drying time to one week. It takes 4–5 days for complete drying.

The use of controlled drying improves the quality of the dried fruits. The best drying temperature is 60–70°C; this gives maximum color values and longest color retention time. The optimum moisture content of fruit is approximately 10%.

Artificial drying of chili gives a quality consistent product. For this type of drying chili is first washed and then dried in trays as whole pods or sliced in 2.5 cm/length. Turnel driers and stainless steel continuous belt driers are commonly used for exposing the fruits to forced current of air at 50–60°C (moisture content reduced to 7–8%). In case of *C. annuum* species it has been observed that there were to losses in color and pungency at 60–75°C even for >2 hrs. Optimum quality product was produced at 65°C where whole fruits were dried at 8% moisture content in 12 hrs and in 6 hrs when the chilies were subjected to slicing/slitting. Since slitting/slicing improved the rate of drying, drying time is reduced and color and luster is enhanced.

3.1.3.6.1 Value Addition

Dehydrated chilies: Dehydrated chilies contain 6% stalk 40% pericarp 54% seeds.

Red Chilies: Good quality dry chilies are generally characterized by bright red medium sized moderately thin pericarp, smooth glossy surface and firm stalk.

Oleoresin

- *Oleoresin capsicum (Bird chilly)*. Oleoresin is prepared form pungent small-fruited bird chilies. Very pungent oleoresin is usually evaluated on its capsicum contents (between 5 lakhs and 18 lakhs scoville units). Color is expressed on ASTA scale of 350 units.
- *Oleoresin red pepper: (Capsicum annuum)*. Obtained from larger, moderately pungent capsicums. Pungency rating is b/n 80,000–5 lac scoville units. Color is expressed on ASTA scale of 20,000 units max.
- *Oleoresin (paprika)*. It is obtained from sweet peppers of paprika. It has intense color but no pungency. Commercial paprika oleoresins are available in color strength ranging from 12,000–1 lac units. 1 kg paprika oleoresin replaces 12–15 kg of paprika powder in color intensity.
- Chili oleoresin. It is extracted by solvent extraction for 5 hrs at 50°C for using ethyl acetate and acetone or ethyl alcohol.

3.2 RHIZOMATOUS SPICES

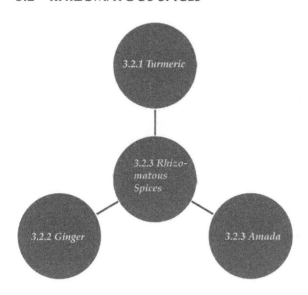

3.2.1 TURMERIC (CURCUMA LONGA L.)

3.2.1.1 Systematic Position

Kingdom: Plantae
Division: Magnoliophyta
Class: Liliopsida
Subclass: Zingiberidae
Order: Zingiberales
Family: Zingiberaceae
Genus: *Curcuma*
Species: *longa*

3.2.1.2 About the Crop, National and International Scenario, Uses, and Composition

Turmeric is an ancient and sacred spice of India. It is the dried rhizome of a herbaceous perennial tropical plant. The primary rhizomes, which are some-what round in shape are called as bulbs while the thin long secondary rhizomes are fingers. It is used as a condiment, dye, drug, and cosmetic in addition to its use in religious ceremonies. India is the largest producer, consumer, and exporter of turmeric across the world. It has an envious monopoly over world turmeric trade. Other producing regions are Pakistan and Bangladesh. On an average India exports around 61,000 tonnes of turmeric annually. The export is growing at a rate of 7.6% (in terms of volume) and 17.6% (in terms of value) respectively. India mainly exports turmeric in the form of fingers and bulbs. It also exports turmeric oleoresin and powder but the quantity is less. USA, UAE, Iran, and Malaysia are the major importers of Indian turmeric.

The curcuminoids, which include curcumin (diferuloylmethane), deme-thoxycurcumin, and bis-demethoxy curcumin, are most important chemical components of turmeric. However, the major compound is curcumin, which generally varies from 3–9%. Curcuminoids consist of curcumin demethoxyc-urcumin, 5'-methoxycurcumin, and dihydrocurcumin, which are found to be natural antioxidants. Volatile oils include d-α-phellandrene, d-sabinene, cinol, borneol, zingiberene, and sesquiterpenes. The components responsi-ble for the aroma of turmeric are turmerone, arturmerone, and zingiberene. The rhizomes are also reported to contain four new polysaccharides-ukonans along with stigmasterole, β-sitosterole, cholesterol, and 2-hydroxymethyl anthraquinone. Nutritional analysis showed that 100 g of turmeric contains 390 kcal, 10 g total fat, 3 g saturated fat, 0 mg cholesterol, 0.2 g calcium, 0.26 g phosphorous, 10 mg sodium, 2500 mg potassium, 47.5 mg iron, 0.9 mg thiamine, 0.19 mg riboflavin, 4.8 mg niacin, 50 mg ascorbic acid, 69.9 g total carbohydrates, 21 g dietary fiber, 3 g sugars, and 8 g protein. Turmeric is also a good source of the ω-3 fatty acid and α-linolenic acid (2.5%).

3.2.1.3   Soil and Climatic Requirements

Turmeric can be grown in diverse tropical conditions from sea level to 1500 m above MSL. It requires temperature ranging from 20–30°C with annual rainfall of 1500 mm or more. It can be grown on various types of soils from light black, ashy loam, clayey loam and red soils. However, it grows well in a well-drained sandy or clay loam soils.

3.2.1.4   Varieties

The improved varieties of turmeric with parentage, average yield (kg/ ha), dry recovery (%), curcumin (%), oleoresin (%), essential oil (%), crop duration (days) and the salient features have been highlighted in Table 3.5.

3.2.1.5 Agrotechniques for Quality Production

The land is prepared with the onset of early monsoon in the month of May. The soil is brought to fine tilth by giving about four deep plough-ing and weeds, stubbles, roots, etc. are removed. Beds of size 1–1.5 m width, 15 cm height and of convenient length are prepared with spacing of 50 cm between beds. Small pits are made with hand hoe in the beds in rows with spacing of 25 × 30 cm and covered with soil or dry powdered cattle manure. Generally ridges and furrows method is used for planting by keeping optimum spacing of 45–60 cm between rows and 25 cm between plants. Turmeric can also be planted during April- May with the receipt of pre monsoon showers. Whole or split mother rhizomes are used for plant-ing. A seed rate of 2500 kg of rhizomes is required for planting in one hectare.

Being an exhaustive crop, turmeric requires heavy manuring for higher yield. Cattle manure or compost at the rate of 40 tonnes per hectare is applied by broadcasting and ploughing at the time of preparation of land or as basal dressing by spreading over the beds to cover the seed pits after planting. Recommended fertilizers dose for irrigated crop is 125 kg N, 80 kg P_2O_5 and 120 kg K_2O per hectare. The beds are earthed up after each top dressing with fertilizers.

The crop is to be mulched with green leaves at the rate of 12–15 tonnes per hectare immediately after planting. It may be repeated again after 50 days with the same quantity of green leaves after weeding and application of fertilizers. Weeding may be done thrice at 60, 120 and 150 days after planting depending upon weed intensity. It could be grown as mixed crop with chilies, colocasia, onion, brinjal, and cereals like maize, ragi, etc. Irrigation should be depends on the weather and the soil conditions so, in case of irrigated crop about 15–20 irrigations are to be given in clayey soils and 40 irrigations in sandy loams. Yield losses of turmeric is mainly affected due to pests like shoot borer, rhizome scales and diseases like leaf blotch, leaf spot and rhizome rot, etc. Necessary control measures are to be adopted.

TABLE 3.5 Improved Varieties of Turmeric

Sl No.	Variety	Pedigree/ Parentage	*Average Yield (kg/ha)	Dry recovery%	Curcumin %	Oleoresin %	Essential oil%	Crop duration (days)	Salient features
1	CO.1	Vegetative mutant by x-ray irradiation of Erode local	30.5	19.5	3.2	6.7	3.7	270	Bold and orange yellow rhizomes, suitable for drought prone areas, water logged, hilly areas saline and alkaline areas
2	BSR.1	Clonal selection from Erode local irradiated with × rays	30.7	20.5	4.2	4	3.7	285	Bright yellow rhizome suitable for problem soils and drought prone areas of Tamil Nadu
3	BSR.2	Induced mutant from Erode local	32.7	–	–	–	–	245	A high yielding short duration variety with bigger rhizomes, resistant to scale insects
4	Krishna	Clonal selection from Tekurpeta collection	9.2	16.4	2.8	3.8	2	240	Plumpy rhizomes, moderately resistant to pests and diseases

Sl No.	Variety	Pedigree/ Parentage	*Average Yield (kg/ ha)	Dry recovery%	Curcumin %	Oleoresin %	Essential oil%	Crop duration (days)	Salient features
5	Sugandham	Clonal selection from germplasm	15.0	23.3	3.1	11	2.7	210	Thick, round rhizomes with short internodes. Moderately tolerant to pest and diseases
6	Roma	Clonal selection from T. Sunder	20.7	31	6.1	13.2	4.2	250	Suitable for both rainfed and irrigated condition. Ideal for hilly areas and late sown season.
7	Suroma	Clonal selection from T. Sunder by x-ray irradiation	20.0	26	6.1	13.1	4.4	253	Round and plumpy rhizome, field tolerance to leaf blotch, leaf spot and rhizome scales.
8	Ranga	Clonal selection fromRajpuri local	29.0	24.8	6.3	13.5	4.4	250	Bold and spindle shaped mother rhizome, suitable for late sown condition and low-lying areas. Moderately resistant to leaf blotch and scales

TABLE 3.5 (Continued)

Sl No.	Variety	Pedigree/ Parentage	*Average Yield (kg/ ha)	Dry recovery%	Curcumin %	Oleoresin %	Essential oil%	Crop duration (days)	Salient features
9	Rasmi	Clonal selection from Rajpuri local	32.0	23	6.4	13.4	4.4	240	Bold rhizomes, suitable for both rainfed and irrigated condition, early, and late sown season
10	Rajendra Sonia	Selection from local germplasm	42.0	18	8.4		5	225	Bold and plumpy rhizome
11	Megha turmeric 1	Selection form Lakadong type	23.0	16.37	6.8	–	–	300–315	High curcumin content and bold rhizomes, suitable for Northeast hill and Northwest Bengal.
12	Pant Peetabh	Clonal selection from local type	29.0	18.5	7.5	–	1	–	Resistant to rhizome rot
13	Suranjana (TCP-2)	Clonal selection from local types of west Bengal suitable for open and shaded condition	29.0	21.2	5.7	10.9	4.1	235	Tolerant to rhizome rot and leaf blotch; resistant to rhizome scales and moderately resistant to shoot borer.

Sl No.	Variety	Pedigree/Parentage	*Average Yield (kg/ha)	Dry recovery%	Curcumin %	Oleoresin %	Essential oil%	Crop duration (days)	Salient features
14	Suvarna	Selection from germplasm, collected from Assam	17.4	20	4.3	13.5	7	200	Bright orange colored rhizome with slender fingers, field tolerant to pest and diseases.
15	Suguna	Selection from germplasm, collected from AP	29.3	20.4	4.9	13.5	6	190	Early maturing, field tolerant to rhizome rot.
16	Sudarsana	Selection from germplasm, collected from Singhat, Manipur	28.8	20.6	5.3	15	7	190	Early maturing, field tolerant to rhizome rot.
17	IISR Prabha	Open pollinated progeny selection	37.0	19.5	6.5	15	6.5	205	High yielding variety
18	IISR Prathibha	Open pollinated progeny selection	39.1	18.5	6.2	16.2	6.2	225	High yielding variety

TABLE 3.5 (Continued)

Sl No.	Variety	Pedigree/ Parentage	*Average Yield (kg/ha)	Dry recovery%	Curcumin %	Oleoresin %	Essential oil%	Crop duration (days)	Salient features
19	IISR Kedaram	Clonal selection from germplasm	34.5	18.9	5.5	13.6	–	210	Resistant to leaf blotch.
20	IISR Alleppey Supreme	Selection from Alleppey Finger turmeric	35.4	19	5.55	16	–	210	Tolerant to leaf blotch.
21	Kanthi	Clonal selection from Mydukur variety of Andhra Pradesh	37.65	20.15	7.18	8.25	5.15	240–270	Erect leaf with broad lamina, big mother rhizomes with medium bold fingers and closer internodes
22	Sobha	Clonal selection from local type	35.88	19.38	7.39	9.65	4.24	240–270	High yielding variety with high curcumin content (7.39%), Erect leaves with narrow lamina. Mother rhizome big with medium bold figures and closer internodes. Inner core of rhizomes is dark orange like Alleppey. More territory rhizomes.

Sl No.	Variety	Pedigree/ Parentage	*Average Yield (kg/ ha)	Dry recovery%	Curcumin %	Oleoresin %	Essential oil%	Crop duration (days)	Salient features
23	Sona	Clonal selection from local germplasm	21.29	18.88	7.12	10.25	4.4	240–270	Orange yellow rhizome, medium bold with low territory fingers. Best suited for central zone of Kerala. Rhizome medium bold. Field tolerant to leaf blotch.
24	Varna	Clonal selection from local germplasm	21.89	19.05	7.87	10.8	4.56	240–270	Bright orange yellow rhizome, medium bold with closer internodes, territory fingers present-suited to central zone of Kerala. Field tolerant to leaf blotch

* Yield tonnes/ha (fresh).

3.2.1.6   Harvest and Post Harvest Technology

The crops become ready for harvest in seven to nine months depending upon the variety. It may extend from January–March. The land is ploughed and the rhizomes are judiciously picked up with a spade. Harvested rhizomes are cleaned of mud and other extraneous matter adhering to them. The average yield comes to 20–25 tonnes of green turmeric per hectare.

After harvesting, turmeric should be processed to improve the quality and storage life of rhizomes. The processing of turmeric consists of four stages.

a) *Curing:* Fingers are separated from mother rhizomes, which are usually kept as seed materials. The freshly harvested turmeric is cured for obtaining dry turmeric. It involves boiling of fresh rhizomes in water and then drying in the sun. Curing is done by two methods.

1. *Traditional method of curing*: In this method, cleaned rhizomes are boiled in copper or galvanized iron or earthern vessels with water just enough to soak them. When froth comes out and white fumes appear jigging out a typical odor, boiling is to be stopped. The whole process of boiling lasts for about 45–60 minutes when the rhizomes are soft. The stage at, which boiling is stopped largely influences the color and the aroma of the final product. Over cooking spoils the color of the final product while under-cooking renders the dried product brittle.

2. *Improved scientific method* (TNAU):
   - Fingers and rhizomes are boiled separately in pans
   - Cleaned rhizome are (50 kg) taken in perforated trough (75 kg) of size with handles for immersing in pan.
   - These troughs are immersed in clean water present in outer tank.
   - The alkaline solution (0.1% of Na_2CO_3/$NaHCO_3$) is added to H_2O.
   - The whole system is boiled till fingers become soft.
   - With the help of needle the extent of cooking to be done can be determined.
   - Cooked rhizomes are taken out of pan by lifting trough and draining.
   - Alkalinity of boiling water helps in implanting tinge to core of turmeric.
   - Salient features: 200 kg/batch Rs. 6000/batch expenditure.

- 25–30 kg Agricultural waste for fuel.
- 2 labors are required.
- Proper utilization of water.

b) *Drying:* After cooking of fingers they are subjected to drying in the sun by spreading in 5–7 cm thick layers on bamboo mats or drying floor. A thinner layer is not desirable as the color of the dried product may be adversely affected. During nighttime, the material should be heaped or covered. It may take 10–15 days for the rhizomes to become completely dry. Artificial drying using cross flow hot air at a maximum temperature of 60°C is also found to give a satisfactory product. The yield of dry turmeric varies from 20–30% depending upon the variety and the location where the crop is grown.

c) *Polishing:* Turmeric, often after drying, has a poor appearance and rough dull color outside the surface with scales and root bits. This appearance can be improved by smoothening and polishing the outer surface through manual or mechanical rubbing.

Manual polishing consists of rubbing the dried turmeric fingers on a hard surface or trampling them under feet wrapped with gunny bags. The improved method is by using hand-operated barrel or drum mounted on a central axis, the sides of which are made of expanded metal mesh. When the drum is filled with turmeric is rotated, polishing is effected by abrasion of the surface against the mesh as well as by mutual rubbing against each other is as they roll inside the drum. The turmeric is also polished in power-operated drums. The yield of polished turmeric from the raw material varies 12–25%.

d) *Coloring:* Attractive color of the dried turmeric always lures the buyers. In order to impart this attractive yellow color, turmeric suspension in water is added to the polishing drum in the last 10 minutes. When the rhizomes are uniformly coated with suspension, they may be dried in the sun. After polishing, it should be kept in clean sacks and stored over wooden pallets in stores. Stores should be clean, free from infestation of pests, spiders, and harborage of rodents. Pesticides should not be applied on the dried/polished turmeric to prevent storage pests.

For better color the emulsion to be used should contain:

Alum – 40 g

Turmeric powder – 2 kg

Castor seed oil – 140 mL

Sodium bisulfate – 30 gm

Concentrated Hydrochloric acid – 3 mL

3.2.1.6.1 Value Addition

Cured turmeric: It is an important item of value added product of turmeric. This product is popular in some parts of the world.

Turmeric powder: Turmeric powder of commerce implies a fine powder of about 60 meshes. The coloring matter is stable to heat but sensitive to sunlight and hence does not pose much problem during grinding ever though heat is generated. In view of this "cryogenic grinding" may not be necessary for powdering turmeric.

Volatile oil: Turmeric is crushed and steam distilled to derive the volatile oil having an orange-yellow liquid occasionally slightly fluorescent with an odor of turmeric. The dried rhizomes yield 5–6% of oil and fresh ones give 0.24% essential oil. About 58% of the oil is composed of turmerones (sesquiterpene ketones) and 9% tertiary alcohol.

Oleoresin: The techniques of manufacturing oleoresin from ground turmeric have been standardized by the CFTRI, Mysore by solvent extraction followed by vacuum concentration. Turmeric generally contains 15% oleoresin depending on cultivars. The semi-liquid viscous stuff contains both volatile aromatic aromatic principles and non-volatile acrid fraction covering the overall aroma and flavor in a concentrated form. Acetone, alcohol, and ethylene dicholoride were found suitable for extraction oleoresin from turmeric. Oleoresin is highly viscous orange brown product containing 30–35% curcumin, 15–20% volatile oil and has characteristic turmeric aroma.

3.2.2   GINGER (GINGIBER OFFICINALE L.)

3.2.2.1   Systematic Position

Kingdom: Plantae

Division: Magnoliophyta

Class: Liliopsida

Order: Zingiberales

Family: Zingiberaceae

Genus: *Zingiber*
Species: *officinale*

3.2.2.2   About the Crop, National and International Scenario, Uses, and Composition

Ginger is a slender herbaceous perennial grown as annual crop belonging to Zingiberaceae family. Ginger of commerce founds the dried rhizome of the plant, which is used as a spice. It is used in different forms such as raw ginger, dry ginger, bleached dry ginger, ginger powder, sliced ginger, ginger oil, ginger oleoresin, ginger in brine, etc. It has usage in foods, beverages, preservatives, medicines, and perfumery industries also.

Indian ginger exports are growing at a rate of 15% and 22% in terms of volume and value respectively. Indian ginger production has increased from 500 thousand tonnes (2004–2005) to 700 thousand tonnes (2015–2016) over the years. It exports ginger mainly in the form of dry ginger, oil, and oleoresin. Spain, USA, and Bangladesh are the major importer of ginger from India.

Ginger contains 4.2 to 10.9% oleoresin, 1.6 to 4.4% crude fiber 1.7 to 6% essential oil. The major functional constituent of ginger is zingibarin, which can aid immunity and is associated with antiinflammatory activity. This is due in part to the fact that ginger acts as an antioxidant with more than twelve constituents superior to vitamin E. Some other chemical includes the well-known ascorbic acid, caffeic acid, capsaicin, β-sitosterol, β-carotene, curcumin, lecithin, limonene, selenium, and tryptophan. The antioxidant components of ginger are polyphenols, vitamin C, β carotene, flavonoids, and tannins. Ash, minerals namely iron, calcium, phosphorous, zinc, copper, chromium, and manganese and vitamin C are about 3.85 (g), 8.0 (mg), 88.4 (mg), 174 (mg), 0.92 (mg), 0.545 (mg), 70 (μg), 9.13 (mg), and 9.33 (mg) per 100 g of sample, respectively.

3.2.2.3   Soil and Climatic Requirements

Ginger requires in warm and humid climate. It is mainly cultivated in the tropics from sea level to an altitude of above 1500 MSL. It thrives best in well-drained soils like sandy or clay loam, red loam or lateritic loam. Friable loamy soil rich in humus is ideal for ginger cultivation.

TABLE 3.6 Improved Varieties of Ginger

Sl No.	Variety	Pedigree/ Parentage	*Average Yield (kg/ ha)	Dry recovery %	Oleoresin %	Crude fiber%	Essential oil%	Duration (days)	Salient features
1	Suprabha	Clonal selection from Kunduli local	16.6	20.5	8.9	4.4	1.9	229	Plumpy rhizome, less fiber, wide adaptability, suitable for both early and late sowing.
2	Suruchi	Clonal selection from Kunduli local	11.6	23.5	10.9	3.8	2.0	218	Profuse tillering, bold rhizome, early maturing, suitable for both rainfed and irrigated condition.
3	Suravi	Induced mutant of Rudrapur local	17.5	23.6	10.2	4.0	2.1	225	Plumpy rhizome, dark skinned yellow fleshed, suitable for both irrigated and rainfed conditions.
4	Himgiri	Clonal selection from Himachal collection	13.5	20.2	4.29	1.6	6.05	230	Best for green ginger less susceptible to rhizome rot disease, suitable for rainfed condition.
5	IISR Varada	Selection from germplasm	22.66	19.5	6.7	3.29–4.50	1.7	200	High yielder, high quality bold rhizome, low fiber content. Wide adaptability and tolerant to diseases.
6	IISR Mahima	Selection from germplasm	23.2	23	4.5	3.26	1.72	200	High yielder, plumpy extra bold rhizomes, resistant to M. incognita and M. javanicapathotype 1
7	IISR Rejatha	Selection from germplasm	22.4	20.8	6.3	4	2.36	200	High yielder, plumpy, and bold rhizome

The improved varieties of ginger with their parentage, average yield (kg/ha), dry recovery (%), oleoresin (%), crude fiber (%), essential oil (%), duration (days) and salient features have been discussed in Table 3.6.

3.2.2.4   Agrotechniques for Quality Production

The land is ploughed 4–5 times to bring the soil to fine tilth and weeds, stubbles, roots, etc. are removed. Beds are prepared with a dimension of about one meter width, 15 cm height and of convenient length and distance between beds should be of 50 cm. In case of irrigated crop, the ridges are formed 40 cm apart. The best time for planting ginger is during the first fortnight of May with the receipt of pre-monsoon showers. Under irrigated conditions, it can be planted well in advance during the middle of February or early March.

Ginger is propagated by portions of rhizomes known as seed rhizomes. Seed rhizomes weighing 20–25 gm, length of 2.5–5 cm and each having one or two good buds are used for planting. For planting one hectare area 1500–1800 kg rhizomes are enough and at high altitudes the seed rate varies from 2000–2500 kg per hectare. The seed rhizome bits are place in shallow planting pits prepared with a hand hoe and covered with well rotten farmyard manure and a thin layer of soil and leveled it.

At the time of planting, 25–30 tonnes well-decomposed and dried cattle manure or compost per hectare is applied. The recommended dosage of fertilizer to ginger is 75 kg N, 50 kg P_2O_5 and 50 kg K_2O. N is applied in two split doses, first dose during 40 days after planting and second dose during 90 days after planting while, whole quantity of P_2O_5 and K_2O are applied as basal dose. Application of neem cake @ 2 tonnes per hectare at the time of planting helps in reducing the incidence of rhizome rot of ginger and increase the yield.

Beds are mulched with green leaves to enhance germination as well as to prevent of washing off soil due to soil conservation. The first mulching is done at the time of planting and repeated at 40th day and 90th day after planting with 10–12 tonnes and 5 tonnes of green leaves per hectare respectively. Weeding is done just before fertilizer application and mulching. Two to three weedings are required depending upon the intensity of weed growth. Proper drainage channels are to be provided to drain off stagnant water. Ginger is grown in rotation with other crops such as tapioca, chilies, paddy, gingelly, ragi, groundnut, maize, vegetables, red gram, castor, etc.

It is also grown as an intercrop with coconut, arecanut, coffee, and orange plantations.

Shoot borer, leaf roller and rhizome scale are major pests infesting ginger as well as diseases like soft rot, bacterial wilt and leaf spot infecting ginger in the field, which affects the quality and yield of the crop.

3.2.2.5   Harvest and Post Harvest Technology

The crop becomes mature harvest within eight months and ready for harvesting when the leaves turn yellow and start drying up gradually. The clumps are picking up carefully with a spade or digging fork and the rhizomes are detached from the dried up leaves, roots, and adhering soil. The average yield of fresh ginger varies from 15–25 tonnes per hectare depending upon the varieties.

For vegetable ginger, harvesting is done from 6th month onwards. The rhizomes are thoroughly washed in water twice or thrice and sun dried for a day. For dry ginger, the produce is soaked in water overnight and then rubbed well to clean them. After cleaning, the rhizomes are removed from the water and the outer skin is removed with bamboo splinters having pointed ends. While scraping, care must be taken not to rupture oleoresin cells lying just below the outer skin. Iron knives should not be used for peeling as they leave black stains on the peeled surfaces affecting the appearance of rhizomes. The peeled rhizomes are washed and dried in sun uniformly for one week. Ginger should be dried on clean surfaces like clean bamboo mats, cemented/concrete drying yards to ensure that the product does not get contaminated by extraneous matter. Ginger should be dried to a safe moisture level of 8–10%. The yield of dry ginger is 16–25% of the fresh ginger depending upon the variety.

Only clean and new gunny bags should be used for packing dried ginger. It is desirable to use polythene laminated gunny bags for packing dried ginger and stored ensuring protecting it from wetness. Dunnage of wooden crates should be used to stack the bags to prevent moisture access from the floor. Care should be taken to stack the bags 50–60 cm away from the walls. No insecticide should be directly used on dried ginger. Insects, rodents, and other animals should be effectively prevented from getting access to the premises where ginger is stored. Stored ginger should be periodically exposed to the sun. Prolonged storage of ginger for long time would result in deterioration of its aroma, flavor, and pungency.

3.2.2.5.1 Value Addition

Ginger oil: On steam distillation dried, cracked ginger yields 0.5 – 2.0% of pale-yellow, visicid volatile oil. Oil is generally extracted from unscraped dried ginger or from ginger scrapping. It is economical and convenient to recover oil and oleoresin from dried ginger than fresh ginger. Ginger oil is greenish to yellowish in color, mobile with the characteristics warm and aromatic odor. The main component of essential oil is ∞-zingiberine. The oil is sparingly soluble in 95% alcohol but it is soluble in 90% alcohol. The oil, oleoresin content varies from cultivar to cultivar (Nybe et al., 1980), which have been presented in Table 4.9.

Ginger oleoresin: Ginger oleoresin is obtained by extraction of powder ginger with suitable solvent like alcohol, acetone or any other solvent like ethylene dicholoride. It contains gingeol, zingerone, shogoal, volatile oil, resin, phenols, etc. the amount of oleoresin extracted by alcohol was much higher than that extracted by acetone or ethylene dicholoride (Natarajan et al., 1972). The oleoresin content of different ginger cultivars vary from 4–10% and it is viscous dark brown in color. The amount of essential oil in oleoresin varies from 7–28%; non-pungent substances nay amount to 30%. The amount of essential oil is an important factor in the evaluation of oleoresins, and the Essential Oil Association of America has specified a content of 18–35 mL volatile oil per 100 g of oleoresin.

Dehydrated ginger: The ginger is sun dried or dehydrated mechanically. Sun dried ginger is available in two forms; unscraped or scraped. Green ginger is placed in a wire gauge cage and dipped in a boiling solution of 20, 25, or 50% sodium hydroxide solution for 5, 1, or 0.5 minutes, respectively. The cage is pulled out of the lye bath and rhizomes are placed in 4% citric acid for 2 hour, then they are washed and dried. The loss in weight due to lye peeling is about 12.5%. Mechanical peeling of 60 seconds gives a product of equal in essential oil content to hand peeled ginger.

However, hand peeling was found to be superior to mechanical peeling in giving a product, which is uniform in appearance, size, and color. Peeling facilities subsequent drying of ginger and drying in a through-flow drier at a temperature not exceeding 60°C gives a satisfactory product. Sreekumar et al. (1980) recorded dry recovery of ginger ranged from 17.7% in China to 28.0% in Tura. They reported the cultivar Maran, Jugigan, Ernad Manjeri, Nadia, Himachal Pradesh, Tura, and Arippa having dry recovery of above 22% were quite suited for conversion of dry ginger. Ratnambal et al. (1987) studied about the correlation matrix of different qualities of ginger and compared with that of dry ginger. The correlation between dry ginger percentage and crude fiber was negatively significant. Mani et al. (2000) studied about the quality characters of dry ginger. According to them, dry ginger contained approximately 6.80% – ash, 0.93% – acid insoluble ash, 7.50% – alcohol soluble ash, 11.25% – water soluble extract, 1.20% – extraneous matter, 1.28% – volatile oil, and 13.40% – moisture.

Bleached ginger: For preparation of bleached ginger properly peeling of the fresh rhizome is done. The peeled rhizome is washed and then repeatedly immersed in milk of lime and allowed to dry in the sun dry until the ginger receives a uniform coating of lime (calcium hydroxide) and assumes a bright color, the final drying takes about 10 days. Finally the product is well rubbed with gunny cloth to remove bits of skin and to provide a smooth finish

Ginger in brine: Five to six months aged fresh ginger should be properly harvested and washed in fresh water. After peeling it is preserved in 16% salt and 1% citric acid solution. Sometimes 0.5% sulphur dioxide may be added with this solution. Ginger in brine is exported in the foreign market. It is mainly used for the preparation of salted ginger, pickles, and preparation some special Japanese dishes.

Salted Ginger: Fresh ginger with low fiber harvested at immature stage, are cleaned and soaked in 30% salt solution containing 1% citric acid. After about 14 days remove the rhizomes from salt solution, clean, and preserve in cold condition. This can be stored in 1–2% brine containing citric acid.

Ginger powder: For preparation of ginger properly peeling of the fresh rhizome is done. The peeled rhizome is washed and then immersed in boiling water for 10 minutes. The dry it properly and thereafter powder is made with special type of machine. The grain size of the ginger powder should be 50–60 mesh.

Ginger paste: It is a novel product, which can meet regular requirement of fresh ginger where it is not readily available. It is also a convenient preparation compared to fresh ginger, which requires peeling and cutting or crushing before domestic use. This product is therefore may compete with fresh ginger as it can be prepared in peak harvesting season when the price of the fresh ginger is very low (George, 1996).

Other value added products of ginger

1. Ginger preserve.
2. Ginger candy.

3. Soft drinks like ginger cocktail.
4. Ginger pickles, salted in vinegar or in vinegar mixed with other materials like lime, green chilies, etc.
5. Alcoholic beverage like ginger wine, ginger brandy and ginger beer.

Ginger Candy: After proper washing of the fresh ginger peeling is done. Then pieces are made in appropriate sizes. This is prepared by crystallizing soaked fresh ginger pieces in sugar solution (60:40). A layer of sugar is spread over the ginger and then dried in sun for 6–8 hours. After proper packaging it is send to the market. Ginger candy is mainly used for chocolate preparation.

Ginger beer: It is prepared by fermenting ginger extract six part dry ginger and three part hops and other spices are mixed and boil for 20–30 minutes. Filter this syrup, mix with sugar and cool up to −10°C. Ferment with yeast and citric acid if required.

Ginger juice: *Ginger Juice* is the perfect timesaving alternative to fresh or ground ginger. The ginger is first to be peeled and sliced into small cubes. Then some water is to be added and the ginger is placed in a mixer grinder. After grinding, it is taken out and the juice is strained in order to get rid of all the lumps into small cubes. Ginger juice contains antioxidants and acts as an antiinflammatory agent. It has certain other excellent medicinal values

Flow-sheet for extraction of Ginger juice

Ginger
↓
Washing
↓
Peeling
↓
Grating
↓
Addition of water (1 part of grated material: 2 parts of water)
↓
Grinding
↓
Straining
↓
Keeping one hour for settling
↓
Siphoning of clear juice

Straining

Juice

Source: Srivastava and Kumar (2002).

3.2.3 AMADA (CURCUMA AMADA ROXB.)

3.2.3.1 Systematic Position

Kingdom: Plantae
Division: Tracheophyta
Class: Magnoliopsida
Order: Zingiberales
Family: Zingiberaceae
Subfamily: Zingiberoideae
Tribe: Zingibereae
Genus: *Cucurma*
Species: *amada*

3.2.3.2 About the Crop, National and International Scenario, Uses, and Composition

Mango ginger (*Curcuma amada* Roxb.) is a unique spice belongs to the family zingiberaceae. Mango ginger is botanically and morphologically related to neither mango nor ginger, but to turmeric (*Curcuma longa*), but has shorter crop duration of six months. The rhizome has pale yellow central core with less fiber and contain traces of essential oil and 0.1% curcumin. It also possesses a unique flavor of a mixture of mango, turmeric, and ginger. It imparts mango flavor mainly due to the presence of car-3-ene and cis-ocimene (Acht and Bandopadhyaya, 1984) and (Rao et al., 1984). Probably due to high usage in the preparations of pickles, sauce, and various other culinary formulations, *C. amada* is second most cultivated curcuma species after *C. longa* (Syamkumar and Sasikumar, 2007). It is mainly cultivated in West Bengal, Kerala, Assam, Tamil Nadu, and Orissa.

3.2.3.2.1 Use

The rhizomes of mango-ginger are used for preparing pickles, chutney, preserve, candy, sauce, and salad and in meat and other culinary preparations. Its use in traditional medicine is well reflected in ethno-botanical studies conducted mainly in Indian subcontinent, Myanmar, and Thailand. Mango ginger has antibacterial, antifungal, insecticidal, aphrodisiac, antipyretic, antiinflammatory, antimycobacterial, antihyper-cholesterolemic, and antioxidant properties (Singh et al., 2010). It is useful in biliousness, itching, skin diseases, bronchitis, asthma, hiccough, and inflammation due to injuries.

3.2.3.3 Soil and Climatic Requirements

It prefers a warm humid climate and well-drained rich loamy soil. The crop comes up well in open conditions, but tolerates low levels of shade and therefore partially shaded situations can also be utilized for its cultivation. It can be well accommodated as an intercrop in coconut gardens and in rotation with other short duration crops like vegetables and also as a crop component in homesteads.

3.2.3.4 Agrotechniques for Quality Production

Whole or split mother rhizomes or well developed, healthy, and disease free finger rhizomes weighing 15–20 g are suitable for planting and optimum time of planting is May–June. Chatterjee and Chattopadhyay (2009) found that among eight planting dates (17 March to 2 July) planting on 2nd June is suggested for highest yield and quality of mango ginger under Nadia, West Bengal condition. Amba is a released variety from High Altitude Research Station, Pottangi, Orissa. A spacing of 30 × 30 cm and at a planting depth of 4–5 cm is adopted for planting of mango ginger. Mulch the crop immediately after planting with green leaves or paddy straw and repeat mulching after 50 days. Remove weeds 45 days after planting and repeat if necessary. Earth up the crop after 60 days of planting. Regarding the application of manure and fertilizer, apply 20–30 tonnes of well decomposed FYM/ha and NPK at the rate of 30: 30: 60 kg/ha. Full dose of FYM and half dose of K should be applied as basal. Half dose of N should be applied at 30 days after planting. Reaming half N and K should be applied at 60 days after planting.

3.2.3.5   Harvest and Post Harvest Technology

A good crop mango ginger may produce 30–40 tonnes/ha fresh rhizome. Regarding pest, shoot borer (*Conogethes punctiferalis*) causes, damage to the crop. If infestation is severe, spray dimethoate or quinalphos at 0.05%. Among the different diseases, appearance of dead heart in the field is the main symptom. To reduce the pest population, pull out the dead hearts with the larvae inside and burn it.

3.2.3.5.1   Value Addition

Mango ginger (*Curcuma amada* Roxb.) is a unique species having mango flavor in its rhizomes and is of high medicinal importance. Its shelf-life and quality is governed by storage temperature and time. Mango ginger rhizome can be stored for 4–5 months at low temp compared with 2–3 months at room temp. Shriveling and sprouting are the limiting factors for further storage at room temp, and the threshold percentage of water loss ranging from 30 to 36% is responsible for commercially objectionable levels of shriveling. Within the range of temperatures, rhizomes exhibits chiling injury symptoms as water-soaked lesions with tissue softening, browning, loss of mango flavor and failure to sprout at the lowest temperature (chiling temp), and rapid deterioration of physical, physiological, and antioxidant properties at room temperature. Moderate low temperature minimizes the biochemical changes, maintained or increased the antioxidant activity and doubles the shelf-life as a function of temperature with storage time.

The following products of mango ginger are popular:

1. Mango Ginger Chews
2. *Blended nectar beverage*: Value added product, nectar was prepared by using three medicinal plant extracts namely, shatavira, aloe, and mango ginger. The nectar having 20% juice concentration possessing extracts of all the three medicinal plants at 50:40:10 ratios respectively and acidity of 0.5% shows good quality attributes for pH, total

soluble solids, acidity, ascorbic acid, reducing sugar and total sugar contents (Ravindra et al., 2012).

3. *Mango Ginger pickle*: It can be prepared with very less oil in short time. Other than pickle, we can use this in Upma, Lemon Rice, Curd Rice, Chutneys, and so on. The pickle is very tasty and the nice flavor is liked by most of the people. The pickle can be consumed after 4 to 5 hours of preparation.

4. *Mango ginger powder*: Wheat flour may be mixed with mango ginger powder to improve the rheological and baking characteristics of the former. Crassina and Sudha (2015) opined that addition of mango ginger powder at different levels (from 0 to 15%) changes the mixing, pasting, and baking characteristics of wheat flour. The incorporation of MGP in formulation markedly increased the total phenolics content and the radical scavenging activity of soup stick extracts. Therefore, such a value added product would be helpful in promoting utilization of underutilized spice, i.e., mango ginger.

3.3 BULBOUS SPICES

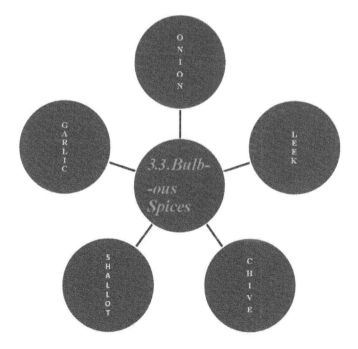

3.3.1   ONION (ALLIUM CEPA L.)

3.3.1.1   Systematic Position

Kingdom:	Plantae
Clade:	Angiosperms
Clade:	Monocots
Order:	Asparagales
Family:	Amaryllidaceae
Subfamily:	Allioideae
Genus:	*Allium*
Species:	*cepa*

3.3.1.2   About the Crop, National and International Scenario, Uses, and Composition

The onion (*Allium cepa* L., from Latin *cepa* "onion"), also known as the bulb onion or common onion, is a vegetable and is the most widely culti-vated species of the genus *Allium*. The onion is most frequently a biennial or a perennial plant, but is usually treated as an annual and harvested in its first growing season. Onions are cultivated and used around the world. As a food item, they are usually served cooked, as a vegetable or part of a prepared savoury dish, but can also be eaten raw or used to make pickles or chutneys. They are pungent when chopped and contain certain chemical substances, which irritate the eyes.

A number of synonyms have appeared in its taxonomic history:

- *Allium cepa* var. *aggregatum* – G. Don
- *Allium cepa* var. *bulbiferum* – Regel
- *Allium cepa* var. *cepa* – Linnaeus
- *Allium cepa* var. *multiplicans* – L.H. Bailey
- *Allium cepa* var. *proliferum* – (Moench) Regel
- *Allium cepa* var. *solaninum* – Alef
- *Allium cepa* var. *viviparum* – (Metz) Mansf

Onions contain 89% water, 1.5% protein, and vitamins B_1, B_2, and C, along with potassium and selenium. Polysaccharides such as fructo-sans, saccharose, and others are also present, as are peptides, flavonoids (mostly quercetin), and essential oil. Onion contains numerous sulfur compounds, including thiosulfinates and thiosulfonates; cepaenes; S-oxides; S, S-dioxides; mono-, di-, and tri-sulfides; and sulfoxides. Mincing or crushing the bulb releases cysteine sulfoxide from cellular compartments, making contact with the enzyme alliinase from the adjacent vacuoles. Hydrolysis results with the release of reactive intermediate sulfenic acid compounds and then to the various sulfur compounds.

3.3.1.3   Soil and Climatic Requirements

Alliums are among the oldest cultivated plant species. References to edible can be found in the Bible, Koran, and in the inscriptions of ancient civilization of Egypt, Rome, Greece, and China. Onion is an important vegetable crop grown in almost all parts of the world. The four major onion growing countries in the world China, India, USA, and Turkey and other important countries for onion are Italy, Egypt, Netherlands, Russia, Thailand, Indonesia, Korea, Japan, and Brazil. It is used in almost all kind of Indian cuisines. The Important onion Growing states in India are Maharashtra, Karnataka, Tamil Nadu, Andhra Pradesh, Gujarat, Punjab, Haryana, Rajasthan, UP, Bihar, and MP.

3.3.1.3.1   Soil

Soils for onion should be rich in humus with good drainage. Sandy soil needs more and frequent irrigation and favors early maturity, whereas heavy soils lead to mishappened bulbs and there is problem in digging of bulbs. The most desirable soils are the one that retains enough moisture and at the same time be favorable enough to be easily cultivated and to allow proper development of bulbs. The optimum pH range is between 5.8–6.5. Highly alkaline and saline soils are not suitable.

3.3.1.3.2   Climate

The onion is a cool season crop. Onion can be grown under wide range of climatic condition. It grows well under mild climates without extreme heat or cold or excessive rainfall. The plant is hardy and in the young stage can withstand freezing temperature it does not thrive well in places where the

TABLE 3.7　Improved Varieties of Onion

Sr. No	Variety	Source	Maturity (days)	Average Yield (q/ha)	TSS (%)	Salient features
1	Punjab Selection	PAU, Ludhiana	-	200	14	For rabi season, bulbs are red, globular, and quite firm with good keeping quality
2	Pusa Ratnar	NBPGR, New Delhi	125	300	11–12	Bulbs dark red color, obviate to flat, globular, less pungent and neck drooping, suitable for rabi season, average storage quality
3	Pusa Red	IARI, New Delhi	140–145	250	12–13	Short day variety, for both late kharif and rabi season in Maharashtra, medium size, bronze red color, flat to globular, less pungent bulbs having good keeping quality
4	N-2-4-1	Dept. Agril, M.S.	140–145	300–350	12–13	Brick red, globe shape and pungent bulbs, good keeping quality. Mainly grown in rabi season but can be grown both in late kharif and rabi season in Maharashtra.
5	Pusa Madhavi	IARI, New Delhi	130–145	300		Best for rabi season, Bulbs are medium to large, light red, flatish round with good keeping quality.
6	Arka Niketan	IIHR, Bangalore	145	340	12–14	Globular light red bulbs with thin neck, high in pungency and dry matter content, suited mainly for rabi season, can also a grown in late kharif in Maharashtra.
7	Agrifound Dark Red	NHRDF, Nashik	95–110	300	12–13	Dark red, globular with tight skin bulbs, moderately pungent, cultivate in kharif season, average keeping quality.

Sr. No	Variety	Source	Maturity (days)	Average Yield (q/ha)	TSS (%)	Salient features
8	Kalyanpur Red Round	CSAU, Kanpur	150–160	250–300	13–14	Good keeping quality, bronze red, globular bulbs with moderately sweet and moderately pungent
9	Agrifound Light Red	NHRDF, Nashik	160–165	300–325	13	Globular light red with tight skin bulbs having good keeping quality, recommended for rabi season, can be grown in late kharif season in Nashik district of MS.
10	Hissar 2	HAU, Hissar	165 days	200–250	11.5–13	Suitable for rabi season in Haryana and Punjab. Bulbs bronze red, globular, tight skin with sweet to pungent taste
11	Punjab Red Round	PAU, Ludhiana	–	300	–	Early maturing and high yielding variety having shining red, medium to large, round with tight neck bulbs
12	Arka Pragati	IIHR, Bangalore	140–145	200	–	Early maturing and highly pungent variety of pink colored, globe shape, uniform size with thin neck bulbs
13	N-53	Dept. of Agril. M.S.	90–100	250	11–12	For kharif season, mildly pungent and poor keeping quality, red, flatish, medium to large bulbs
14	Baswant-780	MPAU, Rahuri (M.S.)	100–110	250	12	Suitable for kharif season in Maharashtra, mildly pungent, less bolting with less twins, average keeping quality. Bulbs are crimson red colored and in globe shape
15	Udaipur-101	RAU, Bikaner Campus, Udaipur	150–160	200–300	12–14	Suitable for rabi season in Rajasthan, dark red flatish globular bulbs, sweet but slightly more pungent.

TABLE 3.7 (Continued)

Sr. No	Variety	Source	Maturity (days)	Average Yield (q/ha)	TSS (%)	Salient features
16	Udaipur-103	RAU, Bikaner Campus, Udaipur	150–165	250–300	10.5–13	Suitable for rabi season in Rajasthan, sweet but slightly more pungent.
17	Line-28	NHRDF	-	-	13–14	For rabi season, dark red, globular, and have reddish thick inner scales, moderately pungent in taste
18	Phule Samarth	MPKV, Rahuri (MS)	-	-	-	For kharif and early rabi (rangda) season, can be grown in northern India as well as central and western India, bulbs have dark red color, globular round with thin neck and moderately pungent taste
19	Early Grano	IARI, Delhi	95–110	500–600	6–7	Suitable for cultivation in plains during kharif and rabi season for salad purpose, very mild pungency and poor in storage
20	Spanish Brown	IARI, RSS, Katrain Kullu	160–180	280–300	13–14	Good in storage at hills and mild pungent.
21	Phule Suvarna	MPKV, Rahuri (MS)	-	240	11.5	Suitable for export to Europe, Australia, and America, excellent keeping quality (4–6 months), suitable for rangda and rabi season
22	Pusa White Round	IARI, New Delhi	130–135	300–325	12–13	White bulbs with round flat in shape, good in storage with drying ratio of 8:1
23	Pusa White Flat	IARI, New Delhi	130–135	-	12–14	White flatish round bulbs, good in storage with drying ratio of 9:1

Sr. No	Variety	Source	Maturity (days)	Average Yield (q/ha)	TSS (%)	Salient features
24	Punjab-48	PAU, Ludhiana	-	300	-	Flatish round white colored bulbs with very good texture and flavor, suitable for dehydration, good in storage
25	Udaipur-102	RAU, Bikaner Campus, Udaipur	120	300–350	12	Bulbs are white in color, round to flat in shape with low percentage of small bulbs
26	N-257-9-1	Dept. of Agril. MS	-	250	-	High yield potential with good keeping quality and suitable for rabi season
27	Phule Safed	MPKVR, Rahuri	-	250–300	13	Suitable for dehydration
28	Agrifound White	NHRDF, Nashik	160–165	200–250	14–15	Suitable for rabi season, good keeping quality and good for dehydration
29	Agrifound Rose	NHRDF, Chickballpur	95–110	190–200	15–16	It is pickling type variety, suitable for growing in kharif season in Cuddapah district and in Karnataka for all seasons
30	Arka Bindu	IIHR, Bangalore	100	250	14–16	Bulbs free from premature bolters and splits and have high pungency
31	Co-1	TNAU, Coimbatore	90	100	-	Medium sizes bulblet of red color, 7–8 bulblets per plant and fairly pungent with medium TSS
32	Co-2	TNAU, Coimbatore	65	120	-	Short duration variety than Co-1, Moderately bigger size bulblet of crimson color, 7–9 bulblets per plant, pungent with high TSS, good storage, moderately resistant to thrips and Alternaria blight

TABLE 3.7 (Continued)

Sr. No	Variety	Source	Maturity (days)	Average Yield (q/ha)	TSS (%)	Salient features
33	Co-3	TNAU, Coimbatore	65	160	13	8–10 bulblets per plant, moderately resistant to thrips, stored upto 120 days without sprouting
34	MDU-1	TNAU, Coimbatore, Campus Madurai	60–75	150	-	10–11 bulblets per plant, better keeping quality, tolerant to lodging due to thick erect leaves
35	Co-4	TNAU, Coimbatore	60–65	180	-	8–10 bulblets per plant, moderately resistant to thrips
36	Agrifound Red	NHRDF, Dindigul	65–67	180–200	15–16	5.79 bulblets per cluster
37	Co-On-5	TNAU, Coimbatore	95–100 (Seed crop)	189 (seed yield 250–300 kg/ha)	-	It is free flowering type with seed setting ability and can be propagated by seeds as well as bulblets

average rainfall exceeds 75–10°C in monsoon period. It requires about 70% relative humidity for good growth. For good vegetative growth 12.8–23°C temperature before bulbing and for bulb development 20–25°C are required. Very low temperature in the beginning results in bolting while sudden rise in temperature favors early maturity of the crop in rabi and results in small sized bulbs.

The improved varieties of onion with their source, maturity (days), average yield (q/ha), TSS (%) and salient features have been discussed in Table 3.7.

3.3.1.4   Agrotechniques for Quality Production

Onion is grown in two distinct season viz; kharif and rabi in the plains of India although sowing and transplanting with the reason.

- Early kharif crop: April–May (TN, Karnataka, AP).
- Kharif crop: May–June (TN, Karnataka, AP, Maharashtra, Parts of Gujarat, Rajasthan, Punjab, and other parts).
- Late kharif crop: August–September (Maharashtra and some parts of Gujarat).
- Early rabi crop: September (West Bengal and Orissa).
- Rabi crop: September–October (TN, Karnataka, AP).
- Rabi crop: Oct–Dec (WB, Orissa, and Northeastern Parts).
- Rabi crop: November–December (Maharashtra and Part of Gujarat).

Sowing for rabi crop in the hills is done in September–October and for summer crop with long verities sowing is done October–December.

3.3.1.4.1   Land Preparation

In preparing the land the field is ploughed to a fine tilth by giving 4–5 ploughings with sufficient interval. The ploughing should be shallow. The planking should be done for proper leveling. The field is divided into beds and channels. The normal width of a bed should be about 1.8 meter and length may vary according to level of land.

3.3.1.4.2 Nursery Management

Seed beds are prepared finely, well drained, 15–20 cm raised, 1.0 m wide and of convenient length. Fine and fully decomposed FYM or compost @ 3–4 kg/ms should be well mixed to the bed. Drench the bed with Formaldehyde (4%) or Captan @ 2–3 g/L and cover with polyethene sheet for 7–10 days to avoid damping off disease. 8–10 kg seeds/ha is required for open pollinated varieties of common onion. About 1000 m^2 area is required to produce seedlings for 1 hectare. Before sowing treat the seed with Thiram or Captan @ 2–3 kg of seed to avoid damping off. Sowing of seed is done in lines spaced at 5–7 cm distance recommended. The seed after sowing should be covered with finely sieved FYM or compost or vermin compost followed by light watering by rose can. Cover the seed bed with straw or dry grass till seed germination and sprinkle water regularly. Seedlings should be drenched with 0.2–0.3% Thiram and Captan twice, 15 and 30 DAS. Side dressing with SSP and MOP and spraying with 0.5% Urea produce healthy seedlings. Time to time hoeing, weeding, and irrigation are required to raise healthy seedlings. The seedlings are ready within 6–7 weeks after sowing for kharif crop and 8–9 weeks for rabi crop. Under-aged seedlings do not stand well in the field and over-aged seedlings may bolt easily.

3.3.1.4.3 Transplanting

Seedlings are transplanted in flat beds. Transplanting on both sides of ridges is, however better for kharif or rainy season crops. The variety with small bulb like Agrifound Rose can be directly seeded by broadcasting or in rows in the main field 20–25 seed kg per hectare. Multiplier onions are directly planted in the main field with 1.0–1.2 tonnes of bulblets of 1.5–2.0 cm size per hectare. Transplanting is done when they are 6–7 weeks old for kharif crop and 8–9 weeks old for rabi crops. Irrigating the seed beds before lifting of seedlings is advisable to facilitate easy pulling of seedlings. Late afternoon hours are best for transplanting the seedlings in the main field for better establishment. Irrigate the field lightly just after transplanting the seedling.

3.3.1.4.4 Spacing

- Common big size onion: 15 cm × 10 cm.
- Small pickling onion: 8 cm × 5 cm.
- Multiplier onion: 30 cm × 15 cm.

3.3.1.4.5 Nutrient Management

Soil should be liberally manured and fertilized for getting good yield. A basal dose of 20–30 tonnes well rotten FYM per hectare is recommended. It should be applied 1 month before transplanting and mixed in the soil. The recommended range of nutrient dose by different state is 60–150 kg N, 30–150 kg P, and 25–125 kg K per hectare. The whole quantity of P and K and half of N should be mixed in soil before transplanting. The remaining half N should be given in top dressing equal quantity in 2 split doses at 30 and 45 DAT. The top dressing must be completed before initiation of bulbing.

3.3.1.4.6 Irrigation

The number of irrigations will depend on several factors such as soil, and climatic conditions. Any one of the irrigation is necessary immediately after transplanting. In rainy (kharif) season depending upon the rains 8–10 irrigations are enough. Late kharif crop requires 12–15 irrigations and in winter (rabi) season 15–20 irrigations are given. At bulb formation irrigation is necessary and moisture stress at the stage result in low yield. Frequent and light irrigation at weekly intervals in general results in good bulb development and increased yield. Irrigation should be stopped when tops mature and start falling in rabi season. In kharif season it should be stopped 10 days before harvesting.

3.3.1.4.7 Interculture and Weed Control

Onion is a poor competitor of weeds hence, weed must be kept down to get good yield. Two lights hoeing at the early stages of growth give good weed control. Apply Basalin @ 1 L/ha or Stomp @ 3.35 L/ha immediately after and before 1st irrigation along with 1 hand weeding is the best.

3.3.1.4.8   Harvesting

The best time to harvest rabi onion is 1 week after 50–70% neck fall. In winter (*rabi*) season since tops do not fall soon after the color of leaves changes to slightly yellow and tops start drying, the bulbs are harvested.

3.3.1.4.9   Yield

The average yield of big size common onion is 25–30 tonnes per hectare, small size common onion is 16–20 tonnes per hectare and multiplier onion is 15–18 tonnes per hectare.

3.3.1.5   Plant Protection

Damping off (*Pythium, Rhizoctonia, Phytophthora* etc.): The seed is treated with Thiram/Captan/Mancozeb @ 3 g/kg seed. Drenching the nursery bed 7 days before sowing with Thiram/Captan or any copper fungicide @ 3 g/L of water and provisions of proper drainage to the bed is advisable.

Purple blotch (*Alternariaporri*): At least 3 summer ploughing are to be given to reduce disease severity. The crop is sprayed with mancozeb @ 2.5 g/L or Chlorothalonil @ 2 g/L with a sticker 4–5 times starting from 15 days after transplanting.

Downy mildew (*Perenospora destructor*): Crop rotation is adopted with 4 years break in onion or garlic cultivation. Weeds are controlled and the field is kept clean. The young seedlings are sprayed with 0.25% Mancozeb or Metalaxyl (0.25%) or Cymoxanil/Mancozeb (0.25%) with a sticker 4–5 times interval at 10 days.

Thrips (*Thripstabaci*): Crop rotation is to be adopted, the weeds and plant debris on, which adults hibernate are to be destroyed and burnt. Insecticides like Acetamiprid @ 0.5 g/L are to be applied observing the infestation of thrips after transplanting.

Onion fly (*Hylemiaantigua*): Crop rotation is followed. Granular insecticide like Phoret 10 g @ 25 kg/ha is applied. The crop is sprayed with Malathion with 1 mL/L or Thiodan 1.5 mL/L.

3.3.1.6   Harvest and Post Harvest Technology

Generally bulbing of onion occurs after 12 to 18 weeks. For eating fresh, the bulbs can be gathered when they attain a ma definite size or as per the need, but for the purpose of keeping them in storage, they should be harvested after the leaves have died back naturally. Allowing the bulbs to keep on the surface of the soil for a few days after harvest is advisable to dry out properly particularly when dry weather prevails. Afterwards they can be placed in nets, roped into strings, or laid in layers in shallow boxes. Storing should be done in a well-ventilated, cool place such as a shed.

3.3.1.6.1   Value Addition

Dehydrated onion: For dehydrated onion and subsequent preparation of onion powder the onion must have some desirable traits viz., (i) white colored flesh, (ii) full globe to tall global shape of bulbs with 5–6 cm diameter, (iii) high TSS content, above 15%, preferably 20%, (iv) high degree of pungency and (v) high yield good, and (vi) keeping quality. The methods of production of dehydrated onion involve removal of water from the raw onions to a maximum level of 4.25% followed by milling it to a specific particle size. Prior to drying, onion is cleaned, peeled, and top removed. The peeled onion is then subjected to washing and slicing. No blanching or sulfiting is advisable for dehydrated onion in order to protect enzyme system, which develops onion flavor when onion cells are cut or broken. For drying of onions a stainless steel belt may be used, followed by a carefully controlled time/temperature programmed by circulation of hot air. The conditions can be controlled to deliver onion with a maximum of 5% moisture taking into account that no heat damage occurs to the product.

Onion powder: Dehydrated onion slices are ground in a hammer mill to a suitable mesh to produce onion powder of desired texture. The powder being highly hygroscopic, due precaution is needed during its storage so as to keep it in air tight containers in cool, dark places, failing, which it may absorb moisture from air and as a result onion powder turn into granular, cake, pastry, and ultimately getting mould attack.

Onion salt: It is also one of the popular value added product of onion. It is prepared by mixing 19–20% of onion powder with 78% free-flowing pulverized refined iodized table salt. Anticaking agent like anhydrous sodium sulphate is to be added, which prevents absorption and caking of onion powder during storage.

3.3.2 *GARLIC* (ALLIUM SATIVUM *L.*)

3.3.2.1 Systematic position

Kingdom: Plantae
Division: Tracheophyta
Class: Magnoliopsida
Order: Aspagales
Family: Amaryllidaceae/alliaceae
Genus: *Allium*
Species: *sativum*

3.3.2.2 About the Crop, National and International Scenario, Uses, and Composition

Garlic is native to central Asia, and has long been a staple in the Mediterranean region, as well as a frequent seasoning in Asia, Africa, and Europe. It was known to Ancient Egyptians, and has been used both as a food flavoring and as a traditional medicine.

India exports about 14,000 tonnes of garlic annually (April–March). In the recent years exports of garlic has gone up from 12,000 tonnes to 21,000 tonnes. Exports have increased because of decline in production in other major growing regions of the World (China, Argentina). Bangladesh, Malaysia, and Indonesia are the major importers of Indian Garlic.

There are two subspecies of *A. sativum,* ten major groups of varieties, and hundreds of varieties or cultivars.

- *A. sativum* var. *ophioscorodon* (Link) Döll, called *Ophioscorodon*, or hard-necked garlic, includes porcelain garlics, rocambole garlic, and purple stripe garlics. It is sometimes considered to be a separate species, *Allium ophioscorodon* G.Don.
- *A. sativum* var. *sativum*, or soft-necked garlic, includes artichoke garlic, silverskin garlic, and creole garlic.

Garlic essential oil represents more than 94.63% of the total essential oil. The major components were diallyl disulfide (37.90%), diallyl trisulfide (28.06%), allyl methyl trisulfide (7.26%), diallyl sulfide (6.59%), diallyl tetrasulfide (4.14%) and allyl methyl disulfide (3.69%). Garlic contains a wider variety of organosulfur volatiles than the intact garlic clove. Typical volatiles that have been identified in crushed garlic and garlic essential oil include diallyl sulfide, diallyl disulfide, diallyl trisulfide, methylallyl disulfide, methylallyl trisulfide, 2-vinyl-4H-1, 3-dithiin, 3-vinyl-4H-1, 2-dithiin, and (E, Z)-ajoenes.

3.3.2.3 Soil and Climatic Requirements

Alliums are among the oldest cultivated plant species. The most widely cultivated Allium next to common onion is garlic. Garlic is one of the important bulb crops grown and used as a spice or condiment throughout India. China is the largest producer followed by India. Although it is grown in all states in India, Madhya Pradesh and Gujarat have more under garlic than other state.

It possesses a high nutritive value; its preparations are administered as a cure against stomach diseases, sore eyes and earache. It is commonly used in the preparation of various dishes. Allicin, the principle of garlic, has antibacterial properties. It is a powerful drug against amoebic dysentery and is also having many other medicinal properties.

3.3.2.3.1 Climate and Soil

It is grown under a wide range of climatic condition; however, it cannot stand too hot or too cold weather. It prefers moderate temperature in summer as well as in winter. Short days are very favorable for the formation of bulbs. It can be grown well at elevations of 1000 to 1300 m above MSL. Garlic required well-drained loamy soils, rich in humus, with fairly good content of potash. The crop raised on sandy or loose soil does not keep for long and the bulbs too are highly lighter in weight. In heavy soils, the bulbs produce are deformed and during harvesting, many bulbs are broken and bruised and so they do not keep well in storage.

TABLE 3.8 Improved Varieties of Garlic

Sr. No	Variety	Parentage	No of cloves	Average Yield (q/ha)	TSS (%)	Silent features
1	Agrifound White (G-41)	Mass selection from a local collection from Biharsharif area in Bihar	20–25	130	41	Early crop, susceptible to purple blotch and stemphylium blight, good storage ability
2	Yamuna Safed (G-1)	Mass selection from a local collection from Delhi (Azadpur) Market	25–30	150–175	38	Tolerant to insect pests and diseases like purple blotch, stemphylium blight and thrips, good storage ability
3	Yamuna Safed-2 (G-50)	Mass selection from a local collection from Karnal area in Haryana	35–40	150–200	38–40	Recommended for Northern India
4	Yamuna Safed-3 (G-282)	Mass selection from a local collection from Tamilnadu	15–16	175–200	38–42	Recommended for cultivation in Northern and Central parts of India, medium storer and suitable for export

5	Yamuna Safed-4 (G-323)	Mass selection from a local collection from Janupur District of U.P.	20–25	175–200	40–42	Recommended for growing in North India
6	Agrifound Parvati (G-313)	Selection from an exotic collection from Hong Kong market	0–16	175–225	-	Long day variety, suitable for only in mid and high hill of Northern states, medium store, suitable for export and tolerant to common diseases
7	Godavari	Selection from Jamnagar collection	-	130–140	-	Pink colored, medium size bulbs
8	Sweta	Selection from Gujarat collection	-	120–130	-	White colored bulbs

The improved varieties of garlic with their parentage, number of cloves, average yield (q/ha), TSS (%) and salient features have been discussed in Table 3.8.

3.3.2.4   Agrotechniques for Quality Production

Garlic is propagated by cloves. All the cloves are planted except the long slender one in the center of the bulb. Healthy cloves or bulbils free from disease and injuries should be used for sowing and about 150 to 200 kg cloves are required to plant one hectare. They are sown by dibbling or furrow planting. Farmyard manure @ 25 tonnes is applied as a basal dose along with 60 kg N and 50 kg in each of P and K. 45 days after planting 60 kg N is applied again as top dressing. Under Karnataka condition, INM with 75:40:40:40 Kg NPKS + 7.5 t/ha of vermicompost is recommended. First, irrigation is given after sowing and then field is irrigated every 10 to 15 days depending upon the soil moisture availability. There should not be any scarcity of moisture in the growing season; otherwise the development of the bulb will be affected. The last irrigation should be given 2 to 3 days before harvesting for making it easy without damaging the bulbs. In south India hills, they are mostly grown as a rainfed crop. First, interculture is given with hand hoe one month of after sowing. Second weeding is given one month after the first weeding and hoeing. Hoeing the crop just before the formation of bulbs (about two and half months from sowing) loosens the soil and helps in setting of bigger and well-filled bulbs. The crop should not be weeded out or hoed at a later stage because this may damage the stem and impair the keeping quality.

3.3.2.4.1   *Harvesting*

Garlic is a crop of four and half to five month's duration. When the leaves turning yellowish or brownish and shows sign of drying up, the crop is ready for harvest. The plants are then pulled out or uprooted with a country plough and are tied into small bundles, which are then kept in the field or in the shade for 2 to 3 days for curing and drying so that the bulbs become hard and their keeping quality is improved. The bulbs may be stored by hanging them on bamboo sticks or by keeping them on dry sand on the floor in a well-ventilated room on dry floor. For taking the bulbs to the market, dried stalks are removed and bulbs are cleaned.

Well-cured garlic bulbs can be kept for 1 to 9 months in an ordinary well-ventilated room. If dust smoke is given to it, the bulbs can be stored for 8–10 months. They can also be stored at 32°F with 60% relative humidity. Average yield level is 6–8 t/ha.

Garlic exhibits certain physiological disorders such as:

1. 'Sprouting of bulbs in the field' – excessive soil moisture and nitrogen supply.
2. 'Splitting' – delayed harvesting or irrigation after long spell of drought.
3. 'Rubberization' – controlled by application of micronutrients $ZnSO_4$ and ammonium molybdate or neem cake application and growth regulator like GA.

3.3.2.4.2   *Yield*

The average yield is 10–20 tonnes/ha depending upon the variety.

3.3.2.5   Plant Protection

Thrips cause withering of the leaves application of chlorantraniliprol 25 18.5% SC(0.25 mL/L) or dimethoate 30 EC 1 mL/L will check the incidence.

Leaf spot is the most importance diseases. Spraying Dithane M-45 at fortnightly interval at 2.5 g in 1 L of water is recommended.

3.3.2.6   Harvest and Post Harvest Technology

To maintain continued vegetative growth, any flower stalk appears during growth stage should be removed. When the leaves are one-third brown and the bulbs reach proper size, it is time to harvest the crop. Putting off harvesting garlic until after the leaves are completely brown will only result in an inedible bulb. Digging the bulbs out of the ground is advisable instead of pulling them out.

Immediately after harvest, the freshly dug unwashed bulbs are kept in a dark, dry place as soon as possible to avoid balancing and burning in the sun.

Domestically, garlic is stored warm (above 18°C) and dry to keep it dormant (otherwise it may sprout). It is traditionally hung; softneck varieties are often braided in strands called plaits or *grappes*. Peeled cloves may be stored in wine or vinegar in the refrigerator. Commercially, garlic is stored at 0°C, in a dry, low-humidity environment. Garlic will keep longer if the tops remain attached.

Garlic bulbs should be clean and white with a dried neck and outer skin and quite firm under pressure. They should be discarded if they are soft or spongy or show signs of mold.

3.3.2.6.1   *Value Addition*

Garlic powder: Powder is always a preferred form of garlic export instead of raw bulbs as the former takes less space, safer to store and most importantly can realize a higher unit value. Adoption of CFTRI technology can also reduce the wastage occurring during storage. Garlic powder is liked by the people in Europe and USA, primarily due to standardization of flavor levels possible in finished product. Therefore, garlic powder is perhaps the most important amongst the various products of garlic viz., garlic oil, dehydrated garlic powder, garlic juice and extract, pickled garlic, etc. However, it suffers from the disadvantage of browning and caking on storage besides losing the flavor. This was obviated by the development of 'stabilized' garlic powder.

Garlic essential oil: Garlic oil, recovered by steam distillation of fleshy ground cloves is a reddish brown liquid commonly used after dilution with vegetable oil or microencapsulated dextrin to alleviate the extreme pungency. The oil constitutes various sulphur compounds like diallyl disulphide (60%), diallyl trisulphide (20%), allyl propyl sulphide (6%) and probably a small quantity of diethyl sulphide. Diallyl polysulphide and diallyl disulphide are said to posses the true garlic odor. One-gram oil is equivalent to in flavoring terms to 900 g fresh garlic or 200 g dehydrated powder. It is being used in beverages, ice cream and ices, confectionary, baked goods, chewing gum and condiments.

Garlic juices: Repeated extrusion of garlic tissue, flash heating (140–160°C) followed by cooling to 40°C results in the production of garlic juice. It is rich in many bioactive compounds and their precursors as well as sugars contributing both flavor and aroma. This kind of juice pressed juice is used in the food industry for seasoning.

Dehydrated garlic and garlic flakes: Processed garlic is one of the highly competitive commodities in the International market as it reduces the transportation cost during transit and avoid losses of bulb during storage. Dehydrated product such as flakes, granules, powder, and other products from the important by-product are being prepared and marketed worldwide. Garlic dehydration is carried out by using tunnel dehydrator. Sliced garlic is dried in series of hot air tunnels, and is prepared in the form of flakes and powder. Garlic is commercially dried upto 6.5% moisture content. To produce one kg of dried product approximately 3.2–4.5 kg of fresh garlic are required.

3.3.2.6.2   Flow Chart for Preparation Garlic Flakes

Garlic bulb-storing and washing
  ↓
Slicing
  ↓
Drying of the sliced garlic (6.5% moisture)
  ↓
Garlic flakes
  ↓
Further processing, packing, and marketed.

Source: Sankar et al. (2005).

Garlic salt: Like onion salt garlic salt is also a preferred value added product of garlic. It should never contain more than 81% salt and the moisture free white garlic powder should be between 18–19%. Garlic salt has much wider culinary potential than the powder, and one tablespoon of the product is considered to be equivalent to a clove of fresh garlic.

Garlic paste: In market garlic paste is available in pure or in mixed forms with ginger or onion. Selected, peeled fresh garlic cloves are cut in to chunks. All ingredients are put in a blender with addition of little quantity of water just to help grinding and packing in airtight container before storing them in refrigerated condition. Instead of water, edible oil can also be used during blending process to enhance the shelf life of the produce. Garlic paste was successfully exported first to London.

Garlic bulbs are subjected to mild pressure by hand to separate into cloves. Cloves were dried in a tray drier at 40°C for 30 min to facilitate manual peeling. After peeling, cloves were blanched at 90°C for 15 min in hot water followed by grinding in a laboratory grinder. This ground material was passed through a 14-mesh sieve to obtain the product of uniform consistency. Sodium chloride (10%, w/w) was added to increase the total soluble solids (TSS). The final pH was maintained to about 4.0 by addition of citric acid (30%, w/v) solution. The prepared paste is filled in glass bottles and processed thermally (Ahmed et al., 2003).

Garlic pickles: This is another attractive example of adding value to garlic. For garlic pickles, cloves are first fermented in brine solution as per requirements and taste. Then ground spices, oil, salt, citric acid/lactic acid and other permitted preservatives are added to enhance the taste to make a sour pickle. The product is packed in sterilized bottle and after through mixing of all ingredients, the bottle is topped with lot of edible oil to plug any air gaps among the cloves.

3.3.3   LEEK (ALLIUM AMPELOPRASUM L.)

3.3.3.1   Systematic Position

Kingdom:	Plantae
Order:	Asparagales
Family:	Amaryllidaceae
Subfamily:	Allioideae
Genus:	*Allium*
Species:	*ampeloprasum*

Leek is a popular vegetable and its edible portions are the white base of the leaves (above the roots and stem base), the light green parts, and to a lesser extent the dark green parts of the leaves. The dark green portion is usually discarded because it has a tough texture. Raw leeks can be used in salads. It is often used by boiling, frying, and even mixing the boiled leaves with rice, herbs (generally parsley and dill), onion, and black pepper (as in Turkish cuisine).

The leek or garden leek is variously classified as *Allium porrum, Allium ampeloprasum, Allium ampeloprasum* var. *porrum, Allium ampeloprasum porrum,* or *Allium ampeloprasum* (Dey and Khaled, 2013). The medicinal property of Allium ampeloprasum is mainly due to the presence of many sulphur containing bioactive constituents, which include dimethyl disulfide, methyl propenyl disulfide, propyl propenyl disulfide, dimethyl trisulfide,

TABLE 3.9 Proximate Composition of the Chemical Constituents of Leek

Proximate composition (units in bracket)		Global average value
Moisture (%)	:	78.32
Total available carbohydrates (%)	:	16.60
Proteins (%)	:	1.67
Lipids (%)	:	0.18
Fiber (%)	:	4.23
Ashes (%)	:	0.79
K (mg/100 g)	:	309.37
Na (mg/100 g)	:	54.60
Ca (mg/100 g)	:	70.16
Mg (mg/100 g)	:	14.03
Mn (mg/100 g)	:	0.11
Fe (mg/100 g)	:	0.60
Zn (mg/100 g)	:	0.75
Cu (mg/100 g)	:	0.11
Ascorbic acid (mg/100 g)	:	4.30
Dehydroascorbic acid (mg/100 g)	:	2.14
Total vitamin C (mg/100 g)	:	6.69
Oxalic acid (mg/100 g)	:	91.65
Glutamic acid (mg/100 g)	:	51.67
Malic acid (mg/100 g)	:	132.86
Citric acid (mg/100 g)	:	38.86
Succinic acid (mg/100 g)	:	2.14
pH	:	5.76
Tritable acidity (mL NaOH/100 g FW)	:	14.35
Energy (Kcal/100 g)	:	78.92

methyl propyl trisulfide, methyl propenyl trisulfide, S-methyl cysteine sulfoxide, S-propyl cysteine sulfoxide, S-propenyl cysteine sulfoxide, N-(γ-glutamyl)-S-(E-1-propenyl) cysteine. *Allium ampeloprasum* has higher amount of methiin and propiin, respectively. This is an important component of flavors or the precursors of flavors and odors of Allium vegetables. Proximate composition of the chemical constituents of leek are given below (Table 3.9)

3.3.3.2   Soil and Climatic Requirements

Rapid leaf growth reported at 21–24°C. High temperature during sowing time inhibits the seed germination. The growing soil should be rich in humus.

3.3.3.3   Agrotechniques for Quality Production

It is either grown directly by seed or by transplanting. Spacing is 20–30 cm × 15–30 cm. Seeds are sown in nursery during October–November. Transplanting is done in 30–40 cm deep trenches. London Flag, Mussel Burg, American Flag Elephant, Goliah, Winterens, Titan, and Ficus are some of the varieties of Leek. High rate of K (potassium) appears to stiffen the edible stalk where as high rate of N (nitrogen) produce more tender stalk. Irrigation is beneficial because it requires large amount of water during growth periods. Irrigation is required during the dried periods of the year. Along with growth and development soil placed around the stem and it is to be done at fortnight intervals.

3.3.3.4   Harvest and Post Harvest Technology

Generally leek is harvested at 100 to 120 days after sowing the seeds, but in case early variety matures within 60–70 days. The harvest starts when the stalks are about an inch across. It is normally harvested from late summer until early spring. Leeks are best used fresh. Owing to its characteristic shape, structure, and soft texture, associated with high moisture content, leeks is more susceptible to physical damages and during transportation losses will be accelerated due to improper infrastructure facilities, poor road accessibility and high climatic variation. Environmental factors such as relative humidity, O_2 concentration, increasing temperature can trigger the several reaction mechanisms that may lead to produce degradation. In case of home storage, wrapping them in a damp paper towel and placing them in a plastic bag in the refrigerator for seven to 10 days is essential.

3.3.3.4.1   Value Addition

Leek Oil: Distillation of the entire leek plant produces an essential oil with a distinctive sulphur odor. It is used in aromatherapy principally as a diuretic, like its close relative, the onion.

3.3.4   SHALLOT (ALLIUM ASCALONICUM L./A. CEPA VAR. AGGREGATM)

3.3.4.1 Systematic Position

Kingdom: Plantae
Division: Tracheophyta
Class: Magnoliopsida
Order: Asparaales
Family: Amaryllidaceae/alliaceae
Genus: *Allium*
Species: *ascalonicum*

3.3.4.2   About the Crop, National and International Scenario, Uses, and Composition

Shallots are a member of the onion family. It is closely related to multiplier onions, but smaller, and has unique culinary value. At maturity, shallot bulbs look like small onions. The edible bulb has a delicate flavor resembles with onion. The small bulbed onions have long, green leaves and are sometimes called scallions or spring onions. It is mainly used for cooking. Shallots have long been associated with fine French cookery. They are eaten as fresh or cooked, chopped or boiled. Shallots have a delicate onion flavor when cooked that adds to but it does not overpower other flavors. It can be successfully cultivated wherever onions are grown. However, most shallots are produced in Europe, particularly France. Most shallots consumed in the USA and Canada are imported chopped and dried from Europe.

Shallot contains 13% carbohydrate, 5% protein, 0.5% fat and different vitamins like folic acid, niacin, pantothenic acid, pyridoxine, riboflavin, thiamine, vitamin A, and ascorbic acid. It also contains good amount of minerals like calcium, iron, copper, manganese, magnesium, selenium, zinc, etc.

3.3.4.3   Soil and Climatic Requirements

The best soils for growing shallots are deep, well-drained clay loams that have high levels of organic matter. Shallots can be successfully grown on a wide range of soils, from relatively shallow, low-pH sandy soils to well-structured red volcanic soils and alkaline alluvial soils. It can grow in a wide range of soil pH but prefers a soil pH of 6.0 to 6.8. The looser the soil, the bigger your shallots will grow.

It is a crop of cool mild-to-mild tropical climates. Shallots can be grown over a wider climatic range than the common onion because they do not have specific day length or temperature requirements. Normally, seeds germinate at a temperature range of about 10°C to 30°C, with an optimal germination temperature of ranges between 18°C and 24°C. The ideal growing temperature is in the range of 13°C to 24°C. Plantings should be timed to avoid periods when daytime temperatures exceed 27°C.

3.3.4.4 Agrotechniques for Quality Production

It is propagated through cloves. Cloves are separated from the healthy and disease free bulbs. Overnight socking of cloves in water is recommended for faster sprouting. Socking in water containing a tablespoon of bicarbonate of soda (baking soda) is also recommended to protect them from fungal diseases. In colder climates, it is normally planted during September to early October) for harvesting them during May–June. A spacing of 15–20 cm × 15 cm is adopted for planting of shallot.

Well-decomposed farmyard manure or compost should be applied @ 15–20 t/ha at the time of final land preparation. Shallots removing about 130 kg of nitrogen, 30 kg of phosphorus and 60 kg of potassium/ha. Application N:P$_2$O$_5$:K$_2$O @ 120:60:60 kg/ha is beneficial for getting higher yield. Apart from major nutrient, application of trace elements like zinc, manganese, and boron as a foliar application increased the productivity as well as improved the quality of the produce. Direct seeded shallots are planted dry, then followed by a light irrigation. Frequent light irrigations are required until plants have emerged and developed a good root system. Frequency of irrigations mainly depends on the weather conditions, soil type and growing season. In case of the established crop irrigation should be given as and when required.

3.3.4.5 Harvest and Post Harvest Technology

Shallots are ready for harvest at 10 to 12 weeks after planting. Hybrid varieties require somewhat higher crop duration. Shallots are ready when the foliage becomes partly withered and bulbs have reached marketable size. Shallots are harvested manually by hand pulling and thereafter washed and bunched with the roots trimmed. Any damaged or diseased leaves should be removed. They are packaged in bunches of 10 to 12, depending on size. Forced air-cooling is desirable to maintain freshness and shelf life. True shallots are hand-pulled and should be stored in a cool, dry environment for about two weeks to allow curing.

3.3.4.5.1 Value Addition

Garlic, ginger, and shallot oil: It is a delicately fragrant and deliciously rich condiment with a multitude of uses made simply by gently frying and stirring the three ingredients in cooking oil of choice on low-medium heat.

3.3.5 CHIVE (ALLIUM SCHOENOPRASUM L.)

3.3.5.1 Systematic Position

Kingdom: Plantae
Division: Tracheophyta
Class: Magnoliopsida
Order: Asparagales
Family: Amaryllidaceae/alliaceae
Genus: *Allium*
Species: *schoenoprasum*

3.3.5.2 About the Crop, National and International Scenario, Uses, and Composition

Chive (*Allium schoenoprasum* L.) is a member of the onion family, but unlike most onions, the greens are harvested instead of the bulb. It is grown for its scapes. It is used for culinary purposes as a flavoring herb, and provides a somewhat milder flavor than those of other Allium crops. Chives are native to Asia and Eastern Europe. It is a common garden plant herb gardens throughout Europe. Chives are hardy, drought tolerant perennials 20 to 50 cm tall that grow in clumps from underground bulbs. The leaves are round and hollow, similar like onions, but smaller in diameter. In comparison to standard onions, chives have a much milder taste. The small grass-like herb is often added to soups, salads, and sauces for its light flavor and having aesthetic appeal. Chives can be eaten either freshly cut or in the dehydrated form. Frozen chives maintain their flavor for several months. It is also used for flavoring baked potatoes, soups, and is often used as a garnish. Chives' constituents equal those of the close relatives, onion, and garlic. The following volatile components have been identified: dipropyl disulfide, methyl pentyl disulfide, pentanethiol, pentyl-hydrodisulfid and *cis/trans*-3, 5-diethyl-1, 2, 4-trithiolane. Chives contain significant amounts of the vitamins A and C.

3.3.5.3   Soil and Climatic Requirements

Chives require well-drained soil, rich in organic matter, with a pH range of 6 to 7 and full sun. Typically, chives need a temperature of 15°C to 20°C for seed germination.

3.3.5.4   Agrotechniques for Quality Production

Before planting of chives, land should be prepared well to produce a suitable tilth for planting. The easiest and most successful means of propagating chives is planting rooted clumps in spring, after frost danger has passed. Established plants usually need to be divided every 3 to 4 years after planting. Division is best done in spring. It can be grown as an annual or a perennial. A spacing of 30 cm × 10 cm is ideal for planting of chives. It is important to give chives consistent watering throughout the growing season for getting high yields. Use mulch material to conserve soil moisture and keep the weeds down. For good production, application of organic manures and inorganic NPK fertilizer is essential for higher yield of the crop.

3.3.5.5   Harvest and Post Harvest Technology

Leaves and flowers typically develop in abundance during the spring and in early summer, and during this time harvesting of leaves and flowers should be done. It may be harvested over successive four- to five-week intervals throughout the summer and autumn as chives will continue producing clusters of leaves and flowers well after spring harvest.

3.3.5.5.1 Value Addition

Chive Powder: Ground chives are a delicate powder having a unique aroma.
  Chive Oil: It is not the oil extracted from chive. It is prepared by gently heating chives in any edible oil to create a concentrated elixir.

3.4 TREE SPICES

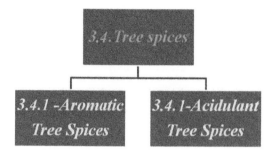

3.4.1 AROMATIC TREE SPICES

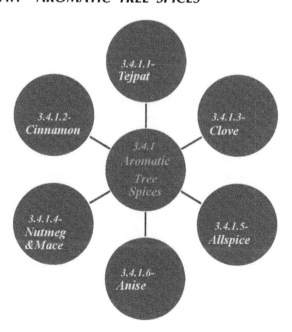

3.4.1.1 Indian Cassia or Tejpat (*Cinnamonum tamala* Nees and Eberm)

3.4.1.1.1 Systematic Position

Kingdom: Plantae
Division: Tracheophyta
Class: Magnoliopsida
Order: Laurales
Family: Lauraceae
Genus: *Cinnamonum*
Species: *tamala*

3.4.1.1.2 About the Crop, National and International Scenario, Uses, and Composition

Tezpat or Indian Cassia (*Cinnamonum tamala* (Buch Ham) Nees and Eberum) belongs to the family lauraeae is a tree spice. It is moderate size evergreen tree, leaves are glabrous and three nerved from the base. Its dry leaves are used as spice having a warm, powerful taste and odor. It is originated in India. Mainly found in tropical and subtropical Himalayan region upto an altitude of 2400 m. It is found frequently in the Northeastern parts of West Bengal, Khasia hills, Jayanti hills and also in homestead gardens of the Northeastern states of India. It is used as flavoring agent in the preparation of many vegetarian or non-vegetarian dishes. On steam distillation of leaves produces essential oil. The main composition of the essential oil is eugenol. Its leaf essential oil is used in pharmaceutical preparation, soap, and cosmetic industries. It is also used in flavoring all kinds of foods, meat, sausages, and sauces.

The yield of the oil on a dry weight basis ranges from 1.2% to 3.9% (w/w). Phenyl propanoids constitute the major portion (88.9–95.0%) of the oils. Fifty-four compounds are identified from the oils. Eugenol (91.4–41.8%) is the main compound, followed by eugenyl acetate (0.0–47.1%) and α-phellandrene (0.6–2.5%) in the oils.

3.4.1.1.3 Soil and Climatic Requirements

It can be grown all types of soil but light soil with having plenty of organic matter is best suited. It prefers humid climate.

3.4.1.1.4 Agrotechniques for Quality Production

The seed are used for propagation. Immediately after harvesting of the seed they are sown in nursery bed because the seeds are recalcitrant in nature. Seedlings are raised in the beds before the onset of monsoon. Usually seeds are sown during March–April and they germinate within one or one and half months. Seedlings are planted in the main field when they are Four- to five-years age. Seedlings are planted at a spacing of 3.0–4.5 m × 3.0–4.5 m.

3.4.1.1.5 Harvest and Post Harvest Technology

It takes another 5–6 years to grow and starts bearing 10th years onwards. Productivity of Tez pat remains up to 50–100 years. Natural regeneration will helps the farmer to raise the new plants at cheap rate. In early stages of growth manuring can be done. No manuring is followed in established garden. Pruning of leaves are done during the harvesting stage. The season of harvest is dry days of October to December and some areas collection extends up to March. Usually leaves are collected once in a year from young trees and every other year from old and weak trees. Normally yield varies from 8–20 kg dry leaves/plant/year and the average yield of tezpat is 12 kg dry leaves/plant/year.

3.4.1.1.5.1 Value Addition

It is used as a spice in local and regional culinary, and for production of essential oils (potential for local value addition) that are used in the food, flavor, and pharmaceutical industries. The following three value added products of tejpat are known:

- Tejpat oil
- Tejpat powder
- Tejpat oleoresin

Processing of leaves into essential oil and powder by wholesale traders takes place in Nepal and India respectively, but no regional trade in essential oil was observed. Three quality grades are sold in markets, which are not known to the farmers. The bay leaf trade is increasing, creating an opportunity for farmers to engage in its cultivation.

3.4.1.2   Cinnamon or True Cinnamon (*Cinnamonum verum*)

3.4.1.2.1   Systematic Position

Kingdom: Plantae
Division: Tracheophyta
Class: Magnoliopsida
Order: Laurales
Family: Lauraceae
Genus: *Cinnamonum*
Species: *zeylanicum*

3.4.1.2.2   About the Crop, National and International Scenario, Uses, and Composition

Cinnamon is an important tree spice of India. Cinnamon, a native of Sri Lanka was introduced into India by Robert Brown in 1835 in Angarakandy Estate of Kannur. Dried inner bark of the tree is the cinnamon of commerce. Sri Lanka, India, Seychelles, Myanmar, South America, Malay Peninsula are cultivating cinnamon. Sri Lanka followed by Seychelles is the largest producer of cinnamon bark with best quality. In India it is commercially cultivated in Kerala, Karnataka, and Tamil Nadu. There are some prominent differences between Ceylon Cinnamon and Cassia Cinnamon as listed in Table 3.10.

3.4.1.2.2.1   Uses

Bark has a delicate fragrance and a warm sweet agreeable taste. It is extensively used as a spice in the forms of small pieces or powder. Cinnamon bark and cinnamon bark oil also used for flavoring confectionary, liquors, soap,

TABLE 3.10 Difference Between Ceylon Cinnamon and Cassia Cinnamon

Ceylon Cinnamon	Cassia Cinnamon
Soft in Texture, easily broken	Hard texture not easily broken
Soft and Sweet Aromatic flavor	Pungent and very Spicy flavor
Light Brown in color	Dark Brown or Reddish in color
Soft in appearance	Rough in Appearance
Number of folders/layers in quill	One inward folded, empty cavity
Native to Sri Lanka	Nativeto, China, Indonesia, Vietnam
Low Cinnamaldehyde	High Cinnamaldehyde
Coumarine content 0.004%	Coumarine content 5%
Generally safe	Toxic in prolonged use

pharmaceuticals, and dental preparations. Leaf oil is a common adulterant for the bark oil. Unripe fruits are used as adulterant in cloves.

Cinnamon consists of a variety of resinous compounds, including cinnamaldehyde, cinnamate, cinnamic acid, and numerous essential oils. The spicy taste and fragrance are due to the presence of cinnamaldehyde and occur due to the absorption of oxygen. The presence of a wide range of essential oils, such as trans-cinnamaldehyde, cinnamyl acetate, eugenol, L-borneol, caryophyllene oxide, β-caryophyllene, L-bornyl acetate, E-nerolidol, α-cubebene, α-terpineol, terpinolene, and α-thujene, has been reported. The most important constituents of cinnamon are cinnamaldehyde and trans-cinnamaldehyde (Cin), which are present in the essential oil, thus contributing to the fragrance and to the various biological activities observed with cinnamon. One of the major constituents of essential oil extracted from *C. zeylanicum* named (E)-cinnamaldehyde has an antityrosinase activity, while cinnamaldehyde is the principal compound responsible for this activity. Cinnamon bark contains procyanidins and catechins. The oil is characterized by linalool (36.0%), methyl eugenol (12.8%), limonene (8.3%), α-terpineol (7.8%) and terpinen-4-ol (6.4%).

3.4.1.2.3 Soil and Climatic Requirements

Cinnamon flourishes in wide range of soils, even in marginal soil with poor nutrient status. Sandy loam soil rich in organic matter is the best.

The quality of the bark is influenced by soil and ecological factors. The best economic produce is obtained when grown in silicaceous sandy soils. Water logged and marshy soil should be avoided as they give an undesirable product.

Cinnamon is a hardy plant. It tolerates wide range of climatic condition. It comes up well from sea level up to an elevation of 1000 m. Annual rainfall of 150–20 cm with an average of temperature of 27°C is ideal for its cultivation. A hot and moist climate is highly suited for its cultivation of cinnamon and prolonged spell of dry weather are not concussive for its growth.

3.4.1.2.4 Botany

Cinnamonum verum is a medium sized tree having a tendency to branch profusely at low level with strongly aromatic bark and leaves. Flowers are small, numerous in panicle, bisexual, and pollinated by insects. The fruit (berry) consists of pulpy pericarp and the seed have the hardy endocarp. The important varieties of cinnamon with parentage, average yield (kg/ha), bark recovery (%), bark oil (%), leaf oil (%) along with salient features are given in Table 3.11.

3.4.1.2.5 Agrotechniques for Quality Production

3.4.1.2.5.1 Propagation

The most widely method of propagation is through seed. Seeds are extracted from ripe fruits from selected mother trees with desirable characters like (i) smooth bark, (ii) erect stem, (iii) vigorous growth, (iv) free from disease and pest, and (v) having good qualities like sweetness, pungency, and flavor. The fully ripened fruits are either picked up from trees or fallen ones are collected from the ground. Seeds are removed from fruits, washed free of pulp and sown immediately. The seeds are sown in bed or polythene bags containing a mixture of sand, well rotten manure and soil (2:1:1). Germination occurs within in 15–20 days. Cinnamon is also propagated by cutting and layering.

TABLE 3.11 Varieties of Cinnamon

Sl No.	Variety	Pedigree/ Parentage	Average Yield* (kg/ha)	Bark recovery%	Bark oil%	Leaf oil%	Salient features
1	YCD.1	Clonal selection from op seedings progenies of Sri Lankan type	360	35.3	2.8	3	Good bark recovery adopted to wide range of soil and rainfed conditions. Recommended for high ranges at 500–1000 m above MSL.
2	PPI (C)-1	Selection from OP seedlings progeny introduced from Sri Lanka	980	34.22	2.9	3.3	Suitable for cultivation in high rainfall zones and hill regions of Tamil Nadu at an altitude range of 100–500 m MSL.
3	Konkan Tej	Seedling selection from progenies of Sri Lankan accessions	334	29.16 51.78	3.2	2.28	Superior qualities with 3.2% bark oil with bark recovery 29.16%, cinnamaldehyde in bark oil 70.23, eugenol in bark oil 6.93%, eugenol in leaf oil 75.5%, yields 4.10 kg fresh bark.
4	Sugandhini (ODC-130)	Single tree selection from Wayanadu local collection. A Sri Lankan type	640	51.0	0.94	1.6	Recommended for cultivation for leaf oil production, cinnamaldehyde in bark oil 45%, eugenol in leaf oil 93.7%; released mainly for leaf oil purpose.
5	RRL(B) C-6	Selection from germplasm collection- OP seedling progenies	250	–	–	–	High quality, sweet, and pungent bark with 83% cinmaldehyde content in bark oil, 94.0% eugenol in leaf oil, leaf oil 0.8%. Spreading, branching, nature.

TABLE 3.11 (Continued)

Sl No.	Variety	Pedigree/ Parentage	Average Yield* (kg/ha)	Bark recovery%	Bark oil%	Leaf oil%	Salient features
6	IISR Nithyashree	Clonal selection from OP seedling progeny	200	–	2.7	3	Good regeneration capacity, bark, and leaf oleoresin contents are high. Good bark recovery with good aroma and taste. Bark oleoresin-10.0%.
7	IISR Navashree	Selection from Op seedling progeny of Sri Lankan collection	200	40.6	2.7	2.8	High quality line, good bark recovery with good aroma and taste, grow well in plains and high elevations. High cinnamaldehyde content (73%) in bark oil, medium quality. High shoot regeneration. Bark oleoresin 8%, cinnamaldehyde in leaf oil 15%, eugenol in bark oil 6%, eugenol in leaf oil 62%.

3.4.1.2.5.2 Planting

1–2 year old seedlings are ready for planting in the main field. Planting starts with the commencement of South-western monsoon. Planting is done in a pit of 60 cm × 60 cm × 60 cm at a spacing of 2 m × 2 m. Pits are prepared in advance and filled with leaf mould and top soil before planting.

3.4.1.2.5.3 Aftercare

Shade and irrigation are essential immediately after planting if dry weather prevails. Three or four weeding are required for first two years depending upon weeding intensity, thereafter two weeding in a year is sufficient.

The fertilizer dose of 20:20:25 g NPK/seedlings along with 20 kg FYM or compost are applied during the 1st year and it is gradually increased to 200:180:200 g/tree/year with 50 kg FYM or compost for grown up plants of 10 years and above. Fertilizers are applied in two equal splits, first in May–June and second in September–October.

3.4.1.2.6 Plant Protection

Stripe Canker: It is a fungal disease and caused by *Phytophtora cinnamoni.* It is mostly found on trunk and branches of young trees grown under poor drainage condition. Vertical stripes of the dead bark occur near the ground level is the common symptom of this disease. Spraying BM (1%) at interval of 40–45 days is a good preventive measure

Gray blight: It is a fungal disease and caused by *Pestalotiopsis palmarum.* Blightening symptoms can be seen on leaves. This disease can be managed by spraying of Bordeaux Mixture (1%), Benomyl (500 ppm) or Captafol (100 ppm).

Pink disease: It is caused by *Pestalotiopsis palmarum.* Pink encrusted areas on stem causing death of smaller shoots are the most common symptom of this disease. This disease is also managed by spraying of Bordeaux mixture (1%), Benomyl (500 ppm) or Captafol (100 ppm).

Pink shoot borer: Spraying carbaryl (0.25%) is the effective way for management of his disease.

Leaf eating caterpillar: Spraying quinolphos (0.05%)is the effective way for management of his disease.

Leaf minor (*Phyllocnistic lhrysophalema*): Larvae of insect mine the leaves of cinnamon is the common symptom of the pest infestation. Pest incidence can be controlled by spraying monocrotophos (0.05%).

Tussock caterpillar: Caterpillar attack the leaves and it can be managed by spraying of dimethoate (0.03%), dimecron (0.04%).

3.4.1.2.7   Harvest and Post Harvest Technology

3.4.1.2.7.1   Coppicing

When the trees are 2–3 years old, it may be pruned to a height of 15 cm above ground and covered by with loose soil. Side shoots are cut to encourage more auxiliary branches. It should be practiced till the whole plant assumes the shape of low bush.

3.4.1.2.7.2   Harvesting and Yield

The first harvest (coppicing) is made after 2nd or 3rd year of transplanting and subsequent harvest is made 12–18 months after previous harvest. The shoot selected for harvest is usually 1–2 m long with 1–2 cm thick. Ideal time for cutting is change of red flush of the young leaves to green and this is the indication of the free flow of sap between the bark and wood. The cut stems are collected, tied, bundled, and carried to the shed.

Quills of 65–125 kg/ha are obtained from the first harvest. The plants with an age of 10–11 years yield about 225–300 kg quills/ha. In addition about 1 tones of leaves are also obtained out of which 2.5–2.6 kg of cinnamon leaf oil can be obtained.

3.4.1.2.7.3   Processing

The processing is done by following these steps:

Peeling: Peeling is done with a help of knife.

Rolling: Bark are packed together, placed one above another and pressed and covered, kept in this way for one day for fermentation.

Pipping: Scarp off the outer skin with a small knife.

Grading: Grading of quills according to the thickness of bark.

Fine or continental grade: ranges from 10–19 mm.

Humburg grade: ranges from 23–32 mm.

Mexican grade: Intermediate in quality between fine and Humburg.

Quillins: Broken length and fragment of quills sold as quilling.

Feathering: Inner barks of the twigs and twisted shoots.

Chips: Includes terminated shoots before they are peeled. They invariably contain more or less woody materials.

3.4.1.2.7.4   Value Addition

- Cinnamon oil
- Cinnamon powder
- Cinnamon tablets

3.4.1.3   Clove (*Eugenia caryophyllus* Bullock et Harrison/*Syzygium aromaticum*)

3.4.1.3.1   Systematic Position

Kingdom: Plantae

Division: Tracheophyta

Class: Magnoliopsida

Order: Myrtales

Family: Myrtaceae

Genus: *Eugenia*

Species: *caryophyllus*

3.4.1.3.2   About the Crop, National and International Scenario, Uses, and Composition

Clove, *Syzygium aromaticum* (Syn. *Eugenia caryophyllus)* is an important spice well-known for its flavor and medicinal values, which is the dried unopened flower buds of the evergreen tropical tree. It is indigenous to Moluccas Island (Indonesia). The major producers of this spice are

Indonesia, Zanzibar, and Madagascar. In India, it is mainly cultivated in the Nilgiris, Tirunelveli, and Kanyakumari districts of Tamil Nadu, Calicut, Kottayam, Kollam, and Trivandrum districts of Kerala and South Kanara district of Karnataka.

Essential oil of clove comprises in total 23 identified constituents, among them the main constituents of the essential oil are phenylpropanoids such as eugenol (87.00%), eugenyl acetate (8.01%) and β-Caryophyllene (3.56%), carvacrol, thymol, and cinnamaldehyde.

3.4.1.3.3 Soil and Climatic Requirements

Clove is strictly tropical evergreen plant. It requires a warm humid climate having a temperature of 20 to 30°C. Humid atmospheric condition and a well-distributed annual rainfall of 150 to 250 cm are necessary. It grows well in all situations ranging from sea level up to an altitude of 1500 meters and also in places proximal to and away from sea.

It grows satisfactorily on well-drained laterite soils, clay loams and rich black soils. But deep black loam soil with high humus content found in the forest region is best suited for its cultivation. Sandy soil is not suitable.

3.4.1.3.4 Agrotechniques for Quality Production

Clove is propagated through seed called mother clove, which become available from June to October. Fruits are allowed to ripe on the tree itself and drop down naturally. Such fruits are collected from the ground and sown directly in nursery or soaked in water overnight to remove pericarp before sowing, which gives quicker and higher germination. They must be sown immediately as they lose their viability within one week after harvest under normal conditions. Generally big sized seeds give higher% of germination. Seeds are sown in nursery at 2–3 cm spacing and in depth of about 2 cm. The seedbeds must be protected from direct sunlight. The germination originates within 10 to 15 days and may last for about 40 days. These germinated seedlings are transplanted in polythene bags (30 cm × 15 cm) containing a mixture of good soil, sand, and well-decomposed cow dung in the proportion of 3:3:1. The nurseries are regularly shaded and irrigated daily to ensure uniform growth. The seedlings are ready for transplanting in the field when they are 18 to 24 months old. Pits of 60 to 75 cm^3 dimension are dug out at a spacing of 6–7 meters for transplanting. These pits are partially filled with compost, green leaf or cattle manure and covered with topsoil. The seedlings are transplanted in the main field with the commencement of rainy season in month of June–July and in low-lying areas, towards the end of the monsoon, in September–October. A clove is shade-loving plant; it requires partial shade for better growth. Hence, it can conveniently be grown as mixed crop with commercial crops like arecanut, coconut etc.

For proper growth and flowering clove trees should be manured regularly and cautiously. During the first year, only 15 kg compost per tree is necessary. In the second year, 20 kg compost along with 80 g nitrogen, 220 g phosphorus and 160 g potassium per tree is desirable. After that there should be an annual increase of 5 kg compost and 40:110:80 g NPK per tree thus reaching a dose of 600:1560:1250 NPK per tree for a 15 year and above tree. After 15 year onwards compost is not required.

No inter cultivation, except weeding is usually done for clove. As the branches of full-grown trees have a tendency to overcrowd, thinning them occasionally may keep the growth within manageable proportion. Dead and diseased shoots should be removed periodically. Irrigation is necessary in the initial stages.

3.4.1.3.5 *Plant Protection*

Sudden death: It is a fungal disease and cause by *Valsa euginae*. This disease is more common at immature stage of the plant. In case of the affected plant chlorosis of the leaves is the initial symptom and thereafter followed by sudden and very rapid leaf fall accompanied by wilt. Considerable proportion of leaves dry up on the tree without abscission becoming a bright russet red. It takes only a few days from onset of leaf fall to the final death of the tree and it is from the apparent suddenness that the disease derives its name sudden death. Finer ramification of the root system of the diseased plant completely disappear. Replacement of diseased tree is the only control measure for this disease.

Acute Blight: This fungal disease and cause by *Cryptosporella* eugeniae. This disease is characterized by sudden dieback. Leaves dry up upon the tree to a conspicuous russet red color. In the severe case attack may be extent to main trunk and tress dies in a manner similar to one caused by sudden death. Careless cultivation of young saplings, sometimes cause injury to the collar and roots. Reddish brown coloration can be seen after splitting the diseased branch. Once the pathogen has entered, its progresses down the branch often girdling and reaching the main trunk. Rotting of individual branches is often noticed. Remove infected branches and all cut surfaces should be covered with protective fungicidal point for management of this fungal disease.

Leaf blight/twig blight: It is caused by *Colletotrichum gleosporoides*. This disease is characterized by small, circular or oral lesions specks scattered on leaf lamina. Specks gradually enlarge and develop into distinct sopt with an ashy gray center surrounded by dark margins. Spots collease and form irregular necrotic patches. It spread from twig to petiole where twigs show dieback symptoms resulting in defoliation. It is severe during rainy season may Cause heavy loss if spread to panicle and flower. Blackening and shedding of flower buds, shriveling, and drying of pedicel. White growth of fungus can be seen on fallen buds. Spraying Bordeaux mixture (1%)/difolton (0.03%)/zineb (0.2%) twice @ fortnightly interval during rainy season but before flower bud formation till the harvest of bud is the common method for the management of this disease.

Little leaf: it is caused by Mycoplasma like organisms. Affected plant shows complete repression of vegetative growth. The branch grow at an acute angle with the trunk giving the tree a cone shape appearance. Trunk shows numerous protuberance. Internodes are shortened. Clustering of small and crackled leaves also noticed. Disease is not transmitted by either through sap or by grafting. Application of 4–5 sprays of tetracycline (200 ppm) at fortnightly intervals on infested seedlings allow production of normal leaves and internode elongation.

Leaf blight: It is a fungal disease and caused by *Cylindrocladium quinqiseptatum*. Incidence of this disease is more common in nurseries during the month of July–August. Defoliation is the common symptom of this disease and defoliated plants often dry up from tip downward. Affected leaves show dark brown lesions on the margins as well as on the tips, the lesions merge causing leaf rot symptoms leading to defoliation at high humid condition. Infected leaf portions shows whitish growth that contains abundant spores.

To control this disease remove the infected plants from the nursery and plantations. Spray Bordeaux Mixture (1%) as prophylactic spray or carbendazium @ 0.2% or copper fungicides or mancozeb (0.2%-0.4%) on plants is effective for management of this disease.

Seedling Blight: Severe incidence of this disease is noticed in poor drainage in polybags/nursery. This results in degeneration of root system associated with rhizobacteria, phytophtora. Affected seedlings shows dark brown coloration at the base of stem and loose natural cluster and also show drooping symptoms and finally such seedlings dry up.

For management of is disease, drench nursery with COC (0.2%) or mancozeb (0.3%) and transplant seedlings in evening to reduce heat and transplant shock.

Black scale (Saissetia nigra), Marked scales (Mycetaspis personalta), Green scales (Lecaniune spp) and Mealy bug (Pulvinaria spp) are the common pests of clove and these pests can be managed by removal of affected leaves and destruction will prevent further spread infection. Application of dimethoate 2 mL/L or monocrotophos 1.5 mL/L is essential in case of severe infection.

3.4.1.3.6 *Harvest and Post Harvest Technology*

Clove tree gives yield from the seventh or eighth year after planting and full bearing stage is attained after about 15 to 20 years. In plains, flowering starts during September–October and during December–February at high altitudes. Flower buds are produced on young flush and it takes about 4 to 6 months for harvesting. The change in color from green to slightly pinkish tinge is the perfect stage for picking clove buds. The harvested flower buds are separated from the clusters by hand and spread in the drying yard for drying. It takes generally 4 to 5 days for drying. The correct stage of drying is reached when the stem of the bud is dark brown and the rest of the bud lighter brown in color. The average annual yield of 2 kg per tree may be taken after 15th year. A well-maintained full-grown tree under favorable conditions may give 4 to 8 kg of dried buds.

3.4.1.3.6.1 *Value Addition*

The clove consists of:

- Volatile oil (15–20%)
- Eugenol (70–90%)
- Acetyl Eugenol
- α,β-Caryopyllene

- Tannins
- Other substances mainly methyl furfural and dimethyl furfural.
- Critical points to consider during manufacturing and processing of clove:
- The right stage of harvesting clove buds is when flower petals change their color from olive green to yellow pink. Clusters of flowers are harvested together with the stalks.
- The flower buds should be separate from the stalks and both buds and stalks are dried in sun or artificial drier until they become dark brown and hard.
- Well dried good quality cloves are in golden brown color and badly dried cloves are soft and pale brown with a whitish mealy appearance, which are known as "khuker" cloves.
- Green clove buds of the right stage give about 30% dry cloves. Well-dried cloves (8–10% moisture) can be stored in gunny bags without damage by fungus and insects for 1 or 2 years.
- *Clove oil:* It consists of several compounds such as Eugenol, which is the major component (85–90%) and eugenol acetate (9–10%). Eugenol has a boiling temperature of 254°C and can be steam distillated from freshly ground clove.
- Clove oil – Obtain from distillation of flower buds
- Clove leaf oil – Obtain from distillation of leaves

Other value addition products:

- Clove is used in a specific type of cigarette.
- Due to the bioactive chemicals of clove, the spice may be used as an ant repellent.
- They can be used to make a fragrance pomander when combined with an orange. When given as a gift in Victorian England, such a pomander indicated "warmth of feeling."
- Toothpastes
- Chewing Gums
- Shaving soaps
- As organic weed Killer
- Perfumes
- Insects Repellents
- Possible/developed future products
- Dairy Products can be implemented with Clove. Such as

- Clove Yoguts
- Clove Butter
- Clove Milk
- Clove Cheese
- Clove Ice Cream

3.4.1.4   Nutmeg and Mace (*Myristica fragrans* Hout)

3.4.1.4.1   Systematic Position

Kingdom: Plantae
Division: Tracheophyta
Class: Magnoliopsida
Order: Magnoliales
Family: Myristicaceae
Genus: *Myristica*
Species: *fragrans*

3.4.1.4.2   About the Crop, National and International Scenario, Uses, and Composition

Nutmeg is an important tree spice crop in India. It is a unique spice crop in the World as two spices namely nutmeg and mace are obtained from the single plant (*Myristica fragrans* Hourt). Kernel portion of the *Myristica fragrans* Hourt is known as nutmeg and aril portion is known as mace. It is originated from the Moluccas Island of the Indonesia. It is cultivated Indonesia, Grenada, Sri Lanka, India, Malaysia, Tanzania, Mauritius, and China. Indonesia and Grenada is the major producer of the World and contribute more than 80% World production. In India, it is commercially cultivated in Kerala, Karnataka, and Tamil Nadu. Eastern and Western Ghat hilly areas of the India are the highly suitable for cultivation of Nutmeg. India's yearly (April–March) exports of nutmeg and mace are around 2.5 thousand tonnes. Vietnam and UAE are the major importer of Indian spice (nutmeg and mace). Meanwhile India annually imports on an average 1 thousand tonnes of nutmeg and mace.

About 30–55% of the seed consists of oils and 45–60% consists of solid matter including cellulose materials. There are two types of oils: (1) The "essential oil of nutmeg" also called the "volatile oil" accounts for 5–15% of the nutmeg seed; and (2) the "fixed oil of nutmeg" sometimes called "nutmeg butter" or expressed oil of nutmeg accounts for 24–40% of the nutmeg seed. Sabinene or camphene is the major component (50%) of the essential oil followed by d-pinene, dipentene, d-linalool, d-borneol, geraniol, myristicin, etc. Mace contains 10–12% of essential oil of fixed oil. The major components of mace are more or less similar to that of nutmeg. Nutmeg and mace is widely used as condiments and in pharmaceutical industries. They mostly used in meat preparation and in bakery products.

3.4.1.4.3 Soil and Climatic Requirements

Clay loam, sandy loam and red lateritic soil is ideal for its cultivation. Being a tropical plant it prefers a warm and humid climate. It can with an annual rainfall of grows up to an altitude of 1300 m. the optimum temperature range is 27–29°C with an annual rainfall of 150–300 cm per annum with a dry period of 2–3 months.

3.4.1.4.4 Agrotechniques for Quality Production

Usually nutmeg is propagated through seed. The fleshy rind of the fruit as well as mace covering the seed is removed before sowing. In the nursery bed, regular watering is necessary for good germination. Germination begins from about 30 days and continues upto 90 days. About 20 days old sprout are transplanted to polythene bags. About 18–24 months old seedlings are transplanted in the mainfield. Presently, nutmeg is propagated vegetatively by epicotyls grafting.

Planting is normally done during the onset of monsoon. Pits of 60 cm × 60 cm × 60 cm size prepared one month before planting and filled with top soil and well rotten farmyard manure. A spacing of 8 m × 8 m is adopted for cultivation of nutmeg. Young plants are provided with artificial shade to protect the young plants from dry wind and sunshine during summer months by planting shade trees like Erythrina, Bananan, Glyricidia, etc. and artificial irrigation should be provided during the dry months to check the mortality of the young plants. Nutmeg can also be planted in coconut, arecanut, and coffee plantation as mixed crop.

TABLE 3.12 Important Varieties of Nutmeg

Sl No.	Variety	Pedigree/Parentage	Average Yield	Myristicin%	Elimicin%	Salient features
1	Konkan Sugandha	Single plant selection from local seedling population	200- 526 fruits/ tree	–	–	Adaptable in Konkan region. Tree canopy is conical and compact. No incidence of major pests/disease
2	Konkan Swad	Selection from nutmeg seedling from Ratnagiri district	761.38 fruits/ tree	–	–	Adapted in Konkan region with warm, humid condition as well as shade. Canopy is erect, conical shape. Bark contains 39.85 essential oil in seed and 10.9% in mace. No incidence of major pests/disease are noticed
3	IISR Viswasree	Clonal selection from elite germplasm	1000 fruits/tree (1.33 kg mace, 9 kg dry, 3122 kg/ha)	–	–	Bushy and compact canopy, low incidence of fruit rot. Nut recovery 70%, mace recovery 35%, oil, 7.14% mace oil, oleoresin 2.48% and mace 13.85 respectively, butter 30.9%, myristicin 12.48% and mace 20.03% respectively. Especially suitable under mixed cropping system

The important varieties of nutmeg with their parentage, average yield, myristicin (%), elimicin (%) and the salient features are given in Table 3.12.

3.4.1.4.4.1   Nutrient Management

The fertilizer dose of 20:18:50 g NPK/seedlings along with 10 kg FYM or compost are applied during the 1st year and it is gradually increased to 500:250:750 g/tree/year with 50 kg FYM or compost for grown up plants of 15 years and above. Fertilizers are applied in two equal splits, first in May–June and second in September–October.

3.4.1.4.5   Plant Protection

Shoot rot or die back: This fungal disease is caused by *Colletotrichum gloeosporioides*. Sunken spots surrounded by yellow halo appear as the initial symptom of this disease. Later the central portion of the necrotic region drops off resulting in shot hole and Die back appears on mature branches. On young seedlings–drying of leaves and subsequent defoliation is seen. It can be controlled by spraying 1% Bordeaux mixture spray and applying Bordeaux mixture (1%) 2–3 times in the rainy season.

Fruit rot: It is caused by *Colletotrichum gloeosporioides* and *Botrydiplodia thiobromae*. Water soaked lesions appears on fruits, the tissues of which become discolored and disintegrated. Premature splitting of the pericarp and rotting of the mace and seed appear as the main symptom. The internal tissues are found rotten and fallen fruits are enveloped with growth on fungi.

Sparying Bordeaux mixture (1%) or Mancozeb (0.25%) 2–3 times in rainy season is very much effective for management of this particular disease.

Dry Rot and Wet Rot: Dry rot is caused by *Diplodia natelensis* and Wet rot is caused by *Phytophtora spp.* Dry rot usually seen from December onwards whereas wet rot starts during monsoon. Depressed necrotic patches appear on stalk of fruit resulting in dark brown drying. Symptoms for both are similar except that the lesions will not be sunken and enlarged fast leading fruit rot in wet rot. Spraying Bordeaux mixture (1%) 2–3 times in rainy season can effectively control these diseases.

Thread Blight: It is a common disease of nutmeg. Falling leaves are held attached to branches by means of fine filamentous fungus threads is the symptom of this disease. Spraying Bordeaux mixture (1%) effectively controls the disease.

3.4.1.4.6   Harvest and Post Harvest Technology

Generally seedling trees bear in the age of 7–8 years and vegetatively propagated trees normally take 4–5 years for first flowering and fruiting. Nutmeg takes 15–20 years for attaining full bearing age. Though, nutmeg fruits can be harvested throughout the year. However, peak harvesting period is June–August. Splitting of the fruit is indication of the ideal harvesting of the nutmeg fruits. After harvesting the fruits, aril portion and karnel portion is separated. Nutmeg: Mace ratio varies from 20:3. A full grown tree produces 2000–3000 fruits per year. Nutmeg produces a yield of 800 kg nutmeg/ha and aril yield of 100 kg/ha.

3.4.1.4.6.1   Sex Expression

Segregation of seedling into a ratio of 1:1 for male and female trees results in unproductive trees. Keeping one male tree per ten female trees are, therefore, necessary. Rest other male trees are converted into female tree by top working.

3.4.1.4.6.2   Value Addition

There are different grades of nutmeg and mace:
- Sound Unassorted Nutmegs (SUNS)
- Sound Selected Nutmegs
- Grenada Unassorted Nutmegs (GUNS)
- Dry in Shell 'Distillation Nutmegs'
- Grenada Broken and Clean (GBC) Nutmegs

    – Whole Red Mace No. 1
    – Whole Pale Mace No. 2
    – Broken Mace No. 3
    – Broken Pieces and Picking

TABLE 3.13 List of Nutmeg Processors and Products

Processor	Products
GCNA	Nutmeg and mace (commodities). Currently developing other value added products
West India Spices, Inc.	Essential oil, butter, oleoresin, spice mixes
De La Grenade Industries Ltd.	Nutmeg jam, jelly, syrup, nutmeg tamarind sauce, nutmeg ginger Barbecue sauce
Rainbow Products	Nutmeg syrup, nutmeg preserves
Noelville Ltd.	Nutmeg pain relief spray and cream
Grenada Distillers Ltd.	Liqueur, rum punch, spicy rum
Baron Foods (Grenada) Ltd.	Nutmeg ketchup
SACCS	Nutmeg jam, jelly, sauce/nutmeg ginger sauce, nutmeg pepper jelly, nutmeg guava jam
The Grenada Chocolate Co. Ltd.	Chocolate nutmeg truffle
Sugar and Spice Investments	Nutmeg ice cream
G-Links	Nutmeg honey
Pappy's Product	Nutmeg extract
GRENROP (members)	Nutmeg jam, jelly, balls, syrup, snacks, preserve, confectionery, seasoning, sauces, spicy cocoa balls
Belmont Estate	Jams, balls, whole and ground nutmeg
Grenada Distillers	Liqueur, rum punch, spicy rum
Carolus Caribbean	Cosmetics, body care products
Arawak Islands Ltd.	Aloe nutmeg balm, nutmeg soap, body oil, nutmeg syrup, spice bar, spice jar, rum punch, spice basket
Cluster of 20 Plus Small agro-processors	Jams, jellies, confectionery
New products to be developed	
Barguna Enterprise Ltd.	Nutmeg wine
Bowen's Products	Nutmeg oil, traditional rums
Market & Tourist Vendors	Nutmeg with shell, without shell, spice mixes

TABLE 3.14 Main Components of Nutmeg Essential Oils

Compounds	Grenadian Oil	Sri Lankan Oil	Indonesian Oil
α-thujene	4.7	1.2	1.7
α-pinene	11.5	14.8	22.3
sabinene	28.2	41.8	19.5
β-pinene	10.8	12.0	14.8
limonene	4	3.7	3.9
8-cineole	3.2	2.2	2.1
Terpinen-4-ol	9.5	2.2	4.3
Safrole	0.17	0.9	0.17
Myrisiticin	0.52	4.1	10.2
Elemicin	0.94	2.1	0.37

The Table 3.13 will summarize the nutmeg processors and products. The Table 3.14 will summarize the main components of nutmeg essential oils.

3.4.1.5 Allspice (*Pimenta dioica* L./Merr.)

3.4.1.5.1 Systematic Position

Kingdom: Plantae
Division: Tracheophyta
Class: Magnoliopsida
Order: Myrtales
Family: Myrtaceae
Genus: *Pimenta*
Species: *dioica*

3.4.1.5.2 About the Crop, National and International Scenario, Uses, and Composition

It is a minor tree spices, indigenous to West Indies and tropical America. All spice of commerce is the dried immature fruit. The flavor resembles a mixture of cinnamon, clove, and nutmeg and hence the name. The word 'Pimento' is derived from 'Pimenta' the Spanish word for the pepper corn, as the spice resembles pepper corn. The plant is reported to be abundantly grown in Jamaica, which is the major producer. Allspice also occurs in Southern Mexico, Honduras, Guatemala, Cuba, and Costa Rica. In India plant is reported to be grown in garden especially in West Bengal, Bihar, and Orissa. The cultivation is slowly coming in Kerala, Karnataka, and Tamil Nadu.

3.4.1.5.2.1   Composition

Typical analysis of allspice berries (ground) shows the following composition:

Moisture – 8.8%	Ca – 0.8%
Protein – 6%	P – 0.1%
Fat (the extract) – 6.6%	Na – 0.08%
Fiber – 21.6%	K – 1.1%
CH_2O – 52.8%	Fe – 7.5 mg/100 g
Total ash – 4.4%	Vit C (Ascorbic acid) – 39.2 mg/100 g
Niacin – 0.9 mg	Vit B_1 (Thyamine) – 0.1 mg/100 g
Vitamin – 1445IU	Vit B_{12} (Riboflavin) – 0.06 mg
Calorie – 380/100 g	

All spices owes its characteristics flavor due to presence of essential oil (3.3–4.5%) concentrated mainly in the pericarp. It also contains soft resin with a burning text, fixed oil – 5.8%, protein – 5.8%, crude starch – 21.8% and traces of alkaloids. The berries are used as a condiment and flavoring agent in ketchup, soup, sauces, pickle, canned, sausages, grieves, relishes, fish dishes, pies, puddings preservative, etc. used as flavoring ingredient for wines and as perfume in soap making. It is an important ingredient of whole mixed pickling spice and spice mixture. Such as curry powder, poultry dressing, frankfurter, and hamburger. All spice is used as an aromatic stimulant in digestive troubles, powdered fruit is used in flatulence and dyspepsia and diarrhea. Earlier it is used as medicine as adjuvant to stonic. It is considered to be carminative. It aids in rheumatism and neuralgia. Pimenta berry oil is used as flavoring agent in food product, perfumery, and soap preparation.

3.4.1.5.3   Soil and Climatic Requirements

The plant grows in a wide range of soil, it is not suited laterite soil and with inadequate moisture. Pimento grows to a sea level to 1000 m above MSL. However, grows below 300 m and annual rainfall of 100–200 cm or more/annum with a monthly minimum temperature up to 27°C is the best. Regarding soil it prefers well-drained fertile soil. Performance of all spice in plains not so good and fruiting does not occur.

3.4.1.5.4   Agrotechniques for Quality Production

Commercially it is propagated through seed. Ripe fruits are collected from high yielding and regular bearing trees. Seeds- stored as ripe berries after collection without extracting upto 3 weeks. Seeds viability gets reduced slowly after a period of 3 weeks and completely lost after 9 weeks Seeds are extracted after soaking overnight. Drying up of seed is done in shade. Seeds are sown in nursery bed of 1.2 m width are prepared by using light soil and incorporating organic matter and mixture of sand and coir dust and coir dust alone. After sowing nursery beds are mulched to hasten germination. Dried leaves, straw, damp sacks are used as mulch. Watering is done by using fine spray. Germination takes between 9–10 days or in some times 15 days. Three-week aged seedlings are transplanted in polybags and normally 6 months aged seedlings are ready for field planting. Vegetative propagation by chip budding, approach grafting, stooling, air layering, and top working is also possible. Applications of IBA (1000–1500 ppm) enhance rooting in allspice. Best time of air layering is January. Application of IBA (4000 ppm) and NAA (4000 ppm) in charcoal give 73% success during the month of December at IISR, Kozhikode.

6–10 months old seedling with a height of 25–30 cm is ideal for field planting. Normally seedlings are planted at a spacing of 6 m × 6 m spacing. Closer spacing is suggested in poor and infertile soil. To avoid losses instead of one seedling three are planted in a pit of 60 cm × 60 cm × 60 cm in size. In case of grafted plant, planting one plant per is advocated. To ensure pollination it is necessary to plant female: male in the ratio of 8:1. Before planting, pit should be filled with well rotten farmyard manure and top soil. Proper care should be taken in the young stage by providing shade and irrigation to reduce the mortality in the main field. About 10 kg of well rotten compost or farmyard manure is applied along with 20:18:50 g N:P$_2$O$_5$:K$_2$O per pant per year. In the second year the dose should be 40:36:100 g N:P$_2$O$_5$:K$_2$O per pant per year. The dose is gradually increased to 50 kg well rotten compost or farmyard manure and 300:200:750 g N:P$_2$O$_5$:K$_2$O per pant per year upto 15 years after planting.

3.4.1.5.5 Harvest and Post Harvest Technology

Male tree flowers earlier as compare to female flower. Usually flowering occurs during March to June. It takes 5 to 6 years flowering under good management condition. Fruits are ready for harvest 2–3 months after flowering. For seed purpose fruits are during July–August. The berries are grown in cluster and fruits are generally harvested when the fruits are green, fully mature but not ripe. The unripe berries are spicier and somewhat peppery in taste. They are gathered by climbing in ladder. Irregular bearing is a problem in all the spice cultivation. Application of Paclobutrazol ensures 100% flowering and increased number of panicles and thus yield. Application of 50 ppm IAA + 5 ppm BA on flower buds, 20 days after flower bud initiation gave abnormally big seeds, more number of seeds and bold berries. After harvest, ripe berries are separated from green ones. Under good management condition a full-grown trees yield up to 20–25 kg/year. Economic bearing period varied from 15–25 years.

The berries are spread out in the sun and turn over with a wooden rack for uniform drying. Normally drying takes 3–12 days depending upon the intensity of sunlight. Dry wind accelerates the sundrying process. Properly cured berries should be bright brown in color. Curing is complete when the berries become crisp and produce a metallic sound if shaken. Berries are cleaned by winnowing and after removing the dust berries should be stored.

3.4.1.5.5.1 Sex Expression

All spices are dioecious plant (i.e., separate male and female trees). Sex of the trees is known only after they flower, i.e., about 5–6 years after planting. Flowers are structurally hermaphrodite but functionally dioecious.

Since the pistillate trees bear profusely, it is necessary to have a maximum number of them interplanted with a few male trees for pollination. Male trees do not set fruits but are essential for the pollination. To ensure pollination it is necessary to plant female:male in the ratio of 8:1. Proper technology has to be formulated for identification of female trees at an early stage.

3.4.1.5.5.2   Value Addition

It contains essential oils such as *eugenol,* which gives pleasant, sweet aromatic fragrances to this spice. Used mostly in Western cooking and less suitable for Eastern cooking. It has medicinal, antimicrobial, insecticidal, nematicidal, antioxidant, and deodorant properties. It is also being used in the preparation of soups, barbecue sauces, pickling, and as a main ingredient in variety of curry powders. It is also used in liquors in many Caribbean countries and in confectionaries. Berry and leaf essential oil is commercially distilled in Jamaica. Berry oil yield varied from 3.3 to 4.3% and eugenol content from 65 to 66%. Dried leaves under steam distillation extraction produced an essential oil yield of 0.7 to 2.9% and fresh leaves oil yield varied from 0.35 to 1.25%.

3.4.1.5.6   Plant Protection

Leaf Rot: It is caused by *Pestalopsis spp.* During monsoon, the disease appears in severe from and results in heavy defoliation. Leaf blight is noticed to result in dieback. Leaf rot starts from margins of the leaves and along with this brown lesions also appear on leaves as a combined infection. Spraying Bordeaux mixture (1%) or Carbendazim (0.2%) is advocated as prophylactic measures for management of this disease.

Pimenta Rust (Puccinia psidii): Rust can be managed by spraying copper fungicides.

Dieback: Pruning and removal of affected branches together with application of white lead paint

Tea mosquito bug (Helopeltis antonii): Occasional attack by this pest can be controlled by spraying plants with 0.05% endosulfan.

3.4.1.6   Star-anise (*Illicium verum* Hook)

3.4.1.6.1   Systematic Position

Kingdom: Plantae
Division: Tracheophyta
Class: Magnoliopsida
Order: Astrobaileyales
Family: Schisandraceae
Genus: *Illicium*
Species: *verum*

3.4.1.6.2   About the Crop, National and International Scenario, Uses, and Composition

Star anise (*Illicium verum*) belongs to the family Magnoliaceae is the dried fruit of a small to medium evergreen tree. It is originated from Southern China. China, Vietnam, India, Japan, Philippines are the major producer of the star anise in the World. The tree bears fruit after about 6 years and has a productive life of about 100 years. Each point of the star shaped husk contains a shiny amber colored seed. Both the seed and husk are used as spice. The aroma of star anise resembles with fennel and anise with a warm notes of liquorice. The flavor is pungent, sweet, and particularly like liquorice. The fruits are harvested before they ripen and sun-dried, which hardens and darkens the carpels and develops the aromatic compounds.

Star anise is one of the most important spices in Chinese and South Indian food preparation. In Chinese and Vietnamese cooking, star anise is used for preparation of soups and stocks, and in meat preparations especially chicken and pork. Star anise is used to add flavor in vegetables and meat preparation and also to marinate meat. It is used as a condiment for flavoring curries, confectionaries, spirits, and for pickling. It is also used in perfumery industries. The essential oil of star anise is used as flavoring agent in soft drinks, bakery products and liquors. The fruit is antibacterial, carminative, diuretic, and stomachic. It is considered useful in flatulence and spasmodic. It is the major source of the chemical compound shikimic acid, a primary precursor in the pharmaceutical synthesis of antiinfluenza drug oseltamivir (Tamiflu).

Distinct licorice taste of star anise is due to the presence of anethol. The star anise seeds are an excellent source of many essential B-complex vitamins such as pyridoxine, niacin, riboflavin, and thiamin. The spicy seed is a source of minerals like calcium, iron, copper, potassium, manganese, zinc, and magnesium.

3.4.1.6.3 Soil and Climatic Requirements

Star anise prefers humus and compost rich soil for better growth and development. Soil texture should be loamy and well drained. Slightly acidic to neutral soil is ideal for its commercial cultivation. It is conducive to grow under warm subtropical climate. It is a frost tender perennial. Star anise only grows in areas where the temperature does not fall below −10°C.

3.4.1.6.4 Agrotechniques for Quality Production

Star anise is generally propagated by seeds or cuttings. In case of seed propagation, seed are sown in nursery or directly n the main field. Regular watering is required for getting higher seed germination and better establishment in the main field. Application of inorganic fertile is not common incase of star anise cultivation. Application of organic manure or compost during spring season is common. As fertilizer a 3-inch layer of compost or aged manure is to be sprayed on the ground surrounding the tree in the spring. However, slow release fertilizer may be applied during spring to ensure good yield. Pruning is practiced in star anise cultivation at the young stage to give the appearance of bush. Apart from the above always prune off dead, diseased, and weak branches.

3.4.1.6.5 Harvest and Post Harvest Technology

Star anise tree takes at least 6 years to bear fruit if grown from seeds. Fruits are picked at unripe stage while they are still green, later on these fruits are sun dried until their color change to reddish-brown. After removal of the seeds, fruits are stored for future use.

3.4.1.6.5.1 Value Addition

Five-spice powder: It is a spice mixture of five spices used primarily in Chinese cuisine but also used in other Asian and Arabic cookery. The mixture essentially contains Star anise, cloves, Chinese cinnamon, Sichuan pepper and fennel seeds. Other recipes may contain anise seed or ginger root, nutmeg, turmeric, *Amomum villosum* pods, *Amomum cardamomum* pods, licorice, Mandarin orange peel or galangal. In South China *Cinnamonum loureiroi* and Mandarin orange peel is commonly used as a substitute for *Cinnamonum cassia* and cloves, respectively, producing a different flavor for southern five-spice powders.

Essential oil: It has a beautiful, fresh, sweet, spicy, liquorice-like aroma. Star Anise essential oil has a beautiful, fresh, sweet, spicy, liquorice-like aroma.

3.4.2 ACIDULANT TREE SPICES

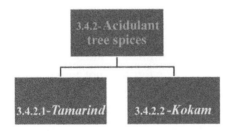

3.4.2.1 Tamarind (*Tamarindus indica* L.)

3.4.2.1.1 Systematic Position

Kingdom: Plantae
Division: Magnoliophyta
Class: Magnoliopsida
Order: Fabales
Family: Fabaceae
Subfamily: Caesalpinioideae
Tribe: Detarieae
Genus: *Tamarindus*
Species: *indica*

3.4.2.1.2 About the Crop, National and International Scenario, Uses, and Composition

Tamarind (*Tamarindus indica* L.) belongs to the family fabaceae is an economically important multi-use acidulant tree spices. It is one the important underutilized multipurpose tree spices in the Indian subcontinent. The tamarind fruit pulp is an important culinary ingredient in India since ancient times. It is ideal for drier-arid regions, especially in the dried prone areas. It can tolerate prolonged drought period upto 5–6 months but cannot tolerate frost or water logged condition. It is considered to be originated in Madagascar. It is now cultivated in semi arid Africa and South Asia. Commercial plantations are reported in Central American and in North Brazil. In India it is grown almost throughout the country, except Himalayan and Western dry region. In India, its cultivation is mainly concentrated in Tamil Nadu, Andhra Pradesh, Kerala, Karnataka, and Orissa. Tamarind can be successfully cultivated with subsistence level of management having minimum disease pest incidence. It has a long life span of 100–200 years and full bearing trees produce a yield of 300–500 kg pod per year. It is a source of timber, fruit, seeds, fodder, medicinal extracts and potential industrial components and potential source of income to the rural farmer.

3.4.2.1.2.1 Use

Fruit pulp is used for the preparation of beverages, canned turmeric, sauce, jam, and jelly. The typical taste of the tamarind is due to the presence of tartaric acid. Tamarind fruit and pulp can be used in fresh or processed form with high export value. Pulp portion of the sweet varieties of tamarind may be eaten fresh and sour fruits and also used for preparation of juice, jams, jellies, and candy. Sour varieties can be used in curries, chutneys, sauces, ice cream, sherbets, and toffees. Roasted or boiled seeds of tamarind used in jams, jellies, confectionery, and condiments. In some countries seeds are eaten alone or mixed with cereal powders, made into flour for bread and cake preparation. Tamarind kernel powder (TKP) is used in textile industry for preparation of adhesive and it is also in paper plywood industry. Turmeric seed powder is used to treat chronic diarrhea, jaundice, and dysentery. Turmeric is used as fodder for cattle and got. Flowers and immature

TABLE 3.15 Composition of Tamarinds, Raw Nutritional Value per 100 g

Energy	239 kcal (1,000 kJ)
Carbohydrates	62.5 g
Sugars	57.4
Dietary fiber	5.1 g
Fat	0.6 g
Protein	2.8 g
Vitamins	
Thiamine (B1)	0.428 mg
Riboflavin (B2)	0.152 mg
Niacin (B3)	1.938 mg
Pantothenic acid (B5)	0.143 mg
Vitamin B6	0.066 mg
Folate (B9)	14 µg
Choline	8.6 mg
Vitamin C	3.5 mg
Vitamin E	0.1 mg
Vitamin K	2.8 µg
Minerals	
Calcium	74 mg
Iron	2.8 mg
Magnesium	92 mg
Phosphorus	113 mg
Potassium	628 mg
Sodium	28 mg
Zinc	0.1 mg

pods of turmeric used in salads and stews. It can also used internally as a remedy to cure jaundice and externally to cure eye diseases.

It is rich in carbohydrate and dietary fiber. Vitamins like thiamine, riboflavin, niacin, folic acid, vitamin C, vitamin K and minerals like calcium, iron, magnesium, phosphorus, potassium, sodium, zinc, etc. are present in tamarind (Table 3.15). The major carotenoid compounds present in tamarind pulp are α-carotene, β-carotene, β-cryptoxanthin and lycopene. Acidity of tamarind is mainly due to the presence of tartaric, citric, and

TABLE 3.16 Important Varieties of Tamarind

Sl. No.	Name	Parentage	Yield	Special features
1.	PKM-1	Selection	-	Dwarf and high yielding type and adopted for cultivation in Tamil Nadu.
2.	DTS-1	Selection	1.5 kg/plant	Sweet red type and having tasty pulp and slightly susceptible to pod borer. It is adopted for cultivation in Karnataka.
3.	Prathisthan	Selection	-	It is adopted for cultivation in Maharashtra.
4.	No. 263	Selection	-	It is adopted for cultivation in Maharashtra.

malic acids, and potassium bitartrate. Acidity level ranges from 1.6 to 3.1%.

3.4.2.1.3 Soil and Climatic Requirements

Tamarind can be grown in all types of soils provided that proper drainage is there. However, tamarind plants grow well in deep loamy or alluvial soils. Tamarind can be grown in poor soils too but soil fertility should be replenished time to time. Tamarind plants are tolerant to saline and alkali soils.

Semi-arid tropical climate is best for the commercial cultivation of tamarind. It grows well in tropical climate where summer months are hot and dry. Tamarind can be grown in any areas where the temperature reaches 46°C maximum and 0°C minimum. Average rainfall requirement is 500–1,500 mm. The optimum altitude required for tamarind cultivation is 1,000 m above MSL (mean sea-level). It is drought resistant but frost tender crop.

The Table 3.16 will describe the following varieties of tamarind with regard to their parentage, average yield (kg/ha), and the special features.

3.4.2.1.4 Agrotechniques for Quality Production

Tamarind is mostly propagated through seeds and it takes about 10 years to come to regular bearing and hence vegetative propagation methods like inarching, patch budding or air layering is commercially used. In most of the cases farmers grow local types available in the locality. But some of the improved varieties are also available. Improved varieties like, Prathisthan, and No.-263 is cultivated in Maharashtra, PKM-1 and Urigam is cultivated in Tamil Nadu. In commercial scale, provided with drip irrigation facilities it can be grown with 5 m × 5 m spacing up to 10–12 years. After that alternate rows are removed to provide 10 m × 10 m spacing. Pits of 1 m × 1 m × 1 m filled with 10–15 kg FYM, 100 g single super phosphate and 1.5 kg of neem cake is ideal for better establishment of the crop. Normally no fertilizer management practices are followed in case of tamarind when it is grown in dry land. However, it responds well to application of nutrient through inorganic fertilizers. In case of first year to fifth year grown plant, apply 20 kg of farmyard manure along with NPK at the rate of 225:160:600 g/plant/year and it is gradually increases and nine years old and above aged plant apply NPK at the rate of 900:500:1800 g/plant/year. Pot watering can is done for establishment of young plants during summer season. It is normally grown as rain fed crop but irrigation helps in getting early yield. Irrigation at flowering and fruit set stage increases the yield. It is essential to keep the plants are free from side branches up to 1 m height. The trees come to flowering from 3rd years of planting. Flowering seen during September–October and fully ripe fruits can be harvested from March–April. At the early stages crop yield is low and increases gradually. A fully bearing trees of 12 years aged produces a yield of 200 kg pod/plant/year. Powdery mildew is the important diseases affecting the crop and tree hopper, mealy bug and scale insect are the important pest of the tamarind that adversely affects the yield.

3.4.2.1.5   Plant Protection

Stony fruit: Causal organism of this disease is *Pestalotia* spp. It is a fungal disease, it also affects seeds, fleshy mesocarp, fruit wall and stalk.

Root rot: It is caused by *Ganoderma lucidum.* Crown of infected trees shows gradual drying by the infection of this disease.

Powdery mildew: This fungal disease is caused by *Odium* spp

Nursery rot: This fungal disease is caused by *Sclerotious rolfsii.* Affected seeds fail to germinate, showing arrested growth.

Bacterial leaf spot: This disease is caused by *Xanthomonas tamarinds*. Small and round spots on leaves, which are initially pale-brown with yellow margins and later to black.

Caeyedon weevil (*Caeyedon serratus*): Infestation range of this pest is 15–17.5%. Adults lay eggs on the fully mature ripened fruits. Grubs on hatching bore into the fruits by cutting a microscopic hole on the pericarp of the fruit and without damaging much of the pulp, enter into seed and feed on the endosperm. Attacked seed must be stored in small container by mixing with dry wood ash and calcium carbonate or precaptulated chalk or stalked lime for management of the pest incidence.

Scale insect (*Aonidiella drientalis*): Scale secrets a thick waxy layer cover the shoot and stem with green barks. In full grown trees symptomps appear as a dieback. Spraying of Diazinone or Monocrotophos (0.04%) has been found effective.

3.4.2.1.6 Harvest and Post Harvest Technology

After harvesting, ripe pods are dried in the sun for 5–7 days. Mechanically it can be dried by using small-scale dehydrators to dry the fresh fruit. Once dried, the shells are hand cracked and separated from the pulp. The fibers, seeds, and shell pieces are removed from the pulp by hand. The pulp is then dried for another 3–4 days before being compressed, ready for storage. The pulp can be mixed with salt or sugar according to preference of the market, prior to storage. Seeds can be used in industrial processing and should be dried for about 2 days.

In rural areas the compressed pulp can be stored in plastic bags, jute bags or closed clay pots depending upon the availability. The pulp is usually stored with the seeds for home consumption. However, when produced commercially, the seeds should be removed before storage. The freshly prepared, dried pulp is light brown in color, but darkens with time in storage. Under cool, dry conditions the pulp remains good for about one year when stored with salt in a clay jar or in polythene, after, which it becomes almost black, soft, and sticky in texture. Various methods can be used to prolong the storage life of the whole fruit and the pulp. Drying is the best and easiest way to store small amounts of tamarind pulp. Freshly harvested pulp can be stored for 4–6 months by packing in high-density polyethylene and storing below 10°C in a dry place.

3.4.2.1.6.1   Value Addition

Tamarind Pulp Powder: Tamarind powder is prepared from half matured tamarind by drying the fruit pulp in the dryer especially in cabinet dryer at 50–60°C till it reaches a moisture content of 5%. After drying, the powder is prepared from the pulp. The color of tamarind is precisely due to the presence of anthcyanin. It is highly soluble in water and used to impart the raspberry red color in the different food products. Tamarind powder is mainly used as red colorant in rassam, sambar, jam, jelly, etc.

Tamarind concentrate: It has the higher self-life than tamarind pulp itself. For making tamarind concentrate, tamarind pulp is heated with boiling water and soluble solids are extracted to the extent of 20% using counter current principle. It is then sieved and concentrated under vacuum in a forced circulation evaporator and directly filled in cans or bottles. CFTRI, Mysore developed the methodology. Tamarind concentrate sets like jam after cooling. The soluble solid content is about 68%. Freshly harvested fruit pulp (not older than 5–6 months) gives good quality products. The composition of the concentrate is tartaric acid (13%), invert sugar (50%), pectin (2%), protein (3%), cellulosic material (2%), and moisture (30%) (Chempakam and Peter, 2000).

Tamarind Kernel powder: Decortication of seed and pulverization of the karrnels are essential for its preparation. Decortication is either done by soaking the seed in water or by roasting the seed. To avoid enzymatic reaction for better storage life sodium metabisulphate is added as preservative. Tamarind kernel powder (TKP) is mostly useful in textile and paper industries as it contains starch. Jellose is the major polysasaccharide present in the karnel powder, which can be effectively used as a remedy against diarrhea, dysentery, and colitis. Bharadwaj et al. (2007) reported that genotype AKT-14 gave maximum TKP however, TKP prepared from AKT-10 was found better with high content of starch.

3.4.2.2   Kokam (*Garcinia indica* Choisy)

3.4.2.2.1   Systematic Position

Kingdom: Plantae
Division: Tracheophyta

Class: Magnoliopsida
Order: Malpighiales
Family: Clusiaceae
Subfamily: Clusioideae
Tribe: Garcinieae
Genus: *Garcinia*
Species: *indica*

3.4.2.2.2 About the Crop, National and International Scenario, Uses, and Composition

Kokam (*Garcinia indica* Choisy) belongs to the family clusiaceae is commonly known as kokan butter tree or mangosteen oil tree or brindonia tallow tree. It has a tremendous commercial potential in the recent times, because of the presence of hydroxyl citric acid (HCA). This chemical can control obesity in man (Mathew et al., 2007). It is a slender evergreen small tree with drooping branches. It is a dioecious tree growing up to 18 m high. The genus *Garcinia* comprises of 200 old world tropical species of which 20 species are found in India. As a part of traditional material and medica of India, *Garcinia indica, Garcinia mangostana* and *Garcinia morella* are documented in Ayurvedic medicine, which is generally known as 'Red Mango.' The fruit is spherical, purple, not grooved having 5–8 seeds compressed in an acid. It is originated from tropical rain forest of the Western Ghats of Kerala and Malaysia. Its commercial cultivation is restricted in the coastal hilly region of Maharashtra and Goa. It is also found in the evergreen forests of Assam, Khasi, Jantia hills, West Bengal and Gujarat. The characteristics aroma and flavor of kokum is due to the presence of garcinol.

3.4.2.2.2.1 Use

Kokum have commercial importance, since the rind of which is used to impart an acid flavor to curries instead of tamarind. The ripened, rind, and juice of Kokkam fruit are commonly used in cooking. The dried and salted rind is used as a condiment in curries. It is also used as a garnish to give an acid flavor to curries and for preparing attractive, red, pleasant flavored cooling syrup. The soft drink and derivative of kokum, popularly known as kokum syrup generally consumed as a natural soft drink to quench thirst during summer. The kernels of *kokum* seed contains about 33 to 44% oil, which is commercially known as "kokum butter." It is considered as nutritive, demulcent, astringent, and emollient. The rind has antioxidant property. Kokum is well known to counteract the heat. Kokum is especially used with fish curries, three or four skins being enough to season an average dish. It is also used in chutneys and pickles.

3.4.2.2.2.2 Composition

Kokum rind contains three important chemicals namely, Garcinol, Hydroxycitric acid and anthocyanin pigment. Garcinol is a fat soluble yellow pigment. Hydroxycitric acid is used as an acidulant and physiologically active compound has been shown to significantly reduce body weight. Anthocyanins are well known for their antioxidant, antiinflammatory, and anticarcinogenic activity.

3.4.2.2.3 Soil and Climatic Requirements

It comes up well in a hot humid climate under partial shade. It thrives under a mean annual temperature of 20–30°C with 60–80% relative humidity and up to an altitude of 800 m above mean sea level. A well distributed rainfall of 250–400 cm is ideal for this crop. It can be grown in a variety of soil from sand loam to clay loam and lateritic soil with having wood water holding capacity and should be deep enough.

The Table 3.17 will describe the following varieties of kokum with regard to their parentage, average yield (kg/ha) and the special features:

TABLE 3.17 Important Varieties of Kokum

Sl. No.	Name	Parentage	Average Yield (kg/ha)	Special features
1.	Konkan Amruta	Seedling selection from ratnagiri and Sinddhudurg districts.	-	Attractive fruit with apple like shape with longer shelf life of 115 days and suitable for processing.

3.4.2.2.4   Agrotechniques for Quality Production

Commercially it is propagated through seed. But kokum is dioecious in nature, as a result upon seed propagation it produces 50% female and 50% male plant. Only female plant produce fruit and male plant can only be identified 7 to 8 years after planting, when produces flower. For this purpose, top working of male plant to convert into female plant after retaining about 10% male plant. Inarching and soft wood grafting is also done to ensure female plant and uniformity on yield. Pits of 60 cm × 60 cm × 60 cm are prepared at a spacing of 6 m × 6 m. In the first year each plant should be applied with 2 kg farmyard manure along with 50 g N, 25 g P, and 25 g K and this dose is increased every year and from the tenth year onward, each plant may be given 20 kg farmyard manure and 500 g N, 250 g P, and 250 g K/plant/year. Konkan Amruta is the improved variety of kokum. Irrigation should be given at 7–10 intervals at early stages of the growth and generally drip irrigation is practiced in kokum. In the 10–12 years, in kokum plantation, vegetable, and flower crop can be grown as inter crops.

3.4.2.2.5   Plant Protection

It is relatively free from devastating pests and diseases attack. However, it is affected by Sooty mould and Blight diseases; and these two diseases can be effectively controlled by spraying Bordeaux mixture (1%).

Among the pests, leaf hopper is important. It sucks sap from leaf as well as twigs resulting in complete leaf fall and final drying of whole plant. It can be controlled by spraying Rogor (0.03%) or Navacron (0.05%) or Imidachlorpid (0.025%).

3.4.2.2.6   Harvest and Post Harvest Technology

Kokum tree starts flowering during the month of November-December and fruits are harvested during April–May. It takes about 7 to 10 years for economical bearing. The fruits are picked when ripe, the rind is then removed and soaked in the juice of the pulp and then sun-dried. The kokum is difficult to cultivate, usually growing as solitary trees in a tropical forest environment. Yield of kokum varies from 20–30 t/ha. There is no serious disease and pest reported to infest kokum plantation. Only leaf minor attack the plant and feed the leaf.

3.4.2.2.6.1   Value Addition

Kokum syrup: Kokam fruits, after washing with clean water, are cut in two halves followed by removal of all seeds and pulp. Then sugar is poured into the opened fruit halves forming one layer. In this way alternate layers of kokum rind halves and sugars are put in cleaned food grade plastic drums for 7 days. When the sugar gets dissolved and the kokum rind gets extracted by osmosis, the syrup is strained through 1 mm sieve or muslin.

Kokum RTS: Kokum juice may be used to prepare an excellent RTS drink after evaluating its TSS and acidity. Then required quantity of citric acid and sugar is added to raise the Brix and acidity levels to 200° and 0.3%, respectively. The normal parameters of kokum RTS are juice – 20%, acidity – 0.3%, water-quite sufficient and sodium benzoate (preservative) – 140 mg/ kg of final product. During preparation, the mixture is boiled till all ingredients including sugar, citric acid and water are dissolved. At the end, preservative is added, the product is filtered, bottled, and sealed followed by pasteurization and cooling of filled bottles.

Kokum squash: It is prepared by maintaining juice – 25%, TSS – 45%, acidity – 1.2%, brix 450° and sufficient water and sugar where all the ingredients are dissolved through boiling. Sodium benzoate @ 610 mg/kg of final product is added as preservative at the end before filling of bottles. Then they are sealed, pasteurized, and cooled.

Kokum rind powder: Dried kokum fruits without any seed and extraneous mater are cut into bits, loaded in percolator and extracted with hot water to obtain kokum concentrate. The concentrate is mixed with a suitable carrier and dried under controlled conditions to obtain kokum powder. It may also be obtained through spray drying technique.

Kokum honey and Kokum wine: Kokum honey may be produced by establishing apiculture unit within the plantation of kokum trees. It is, of course, has not been successfully implemented and still lot more works to be done for validation. However, kokum wine is being produced by fermenting kokum juice (containing 4% sugar, approx.) largely at Goa.

3.5   SEED SPICES

3.5.1   *CORIANDER* (CORIANDRUM SATIVUM *L.)*

3.5.1.1   Systematic Position

Kingdom: Plantae
Division: Magnoliophyta
Class: Magnoliopsida
Order: Apiales
Family: Apiaceae
Genus: *Coriandrum*
Species: *sativum*

3.5.1.2   About the Crop, National and International Scenario, Uses, and Composition

Coriander (*Coriandrum sativum*) commonly known as "Dhania" belongs to Apiaceae family. It is cultivated in Rajasthan, Gujarat, Madhya Pradesh, Tamil Nadu, U.P., etc. It is mainly used as a condiment for its medicinal properties. Green leaves of coriander are also used for culinary purposes. India is the largest producer, consumer, and exporter of coriander with greater share (80%) in world export market. Other major producers are Bulgaria, Romania, and Morocco. On an average India exported around 34 thousand tonnes (in terms of volume) and 38 million dollar (in terms of value) respectively. From 2004 to 2015, there was a 6% rise in term of volume and 20% rise in term of value. The main international markets where Indian coriander sold are: Malaysia, Saudi Arabia, UAE, and UK.

The different parts of this plant contain monoterpenes, α-pinene, limpnene, γ-terpinene, p-cymene, borneol, citronellol, camphor, geraniol, coriandrin, dihydrocoriandrin, coriandrons A-E, flavonoids, and essential oils. Essential oil yields of 0.1%–0.35% (v/w) while the latter has smaller fruits (1.5–3 mm diameter) with essential oil yields of 0.8%–1.8%. Essential oil of coriander contains different alcohol like Linalool (60–80%), geraniol (1.2%–4.6%), terpinen-4-ol (3%), α-terpineol (0.5%); hydrocarbons like γ-terpinene (1–8%), r-cymene (3.5%), limonene (0.5%–4.0%), a-pinene (0.2%–8.5%), camphene (1.4%), myrcene (0.2%–2.0%); ketones like Camphor (0.9%–4.9%) and esters like Geranyl acetate (0.1%–4.7%), linalyl acetate (0%–2.7%).

3.5.1.3   Soil and Climatic Requirements

Coriander requires cool climate during growth stage and warm dry climate at maturity. It can be cultivated in all most all types of soils but well drained loamy soil suits well. It is observed that quality of seed is superior and essential oil content is more when the crop is grown in colder regions and at high altitudes.

The important varieties of coriander with their parentage, average yield (kg/ha), essential oil (%), duration, and the salient features are presented in Table 3.18.

TABLE 3.18 Important Varieties of Coriander

Sl No.	Variety	Pedigree/Parentage	*Average Yield (kg/ha)	Essential oil%	Duration (days)	Salient features
1	Guj. Cor.1	Selection from germplasm	1100	0.35	112	Suitable for early sowing, erect plant, round bold grains, moderately tolerant to wilt and powdery mildew
2	Co.1	Selection from Koilpatti local	440	0.27	110	A variety with small statuted plant, suitable for rainfed areas and for greens and grains, small grain
3	Co.2	Reselection from culture P2 of Gujarat	520	0.40	90–100	A dual purpose variety, suitable for saline, and alkaline and drought prone areas seeds oblong, medium.
4	Co.3	Reselection from Acc.695 of IARI, New Delhi type	650	0.38–0.41	85–95	A dual purpose variety, good yielder, medium sized grains, suitable for both rainfed and irrigated condition, *rabi* as well as *kharif* season. Field tolerant to powdery mildew, wilt, and grain mould.
5	Co (CR).4	Reselection from germplasm ATP77 guntur collection	600	0.4	65–70	Early maturing variety suitable for both rainfed and irrigated condition; grains oblong and medium; field tolerant to wilt and grain mould
6	Guj. Cor.2	Reselection from Co.2	1450	0.40	–	Semi spreading type, suitable for early sowing, moderately tolerant to powdery mildew, grains oblong, lodging, and shattering resistant.
7	Rajendra Swathi	Pureline selection from Muzaffarpur collection	1300	0.65	–	Medium sized plant with fine, aromatic round grains, Suitable for intercropping, field tolerance to aphids

TABLE 3.18 (Continued)

Sl No.	Variety	Pedigree/Parentage	*Average Yield (kg/ha)	Essential oil%	Duration (days)	Salient features
8	Sadhana	Mass selection from local Alur collection	1025	0.20	95–110	A dual purpose, semi-erect variety; Suitable for rainfed condition field tolerance to white fly, mites, and aphids. A mid-late variety withstands moisture stress, responded well to input management under optimum moisture level.
9	Swathi	Mass selection from Nandyal germplasm	855	0.30	82–85	Plants medium size semi-erect type, early maturing variety, suitable for rainfed condition, and late sown season. Field tolerant to white fly, moderately tolerant to disease. Suits well to the areas where the soil moisture retentiveness in comparably less, being early maturity. It escapes powdery mildew disease.
10	CS 287	Reselection from Guntur collection	600	–	–	Early maturing variety, suitable for both rainfed and irrigated condition. Field tolerant to wilt and grain mould.
11	Sindhu	Mass selection germplasm, Warangal local	1000	0.40	100–110	Oval medium breakable grains, suitable for rainfed areas, tolerant to wilt, powdery mildew as well as drought condition, medium duration.

Sl No.	Variety	Pedigree/Parentage	*Average Yield (kg/ha)	Essential oil%	Duration (days)	Salient features
12	Hisar Anand	Mass selection from Haryana collection	1400	0.35	–	A medium tall dual purpose variety, oval medium sized seeds, wider adaptability to different soil conditions. Resistant to lodging due to spreading habit.
13	Hisar Sugandh	Mass selection from indigenous germplasm	1400		–	Suitable for irrigated conditions. Resistant to stem gall diseases.
14	Hisar Surabhi	Mass selection from local germplasm	1800	0.4–0.5	130–140	Bushy erect plant type, seed medium, oblong; tolerant to frost, less susceptible to aphids, medium duration
15	Azad Dhania-1	Mass selection from Kalyanpur germplasm collection	1000	0.29	120–125	Erect, early branching, number of umbellates per umbel 5, tolerant to moisture stress, powdery mildew and aphids.
16	Pant Haritima	Selection from local type Pant Dhania	1200	0.4	150–160	Tall erect plant, a dual purpose type, good yielder of leaves, smaller seeds with high oil. Resistant to stem gall.
17	DWA 3	Pureline selection from Karnataka collection	400	0.27	–	A dual purpose variety and for seed production in rabi crop, moderately tolerant to powdery mildew, black clay soils are best suited
18	CIMPOS-33	Selection from germplasm introduced from Bulgaria	2100	1.3	–	Tall erect, compact, profusely branching and flowering, grains small, and bold. Mainly recommended for oil production.

TABLE 3.18 (Continued)

Sl No.	Variety	Pedigree/Parentage	*Average Yield (kg/ha)	Essential oil%	Duration (days)	Salient features
19	ACR-01–256(NRCSS ACR-1)	Reselection from EC-467683 from Russia	1100	0.35–5	Long duration	Dual purpose variety, long duration, resistant to stem gall and wilt.
20	RCr 20	Recurrent half sib election from Jaipur local	900	0.25	100–110	Medium sized bushy plant suitable for rainfed crop or limited moisture condition and heavy soils of south Rajasthan. Moderately resistant to stem gall, bold grains, early maturity.
21	RCr.41	Recurrent half sib selection from local type from "Kotta"	909	0.25	130–140	A tall erect plant with thick stem. Grows well under irrigated conditions, resistant to stem gall, wilt, and moderately resistant to powdery mildew; small seeds (9.3 g/1000 seed), long duration variety
22	RCr 435	Recurrent selection from local germplasm from Jalore	1000	0.33	110–130	Plants are bushy, erect, bold seeds, medium sized, medium maturing variety, adapted for irrigated condition moderately resistant to root knot and powdery mildew.
23	RCr 436	Recurrent half sib selection from local germplasm from Kotta	1100	0.33	90–100	Plants semi dwarf, bushy type with quick early growth and bold seeds. Resistant to root rot and root knot nematodes most suitable for limited moisture condition and heavy soils of south Rajasthan

Sl No.	Variety	Pedigree/Parentage	*Average Yield (kg/ha)	Essential oil%	Duration (days)	Salient features
24	RCr446	Half sib selection from local type from Jaipur local	1200	0.33	–	Plants tall, are leafy erect with higher number of seeds per umbel. Seeds are medium in size and moderately resistant to stem gall.
25	RCr 684	Mutation breeding of gamma rays. Induced mutant of Rcr-20	990	0.32	110–120	A variety, resistant to stem gall and less susceptible the powdery mildew. Adapted to medium heavy textured soil and sandy loam soil under irrigation. Seeds of the variety are bold. Plants are tolerant and erect with higher number of seeds per umbel, medium maturity.

* Yield kg/ha (Dry)

3.5.1.4   Agrotechniques for Quality Production

Land should be ploughed 2–3 times followed by planking, to bring the soil to fine tilth. Clods should be broken and stubbles of previous crop should be removed. Before land preparation, pre-sowing irrigation should be given, if optimum moisture for seed germination is not available in the soil.

Generally coriander is cultivated during rabi season. The best time for sowing of coriander is 15th October to 15th November. Before sowing, seeds should be rubbed to split into two halves and care should be taken to save the sprouting portion during rubbing. Timely sowing is beneficial as in early sowing germination is affected due to excess sunshine and late sowing leads to paltry growth of plants as well as development of several diseases. Sowing time may be adjusted in frost prone areas so that the frost incidence can be avoided during flowering stage. Sowing should be done in rows at spacing of 30 cm apart and for sowing of one hectare area 15 to 20 Kg of seed is required.

At the time of land preparation 15 to 20 MT FYM per hectare should be incorporated in soil. In case of rain fed crop, apart from FYM 20 Kg nitrogen 30 Kg phosphorous and 20 Kg Potash per hectare should also be applied in the soil as basal dose. For irrigated crop; 20 MT FYM, 60 Kg of nitrogen, 30 Kg phosphorous and 20 Kg Potash per hectare is recommended. Full dose of FYM should be mixed in soil at the time of land preparation and one third dose of nitrogen (20 Kg), full dose of phosphorous and Potash should be applied as basal dose and remaining 40 kg nitrogen should be top dressed in two equal split doses at the time of first irrigation and at flowering stage.

Generally, 4–6 irrigations are given depending on type of soil and climate. First, irrigation should be given within 30–35 days from the date of sowing, second after 50–60 days, third after 70–80 days, fourth after 90–100 days, fifth after 105–110 days and sixth after 115–125 days. Primarily, coriander plants grow very slow, hence weeding during early growth period is very essential to save the plants from weed competition. In rain fed crop, first weeding should be done at about 25–30 days after sowing and in irrigated crop about 40–45 days after sowing. Second weeding should be done 50–60 days after sowing, in case of rain fed crop. Thinning of plants should be done before first irrigation to maintain a spacing of 5–10 cm between plants.

3.5.1.5 Plant Protection

Wilt: The fungus disturbs root system of the plants. To prevent the crop from infection the fungus, deep ploughing should be done during summer season. Crop rotation may also be followed. Avoid cultivation of coriander crop for 2–3 years where disease has been noticed previously. Seed should be treated with Bavistin or Thiram @ 1.5 gm per kg seed before sowing.

Powdery mildew: This disease is mostly seen during cloudy weather condition. White powdery growth appears on the leaves and buds during its primary stage. Seed formation may not take place in affected plants due to this disease. To control this disease, dusting of Sulphur dust @ 20–25 Kg per hectare should be done or spraying of wettable sulphur or Kerathane can also control this disease.

Blight: In this fungal disease, dark brown spots appear on the stem and leaves of infected plants. Spray the crop with 0.2% solution of Mancozeb effectively controls this disease.

Stem gall: Galls appear on the leaves and stems of the plants affected by this disease, which change the shape of seeds and reduced the quality. Disease is more effective with increasing level of humidity. To control the disease, sowing may be done only after treating the seeds with 1.5 g Thiram and 1.5 g Bavistin (1:1)/Kg seeds. Spray 0.1% solution of Bavistin when the symptoms start appearing and repeat the spraying at an interval of 20 days till the disease is completely controlled.

Aphid: Aphid infestation occurs at the time of flowering or after flowering in coriander crop. It sucks the sap from tender parts of the plants resulting in heavy loss. Spraying of 0.3% solution of Malathion or 0.1% solution of Dimethoate should be done to control this insect. Sowing during last week of October to first week of November can minimize the damage caused by aphid.

Leaf eating caterpillar: It is a caterpillar, brownish in color, which cuts the plants from ground level and makes them to fall down. Infestation of this pest starts at the initial stage of plants resulting in heavy loss to the crop. Dusting of 4% Endosulphan dust @ 20–25 kg/ha should be done while ploughing to check the crop from infestation. Before last ploughing land should be drenched with Chlorpyriphos is also recommended.

3.5.1.6   Harvest and Post Harvest Technology

The crop become matures within 110 to 140 days. Seeds turn to yellowish green color at maturity and mature seeds are medium-hard while pressing. After harvest, the crop should be dried under partial shade to retain the green color and its aroma. When the plants are completely dried, the seeds should be separated by thrashing. Seeds should be cleaned by winnowing or with the help of sieve or by vibrator and graded. Average yield of about 500 to 800 kg coriander/ha can be obtained in case of rain fed crop and 1200 to 2000 kg from irrigated crop.

3.5.1.6.1   Value Addition

Coriander essential oil: Coriander essential oil is a colorless or pale yellow liquid. The aroma is pleasant, sweet, and somewhat woody and spicy. It is prepared by steam distillation method. The oil content varied from 0.1 to 1.7%. Generally large fruited types produced lower oil yield as compared to small fruited types. The main composition of the oil is d-linalool (45–70%), α-pinenel (6.5%) β-terpene (10%), camphor (5%), limonene (1.7%) and geranyl acetate (2.6%).

Coriander oleoresin: Coriander oleoresin is a brownish yellow product and the replacement strength of the oleoresin is 1:33. It is prepared through solvent extraction method. In commercial oleoresin, essential oil content varies from 5–40%.

Coriander dal: Coriander dal (dhania dal) is obtained from the seeds of coriander and is mainly used as an adjunct in supari or pan masala. The treated seeds are highly flavored and consumed as a digestive chew. The coriander seeds are dehusked, flaked, and given a mild heat treatment before being salted. The traditional method of preparing dhania dal from the whole seeds of coriander is quite laborious and time consuming, besides being less efficient with high amount of broken ones. In this method, the coriander dal is soaked overnight in salt water and partially dehydrated and husk removed manually. CFTRI has developed a process to get superior quality coriander dal with less broken ones. In this process, indigenous milling equipments are used along with improved conditioning technique.

Coriander fatty oil: Coriander Fatty oil is brownish green color with a similar odor to that of coriander essential oil. Fatty oil has soft consistency and having leathering properties. It is mainly used as flavoring agent for spiritual liquors, chocolate, and cocoa industries. This fatty oil is prepared from the coriander seed containing about 19–21% fatty oil.

Coriander powder: It is a very common but popular value added product of coriander. The powder is prepared by grinding matured dried coriander. Ground coriander is highly hygroscopic in nature. Various changes like loss of volatile oil, caking of the products, microbial spoilage infestation of mould and insect occur during storage and transportation due to absorption of moisture from atmosphere. Hence, proper packaging is essential for protection of above-mentioned hazard. It is generally packed in polythene pouches for better packaging it should be packed in aluminum foil laminated, LDPE, and polycell laminated pouch.

3.5.2 CUMIN, WHITE (*CUMINUM CYMINUM* L.)

3.5.2.1 Systematic Position

Kingdom: Plantae
Division: Tracheophyta
Class: Magnoliopsida
Order: Apiales
Family: Apiaceae
Genus: *Cuminum*
Species: *cyminum*

3.5.2.2 About the Crop, National and International Scenario, Uses, and Composition

Cumin is a flowering plant in the family Apiaceae, native from the east Mediterranean to Pakistan/India. The seeds with its distinctive flavor and aroma are used in the cuisines of many different cultures, in both whole and ground form. It is globally popular and an essential flavoring in many cuisines, particularly South Asian (where it is called *jeera*), Northern African, and Latin American cuisines. Cumin has been in use since ancient times as a traditional medicinal plant. The plant is mostly grown in Pakistan, India, Uzbekistan, Tajikistan, Iran, Turkey, Morocco, Egypt, Syria, Mexico, Chile, and China.

TABLE 3.19 Important Varieties of Cumin

Sl No.	Variety	Pedigree/ Parentage	*Average Yield (kg/ha)	Essential oil	Crude fiber	Duration (days)	Salient features
1	MC.43	Selection from germplasm	580	2.7	15.5	–	Plant semi spreading, grains bold lustering withstand lodging and shattering, moderately tolerant resistant to *Fusarium*wilt, *Alternaria* blight & powdery mildew.
2	Guj. Cumin 1	Selection from local germplasm (Vijaypur-5)	550	3.6	14.25	–	Plants bushy and spreading, grains bold, linear oblong; Withstand shattering and lodging, moderately tolerant to wilt, powdery mildew and blight.
3	RZ-19	Recurrent selection from UC.19	500			–	Erect plant, bold, lustrous grain, tolerant to wilt and blight suitable for late sowing season.
4	Guj Cumin 2	Pure line selection from M2 irradiated seeds from MC-43	620	4	22.1	–	Bushy plant, good branching habit, grains bold, medium sized, lustrous grain, tolerant to wilt and blight suitable for late sowing season
5	Guj. Cumin 3	Recurrent selection derived from W.German entry EC-232689	620	4.4		–	Bushy dwarf plant, fruit medium sized, frost wilt resistant variety suitable for winter season in limited irrigation. Higher essential oil content, seed pungent with good aroma
6	RZ-19	Recurrent single plant progeny selection from Ajmeer	560	–	–	140-150	Erect plant, pink flowers, bold, lustrous grain, gray pubescent, tolerant to wilt and blight suitable for late sowing season.

Sl No.	Variety	Pedigree/ Parentage	*Average Yield (kg/ha)	Essential oil	Crude fiber	Duration (days)	Salient features
7	5–404	Selection from local germplasm	350	2.2	7.7	–	An erect plant, medium sized fruit, moderately tolerant to powdery mildew.
8	RZ-209	Recurrent single plant progeny selection from Jore	650	–	–	120–130	A variety shown some resistance with blight and wilt disease
9	RZ-223	Mutation breeding in UC-216	600	3.0–3.5	–	120–130	Wider adaptability, resistant to wilt, superior in yield and seed quality over RZ-19. Plants bushy, semi-erect, long bold attractive seeds, medium duration.
10	Ac-01–167	Reselection from EC-243373	515	3	–	–	Bold seeds resistant to wilt.

* Yield kg/ha (Dry).

Cumin seeds contain numerous phyto-chemicals that are known to have antioxidant, carminative, and antiflatulent properties. The seeds are an excellent source of dietary fiber. Its seeds contain certain health-benefiting essential oils such as *cuminaldehyde* (4-isopropylbenzaldehyde), *pyrazines, 2-methoxy-3-sec-butylpyrazine, 2-ethoxy-3-isopropylpyrazine,* and *2-methoxy-3-methyl-pyrazine.* The spice is an excellent source of minerals like iron, copper, calcium, potassium, manganese, selenium, zinc, and magnesium. It also contains very good amounts of B-complex vitamins such as thiamin, vitamin B-6, niacin, riboflavin, and other vital antioxidant vitamins like vitamin E, vitamin A, and vitamin C. The seeds are also rich source of many flavonoid phenolic antioxidants such as carotenes, zea-xanthin, and lutein. The major compounds characterized in the cumin oils were cuminaldehyde (19.25–27.02%), p-mentha-l, 3-dien-7-al (4.29–12.26%), p-mentha-l, 4-dien-7-al (24.48–44.91%), α-terpinene (7.06–14.10%), p-cymene (4.61–12.01%) and β-pinene (2.98–8.90%).

3.5.2.3 Soil and Climatic Requirements

Cumin plant grows well in cool dry weather. Heavy rains during growth period damage the plant. Warm weather increases the incidence of disease, which attains an epidemic form at warm flowering and seed formation stage and as a result, seeds become shrivelled. Frost and cool air also blast the flowers. It requires loamy soil, but it grows well on sandy loam and clay loam soils. Heavy clay soil or light sandy soils are not suitable for this crop. Soil pH of 7.5 is quite suitable for its cultivation.

The important varieties of cumin with their parentage, average yield (kg/ha), essential oil (%), crude fiber (%), duration, and the salient features are presented in Table 3.19.

3.5.2.4 Agrotechniques for Quality Production

Cumin requires a very fine seedbed. Loamy soil always give a fine field preparation with 3–4 deep ploughing with mould-board plough, followed by harrowing and planking. Any finely prepared field for rabi crop, which could not be cultivated due to some unavoidable reasons can be sown with late sown variety. Keep the field free from water logging by providing suitable drainage channels. Selection of seeds, seed treatment, sowing time and method of sowing are the important aspects, which influence the production in cumin crop. About 10–12 kg healthy seed per hectare is required. Seed should be free from insect damage. Select only healthy seeds for sowing. For mixed cropping, 5–6 kg seed/ha is generally required seed is treated with any suitable fungicide like Agrosan or Ceresan @ 2 g/kg of seeds, as precaution against the seed borne diseases.

The proper time of sowing cumin seeds is from the middle of October to middle of November. Early sowing in North Indian conditions prolongs vegetative period and incidence of blight increases. Therefore, late sowing in December is preferred and, which also produces high quality seed. The best time of sowing in South India is during November. Cumin seed is very small in size so it requires a very careful sowing for even distribution of seed. It can be sown by any of the following methods.

1. **Broadcasting:** seed (20 kg) is mixed with dry sand to increase the quality of material to be broadcasted, which helps in even spreading of seed. A shallow ploughing followed by planking covers the seed up to a depth of 2–3 cm or seed is mixed with soil by raking with a hand-raker.

2. **Kera method:** seed is mixed with sand and evenly dropped in furrows, made by shallow ploughing and followed by a light planking to cover the seed.

3. **Pora method:** in this method, seed is dropped in manually made furrows through a funnel like pipe called pora. It is followed by light planking.

4. **Seed drill method:** in a very large area, where sowing by manual laborers is a tedious job, seed is dropped in lines by specially designed mechanically operated seed drills.

The recommended space between furrows is 30 cm and between plants is 10 cm. the space between plants is adjusted by thinning operations, after 15–20 days of sowing.

The addition of 10–15 tonnes of well-decomposed farmyard manure, a month before sowing seed in the field, improves the tilth of the soul to a desirable extent for cumin crop. Under rainfed condition, 20 kg of nitrogen is applied after the rain showers as foliar spray with 2% urea. Under irrigated condition, 30 kg nitrogen, 30 kg phosphorus, and 20 kg potash are applied as basal dose and remaining 20 kg nitrogen is applied in two splits doses; one after irrigation and another during the early stage of vegetative growth or as foliar spray with 2% urea. Many weeds like *Chenopodium album*, *Lathyrus* sp and *Cyperus rotundus* are found growing in cumin fields. Care should be taken to remove all the weeds in the early stages of crop growth to avoid competition for the reserve moisture, nutrients, sunlight, etc. Weed control can be achieved by hand hoeing after 15–20 days of sowing or spraying with herbicides such as Isoproturon 1 kg ai/ha in 800 L of water as pre-emergence application. Adequate moisture level of the field is necessary for cumin crop. A light irrigation can be given immediately after sowing, taking care that seeds are not wasted away. Cumin seeds germinate within 9–10 days after sowing. Depending upon soil moisture, frequent irrigations at the interval of 10–15 days are given. Care must be taken to avoid water logging in the field. After flowering, only single light irrigation is given and the crop is left to mature as such.

Cumin crop can be grown well as mixed crop with a many crops like barley, oil seeds, gram, etc. in various proportions. Cumin grown in between lines is easy to be harvested separately. Suitable crop rotations help the crop escape the attack of insect pests (e.g., aphids) and diseases (e.g., blight). The commonly used rotations with cumin crop are given below.

    i.   Maize-cumin-cotton
    ii. Maize-cumin-sugarcane-ratoon
    iii. Maize-cumin-barley
    iv. Maize-cumin-wheat

3.5.2.5   Plant Protection

Cumin is attacked mainly by fungi, which cause two serious diseases viz. blight and powdery mildew.

Blight diseases: This is a fungal disease caused by Alternaria sp. It appears during warm humid weather in the epidemic form at flowering stage and seed cannot mature to full size. Seeds become shriveled and are easily blown away during winnowing. Early sown crop gets high intensity of disease and produces unmarketable seed.

Foliar spray with Difolatan or Dithane M-45 effectively reduces the disease incidence. Spray the crop repeatedly with Cuman @100 g in 100 L of water or spray the crop with Bordeaux mixture 1%. Give the first spray three weeks after sowing and subsequent three sprays at interval of a fortnight.

Powdery mildew: The disease is caused by a fungus. In warm humid climate, a white growth of fungus mycelium is visible on ventral and dorsal surface of leaves and on the stem. Flowers are malformed and covered with white powder. No seeds are formed.

Dust the crop with fine sulphur @ 20 kg/ha in two doses; the first with the onset of flowering and second after 10 days. Give two or three spray with Cosan @ 100 g in 100 L of water at fortnightly intervals from the onset of flowering.

3.5.2.6   Insect Pests

Cumin crop is mainly attacked by aphids and caterpillars.

Aphid: It is a very serious pest of cumin. It sucks the sap of the tender leaves, twigs, and inflorescence and disturbs the normal physiological activities of plants. Spray the crop with Phosphamidon or Dimethoate during the early plant growth period.

Caterpillar (Prodina sp.): These caterpillars eat tender leaves at seedlings stage and inhibit the plant growth. The insect can be controlled by prophylactic (preventive) spray with Folidol (0.03%) or Metasystox (0.1%), or Dimecron-100 (0.02%) at early seedling stages. The spray is repeated till the plants are 50–60 days old.

3.5.2.7   Harvest and Post Harvest Technology

The crop is ready for harvesting in 90 to 130 days. Harvesting is generally done early in the morning to avoid shedding of seeds. Cutting of weeds along with the crop may be avoided to maintain the purity of seeds. Harvesting can also be done by uprooting the plants. Harvested crop is carefully staked for two days and spread out on the threshing floor to dry in the sun. Threshing is done by beating with stick or trampling under the feet of cattle. Cumin crop can yield 5 to 12 q of seed per hectare under good climatic conditions and without any epidemic diseases and damage by insect. A national average of 5–6 quintal is elected. The yield of volatile oil from the mature seed ranges from 2–3% by weight and yield a strong aromatic greenish fixed oil to an extent of about 10%.

3.5.2.7.1   Value Addition

Cumin oil: Colorless or pale yellow colored essential oil (2.5–4.5%) is obtained from ground cumin on steam distillation. The chief constituent of the oil is cuminaldehyde (20–40%), which is essentially used in perfumery. The yield of oil depends upon the quality age of the seed, variety, and methods of extraction. The highest oil recovery was recorded when (35–48) mesh and (48–65) mesh ground cumin used extraction of essential oil (Sagani et al., 2005).

Cumin powder: Cumin powder is an important item of value added product and gaining commercial importance in world trade.

3.5.3 FENNEL (FOENICULUM VULGARE MILL.)

3.5.3.1 Systematic Position

Kingdom: Plantae
Division: Magnoliophyta
Class: Magnoliopsida
Order: Apiales
Family: Apiaceae
Genus: *Foeniculum*
Species: *vulgare*

3.5.3.2 About the Crop, National and International Scenario, Uses, and Composition

Fennel (*Foeniculum vulgarisa*) commonly known as Saunf, which belongs to Apiacae family. It is used as condiment and culinary spice. In India, it occupied about 25, 000 ha area with a production of about 30, 000 MT. Mainly cultivated in Gujarat, Rajasthan, and Uttar Pradesh states of India. India's annual (April–March) Fennel export is around 8 thousand tonnes. USA and Vietnam are the largest importer of fennel.

The main constituents of the oils were: anethole (72.27%-74.18%), fenchone (11.32%–16.35%), estragole (10.42%) and methyl chavicol (3.78–5.29%). Essential oil of fennel also contains 3-caffeoylquinic acid, 4-caffeoylquinic acid, 1,5-O-dicaffeoylquinic acid, rosmarinic acid, eriodictyol-7-O-rutinoside, quercetin-3-O-galactoside, kaempferol-3-O-rutinoside, and kaempferol-3-O-glucoside. Chlorogenic acid, quercetin-3-O-beta-D-glucuronide has also been identified by HPLC-DAD and HPLC-MS as constituents of fennel teas. In addition, minor unidentified flavonol constituents were found in two teas. Major phenolic compounds

TABLE 3.20 Important Varieties of Fennel

Sl No.	Variety	Pedigree/ Parentage	*Average Yield (kg/ha)	Essential oil%	Crude fiber %	Duration (days)	Salient features
1	PF – 35	Selection from local germplasm	1280	–	–	–	Plant tall and spreading moderately tolerant to leaf spot, leaf blight and sugary diseases
2	Co.1	Reselection from PF 35	570	–	–	220	Medium statured, diffuse branching. Suitable for intercropping and border cropping with chili and turmeric. Suitable for drought prone, water logged, saline, and alkaline conditions
3	Guj Fennel 1	Pure line selection from Vijaypur local	1695	–	–	–	Plant tall and bushy, shattering, and lodging, suitable for early sowing and rabi crop, reasonably tolerant to drought, moderately tolerant to surgery disease, oblong, medium bold and dark green seeds.
4	Guj fennel 2	Pedigree selection from local germplasm	1940	2.4	–	–	Plants bushy, bold grains, rich in volatile oil and suitable for both rain fed and irrigated condition
5	RF 101	Recurrent half sib selection from local germplasm collection from Jobner	1400	–	–	150–160	Erect medium tall nature, medium maturity type with long bold grains, most suitable for loamy and black cotton soil.

Sl No.	Variety	Pedigree/ Parentage	*Average Yield (kg/ha)	Essential oil%	Crude fiber %	Duration (days)	Salient features
6	S-7-9	Selection	1100	1.2	24	210	A bushy plant with big umbel, moderately tolerant to blight.
7	Guj Fennel II	Selection based on individual plant progeny performance from local germplasm	2489	1.8	—	148	A medium maturity type adopted to rabi season un der irrigation; seeds medium bold.
8	RF 125	Recurrent half sib selection is an exotic collection EC 243380 from Italy	1700	—	—	110–130	Plants are short statured with compact umbels and long bold seeds when green, there are denser view of plants. Tolerant to sugary disease.
9	Hisar Sawrup	Mass selection from indigenous germplasm of Haryana	1600	1.6	—	175–185	Plants grow up right, spreading, gives a bushy appearance. A late maturity type grain long and bold, resistant to lodging, no shattering of grains.
10	Azad Sanuf-1	Selection from germplasm	1500	2	—	160–170	Medium plants, resistant to blight and root rot diseases. Escapes attack of aphids due to early maturity, seeds are bold green.
11	Pant Madhurika	Pure line selection from local germplasm			180–185		Tall robust eruct plant with big umbels having bold seeds with green fine ridgs sweet in taste, medium maturity.

Sl No.	Variety	Pedigree/ Parentage	*Average Yield (kg/ha)	Essential oil%	Crude fiber %	Duration (days)	Salient features
12	Rajendra Sourabha	NA	NA	NA	NA	NA	NA
13	AF-01– 119NRCSS-AF1	Recurrent selection from individual plant progeny	1950	–	–	–	Medium maturity seed, bold, tolerant to blight
14	RF 143	Recurrent selection from individual plant progeny	1200	1.87	–	–	Medium tall and recommended for loamy and black cotton soils

* Yield kg/ha (Dry).

present in fennel plant material are 3-O-caffeoylquinic acid β-CQA), chlorogenic acid, 4-O-caffeoylquinic acid (4-CQA), eriocitrin, rutin, miquelianin, 1, 3-O-dicaffeoylquinic acid (1, 3-diCQA), 1, 5-O-dicaffeoylquinic acid (1, 5-diCQA), 1, 4-O-dicaffeoylquinic acid (1, 4-diCQA) and rosmarinic acid

3.5.3.3   Soil and Climatic Requirements

Fennel is cultivated as kharif as well as rabi crop. It requires cool, dry climate. Frost prone areas should be avoided as this crop is more susceptible to frost. Rainfall at the time of maturity spoil color and reduce quality of fennel seeds. Well-drained loam soil and black cotton soils are mainly suited for cultivation of this crop.

The important varieties of fennel with their parentage, average yield (kg/ha), essential oil (%), crude fiber (%), duration, and the salient features are tabulated in Table 3.20.

3.5.3.4   Agrotechniques for Quality Production

3.5.3.4.1   Kharif Fennel

Few improved varieties like PF-35, Guj. F-1 and Guj. F-2, etc. give better yield than the old varieties. The times taken to attain maturity stages by the different fennel varieties like PF-35, Guj. F-1 and Guj. F-2 as kharif crops are 216-, 208- and 207-days, respectively.

3.5.3.4.1.1   Raising Nursery

The land should be prepared to fine tilth by ploughing 2–3 times. For raising nursery 100 sq. meter area is enough for transplanting one hectare area. Manures and fertilizers should be mixed in soil properly at the time of land preparation. Beds of 3 m × 1 m size should be prepared for nursery. 15th June is appropriate time for sowing of seeds in nursery. 2–2.5 kg seeds per 100 sq.m area should be broadcasted/sown in the nursery beds. Seed should be treated with organo-mercurial fungicides like Ceresan or Captan @ 3 gm per kg seed before sowing. Beds should be mulched with plant waste for up to 12 days to fasten the germination process. Tender seedlings should be protected from direct sunlight, till they are under danger, by erecting temporary shed over the bed. When seedlings are free from the danger of sunlight, temporary shed should be removed.

Apply 250 kg FYM, 0.6 kg nitrogen and 0.3 kg phosphorous per 100 sq. m nursery area. Whole quantity of FYM should be mixed in soil at the time of land preparation and whole dose of nitrogen and phosphorous should be used applied as basal dose. Supplementary dose of urea can be applied 3 weeks after germination of seeds, if seedlings are very weak. First, light irrigation should be given just after sowing and second after 3 to 4 days after sowing and then light irrigations are given as per requirement to keep the soil moist up to full germination. Later on irrigations are given at and interval of 2–3 days. About 45 days old seedlings are ready for transplanting. For keeping weeds free nursery, 1–2 hand weeding should be done depending on weed growth.

3.5.3.4.1.2 Planting

Land should be brought to fine tilth by one ploughing followed by 2 or 3 harrowing and planking. Suitable time for taking up transplanting of fennel seedling is around 15th August and it should be done during evening time. Seedlings are transplanted in main field at a spacing of 90 × 60 cm or 100 × 60 cm.

For kharif cultivation, 20 MT FYM, 100 kg nitrogen and 60 kg phosphorous per ha is recommended. Whole quantity of FYM should be mixed in soil at the time of land preparation and 40 kg of nitrogen and whole dose of phosphorous should be used applied as basal dose, while remaining 60 kg nitrogen should be top dressed up in two equal splits of 30 kg each at an interval of 30 days and 60 days (at the time of earthing up) after transplanting.

3.5.3.4.1.3 Intercultural Operations

First hoeing should be done at 20–25 days after transplanting. Intercultural operations should be repeated twice based on need at an interval of 20–25 days. Earthing up is done after 60–75 days from transplanting to check the lodging of plants. If necessary 2 to 3 hand weedings should also be done.

Fennel is a long duration crop thus, during dry spell in monsoon and after the Monsoon; it should be irrigated as per the requirement. Generally, 10–12 irrigations are given to fennel crop depending on soil and climate. Irrigation interval is longer in black soils as compared to light soils. First, irrigation should be given on the day of transplanting and subsequent irrigations should be given at an interval of 15–20 days up to February. Due attention should be given at critical stages of crop like seed development however irrigated.

3.5.3.4.1.4   Plant Protection

Damping off: This is a fungal disease mostly affects the nursery and plant shows symptoms of dehydration/wilting. This disease appears in those plots where water stagnation occurs near the plant stem. Collar portion of the plant start decaying and the plants turn to yellow color and die later on. To control the disease, avoid water stagnation in the field. Drenching of nursery beds with Blitox @ 20 g per 10 L water or drenching by 1% Bordeaux mixture (3:3:50) or spray with 0.2% solution of any copper fungicides should be done to control this disease effectively.

Aphid: It sucks the sap of tender parts of plants and affects the growth. Spraying of 0.03% solution of Dimethoate or 0.05% solution of Phosphamidone or any systemic pesticide should be done to control the aphid infestation.

Seed midge is a small dark shining insect having two transparent wings. It affects the fennel seeds. Female of this insect lays eggs on the ovary of flowers, which hatches and small caterpillar comes out of seed. To control this pest spraying of 0.07% solution of Endosulphan at the flowering stage should be done. Spraying of Endosulphan should be repeated twice at an interval of 10 days.

3.5.3.4.2   Rabi Transplanted

3.5.3.4.2.1   Raising Nursery

The land should be ploughed 2–3 times to brought to fine tilth. Nursery should be raised in about 150–200 sq. m land for planting one-hectare area. Manures and fertilizers should be mixed in the soil properly at the time of land preparation. Beds of 3 m × 1 m size should be prepared for raising of nursery. Seeds should be sown in nursery from last week of August to 1st week of September. 2.5–3.0 kg seed are enough for planting one-hectare area. Before sowing, seed should be treated with any organo-mercurial fungicides like Ceresan or Captan @ 3.0 g per kg. To get good germination beds should be covered with plant waste material and tender seedlings should be protected from direct sunlight.

Around 375 to 500 kg FYM, 1.0–1.2 kg nitrogen and 0.5–0.6 kg phosphorous should be applied per 150–200 sq. m nursery area. Whole quantity of FYM should be mixed in soil at the time of land preparation and Nitrogen and phosphorous should be used as basal dose. Supplementary dose of nitrogen can be given 3 weeks after germination of seeds, if seedlings are very weak. 1–2 hand weeding are required to control weed in the nursery depends upon weed intensity. First, light irrigation should be given just after sowing and second after 3 to 4 days after sowing and then light irrigations are given to keep the soil moist up to full germination as per need. Later on irrigations should be given at an interval of 2–3 days. About 40–45 days old seedlings are ready for transplanting.

3.5.3.4.2.2 Planting

Prepared land to fine tilth by one ploughing followed by 2 or 3 harrowing and planking. Transplanting of seedlings in main field should be done during evening time by adopting a spacing of 45 × 10 cm and it should be confirmed that optimum moisture is available in the soil. Suitable time for transplanting of fennel is around 15th October.

For rabi transplanted fennel, 20 MT of FYM, 90 kg nitrogen and 45 kg phosphorous per ha is recommended. Whole quantity of FYM. should be mixed in the soil at the time of land preparation and half dose of nitrogen (45 kg) and full dose of phosphorous (45 kg) should be applied as basal dose. Rest of nitrogen dose (45 kg) should be applied as top dressing in two equal splits of 22.5 kg each at an interval of 30 days and 60 days (at the time of earthing up) after transplanting.

First hoeing should be done 20–25 days after transplanting. It should be repeated twice based on need at an interval of 20–25 days. Earthing up is done after 60–75 days of transplanting to check the lodging of plants. If necessary 2 to 3 hands of weedings should also be done. Rabi fennel crop needs 8–10 irrigations according to soil and climate types. Irrigation at the time of seed development is critical, therefore due attention is needed. First, irrigation should be applied on the day of transplanting, after that subsequent irrigation should be given at an interval of 15–20 days up to February. Plant Protection is similar to kharif crop. Average yield of rabi crop is about 1800 to 2000 kg/ha.

3.5.3.4.3   Rabi Drilled

For land preparation, one ploughing followed two harrowing and planking should be done to bring the soil to fine tilth for growing of drilled fennel crop (Rabi). Stubbles of previous crop should be removed and make field clean. Ideal time for sowing is around 15th October and around 5–6 kg of seed/ha is required for sowing drilled fennel crop. The spacing should be maintained 45 to 60 cm between the line and 10 to 15 cm between the plants.

FYM @ of 25 MT, 90 kg nitrogen and 45 kg phosphorous per ha is commended for rabi drilled fennel. Whole quantity of FYM/ha should be mixed in the soil at the time of land preparation. Half dose of nitrogen (45 kg) and full dose of phosphorous should be applied as basal dose. Remaining nitrogen (45 kg) dose splits in two equal amounts of 22.5 kg each should be top dressed at an interval of 30 days and 60 days after sowing.

Under vapsa condition, two harrowing followed by planking should be done. Suitable beds should be prepared and irrigated. Nitrogen and potash should apply as basal dose by drilling in lines. Seeds also should be sown by drilling in lines 45 cm apart at a depth of about 1.0 cm. Seed should be soaked in water for 8 hours and dried in shade before sowing, to attain good germination. It germinates after 12 to 14 days. 8–10 irrigations should be given depending on soil and climate conditions. First, irrigation should be applied on the day of sowing, second light irrigation after 7 to 8 days from 1st irrigation and subsequent irrigations should be given at an interval of 15–20 days up to February.

For control of weeds; 2 or 3 hand weeding should be done on need basis. First, light intercultural operation should be done 20–25 days after sowing. Second and third intercultural operation should be done after 40 to 60 days from sowing. At the time of third intercultural operation harrow should be adjusted in such a way that it can throw sufficient soil near the base of the plants in addition to the earthing up operation. Plant protection is as similar to kharif fennel.

3.5.3.5 Harvest and Post Harvest Technology

When umbels reach to physiological maturity, i.e., when seeds are fully filled up and green in color should be harvested by cutting down. After cutting the umbels should be dried in partial shade. Umbels should be rotated if desirable at drying process for better aeration. The seeds are separated by thrashing the dry umbels with sticks followed by winnowing. Grading is done with the help of sieve or vibrator. For seed purpose, fully matured umbels are harvested. Yield is 2000 to 2500 kg/ha seed.

Rabi crop is raised for market type produce. Hence, harvest the crop should be done when 70% umbels reach at physiological maturity, by cutting upper half portion of the plants. Harvested plants should be dried and seeds are separated by thrashing followed by winnowing. Yield is 1500 to 1800 kg/ha.

3.5.3.5.1 Value Addition

The processed products of fennel include essential (volatile) oil, powder, fixed oil and oleoresins. Essential oil extracted from fennel fruits is a rich source of bioactive compounds and extensively used in pickes, liqourice, and candy for food flavorant antioxidant and antimicrobial properties. Fennel oleoresin prepared from seeds gives a warm, aromatic, and pleasant flavor to food products like snacks and sauces. Apart from sugar-coated fennel, oleoresin, whole, etc., other fennel based commercial blends are: Fennel tea (from fresh leaves and dried herbs), cough syrup (with honey and other herb based mixtures), absinthe (an alcoholic mixture/medical elixir), Indian *panch foron* (five spices), Chinese five spice blend, etc. Indian five-spices blend contains nigella, fenugreek, cumin, black mustard and fennel usually in equal parts, with ajwain sometimes used instead of cumin and black pepper sometimes added. Whereas, Chinese five spice blend contains anise, black pepper, fennel, cinnamon, and cloves.

3.5.4   FENUGREEK (TRIGONELLA FOENUM-GRAECUM L.)

3.5.4.1   Systematic Position

Kingdom: Plantae
Division: Magnoliophyta
Class: Magnoliopsida
Order: Fabales
Family: Fabaceae
Genus: Trigonella
Species: *foenum-graecum.*

3.5.4.2   About the Crop, National and International Scenario, Uses, and Composition

Fenugreek commonly known as "Methi" (*Trigonella foenum-graecum*) belongs to Fabaceae family. It is cultivated as a leafy vegetable, condiment as well as medicinal plant and is a native of South Eastern Europe and West Asia. It has a high medicinal value as it prevents constipation, removes indigestion, stimulates spleen and liver and is appetizing and diuretic. Major fenugreek producing states are Rajasthan, Madhya Pradesh, Gujarat, Uttar Pradesh, and Tamil Nadu. There are two economically important species of the genus *Trigonella* viz., *T. foenum-graecum,* the common methi and *T. corniculata,* the Kasuri methi. These two differ in their growth habit and yield. The latter one is a slow growing type and remains in rosette condition during most of its vegetative growth period. India exports around 20 thousand tonnes of fenugreek annually (April–March). The major importing countries are Egypt, UAE, and USA.

3.5.4.3   Soil and Climatic Requirements

Fenugreek is a winter crop and requires cool climate during vegetative growth while warm dry climate during maturity. It is tolerant to frost. It can be cultivated in almost all types of soils but well drained loamy soil suits well. For better growth and development of fenugreek, soil pH should be 6.0 to 7.0.

TABLE 3.21 Important Varieties of Fenugreek

Sl No.	Variety	Pedigree/ Parentage	*Average Yield (kg/ha)	Seed protein%	Duration (days)	Salient features
1	Co.1	Reselection from TG-2356 introduced for North India	680	–	80–85	A quick growing, dual purpose, early maturing variety tolerant to root rot disease. Seeds contain 21.7% protein.
2	Co 2	Selection from CF 390	480	–	85–90	Short duration dual purpose variety, field tolerant to *Rhizoctonia* root rot disease, suitable for both kharif and rabi season. Early maturity, short duration.
3	Rajendra Kanti	Pure line selection from Reghunathpur collection	1300	9.5	–	Medium sized bushy plant; early maturity, suitable for intercropping in both kharif and rabi season, field tolerant to *cercospora* leaf spot, powdery mildew and aphids.
4	RMt.1	Pure line selection from Nagpur local	1400	21	–	Vigorous semi erect medium sized, moderately branched growth habit, medium sized, bold, and attractive typically yellow colored grains, moderately resistant to root knot nematode and powdery mildew and aphids
5	Lam Sel.1	Selection from germplasm collection of Uttar Pradesh	740	53	–	Dual purpose varieties, early muturing, bushy type and medium height 94°C, more number of branches and green matter. When cultivated for green leaf purpose it gives an average green yield of 12 tonnes per hectare. Field tolerant to major pests and cases.
6	Hisar Sonali	Pure line selection from germpalsm	1700	–	–	Tall and bushy vigorous growing variety, dual purpose variety, late maturity (140–145 days), suitable for cultivation under irrigated condition. Moderately resistant to root rot and aphids.

TABLE 3.21 (Continued)

Sl No.	Variety	Pedigree/ Parentage	*Average Yield (kg/ha)	Seed protein%	Duration (days)	Salient features
7	Hisar suvarna	Pureline selection from local germplasm	1600	–	–	A quick growing, erect, and tall, dual purpose, medium maturity (130–140 days), moderately resistant to percospora and powdery mildew. Wider adaptability, suitable for cultivation throughout the country.
8	Hisar Madhavi	Pureline selection from local germplasm of UP	1900	–	–	A quick growing, erect, and tall, dual purpose, medium maturity (130–140 days), moderately resistant to powdery mildew and to downy mildew. A variety with under adaptability suitable for both irrigated and rainfed condition.
9	Hisar Muktha	Pureline selection natural green seed coated mutant line from UP	2000	–	–	A quick growing seed type variety, medium maturity (135–140 days), moderately resistant to powdery mildew and to downy mildew. Erect and tall plants. Wide adaptability. Suitable for both irrigated and rainfed condition
10	RMt 303	Mutation breeding from variety RMt 1	1900	–	–	Medium maturity variety (145- 150 days) seeds bold, with typical yellow color, less susceptible to powdery mildew
11	RMt 305	Mutation breeding from variety RMt 1	1300	–	–	First determinant type, multipodant, early maturing, wider adaptability, resistant to powdery mildew and rootknot nematodes. Seeds bold, attractive, and yellow, duration 120–125 days.
12	Guj Methi 1	Recurrent selection based on pure line selection from J. Fenu 102	1864	–	–	The first variety from Gujarat released for the state. Plant dwarf.

Sl No.	Variety	Pedigree/ Parentage	*Average Yield (kg/ha)	Seed protein%	Duration (days)	Salient features
13	RMt 143	RMt143	1600		140–150	Moderately resistant to powdery mildew, seeds bold yellow color, suitable for heavier soils.
14	Rajendra Abha (Kasuri Methi)	NA	NA	NA	NA	NA
15	Pant Ragini	Selection from local germplasm	1200	–	170–175	A dual purpose tall bushy type resistant to downy mildew and root rots, medium maturity. Seed contain 2–2.5% essential oil
16	Rajendra Khushba	NA	NA	NA	NA	NA
17	Pusa Early Bunchy	NA	NA	NA	NA	NA
18	AM-01–35	Selection from local germplasm	1720	–	–	Dual purpose, tolerant to powdery mildew

* Yield kg/ha (Dry).

Some other improved varieties are Gujarat Methi-1, Prabha, Methi No. 14, RM-16, Co.1 UM-34, UM-35, Kasuri, and Kasuri selection.

The important varieties of fenugreek with their parentage, average yield (kg/ha), protein (%), duration, and the salient features are given in Table 3.21.

3.5.4.4   Agrotechniques for Quality Production

Land should be ploughed 2–3 times followed by planking to bring the soil to fine tilth. Soil clods should be broken and stubbles of previous crop should be removed. Sowing may be done from 2nd fortnight of October to 1st fortnight of November, but first fortnight of November is the best time for sowing fenugreek. It required 10–15 kg of seeds for sowing one-hectare area. Seeds should be soaked in water for 6 to 8 hours and dried in shade before sowing to hasten germination. Seeds take about 6–8 days for complete germination. Seed should be sown using a spacing of 30 cm between lines and 8–10 cm plant to plant or by broadcasting. Farmyard manure @10–15 MT should be mixed in soil at the time of land preparation. A dose of 40 kg nitrogen and 20 kg phosphorous per ha is recommended for fenugreek. Half dose of nitrogen and full dose of phosphorous should be applied as basal dose and remaining 20 kg nitrogen at an interval of 30 days after sowing. In broadcasted crop, plants should be thinned to maintain the plant distance about 8–10 cm. 2 to 3 weeding are enough to keep crop weed free. One weeding and hoeing should be done about 20–25 days after sowing and second weeding 45–50 days after sowing. Intercultural operation during the early stage of plant growth minimizes weed competition. About 4–6 irrigations are required depending on soil type and climate. Pre-sowing irrigation should also be given, if moisture level of the soil is not optimum for seed germination. First, irrigation should be given at the time of thinning and subsequent irrigation at an interval of 20–25 days.

3.5.4.5   Plant Protection

Major diseases like powdery mildew, downy mildew and pests like aphid are adversely affect the growth and yield of the fenugreek crop.

Aphid: It sucks the sap of tender parts of plants and affects the growth adversely. Spraying of 0.03% solution of Dimethoate or 0.025% solution of Methyl demetone or 0.04% solution of Monocrotophos is recommended to control the aphid. If the crop is grown for green purpose then spray Malathion is suggested.

Powdery mildew: This disease appears usually in later stage of crop and becomes serious when pod formation takes place. In this disease, white powdery patches appear on the lower and upper surface of leaves and other parts of plant. To control this disease crop should be dusted with 300 mesh Sulphur dust @ 25 kg/ha as the symptoms are noticed. Spraying of wettable Sulphur or Dinocap (Kerathan or Thiowet) can also be used to control the disease @ 20–25 g per 10 L of water at the initial stage of this disease. If needed two more sprays should be given at an interval of 15 days after first spray.

Downy mildew: Disease occurs during February and March. Infected plants show yellow patches on the upper surface of leaves and white cottony mycelium on the lower surface of leaves. This disease can be controlled by spraying of 0.2% solution of Difoltan or any other copper fungicide.

3.5.4.6 Harvest and Post Harvest Technology

For leaf, young shoots are nipped off 4 to 5 cm above ground level in about 25 to 30 days and subsequent cuttings of leaves may be taken after 15 days. It is advisable to take 1 to 2 cuttings before the crop is allowed for flowering and fruiting.

For seed purpose, fruit are ready for harvest in about 120–150 days. At the time of maturity, leaves, and pods become yellowish and leaves falling start. Timely harvesting is very important for this crop as late harvest leads to seed losses due to pod bursting, whereas in early harvest, the grains remain immature and smaller in size as compared to mature ones. Harvesting should be done early in the morning to avoid shattering loss. After harvest, plants should be dried in threshing yard and threshed mechanically by trampling under the feet of bullocks. Seeds should be separated and cleaned by winnowing. Average seed yield is 1800 to 2000 kg/ha and green leaf yield is about 800 to 1200 kg/ha.

3.5.4.6.1 Value Addition

Fixed oil: Fenugreek seed contains is about 7% fixed oil. It has marked drying properties. The dried oil is golden yellow in color and is insoluble in ether. It is recently used in perfumery industries.

Volatile oil: Seed contains less than 0.02% essential oil.

3.5.5 AJWAIN OR BISHOP'S WEED (TRACHYSPERMUM AMMI L./ SPRAGUE)

3.5.5.1 Systematic Position

Kingdom: Plantae
Division: Tracheophyta
Class: Magnoliopsida
Order: Apiales
Family: Apiaceae
Genus: *Trachyspermum*
Species: *ammi*

3.5.5.2 About the Crop, National and International Scenario, Uses, and Composition

Ajwain or Bishop's weed (*Trachyspermum ammi* L.) belongs to the family apiaceae is an annual herbaceous underutilized seed spice. It is originated from Mediterranean region and grown in Iran, Afghanistan, Egypt, Pakistan, North Africa and India. In India it is mainly cultivated in Gujarat, Rajasthan, Madhya Pradesh, Uttar Pradesh, West Bengal, Punjab, Haryana, Bihar, and Tamil Nadu. Essential oil content of the seed varies from 3–4% and the main composition of the essential oil is thymol. The other constituents are pinene, cymene, dipentene, terpinene, and carvacrol. Ajwain is widely used for seasoning various food preparations. Ajwain is mostly found in pulse dishes (dhal), vegetable dishes and for preparation of pickles. It is also used as preservative, antioxidant, and essential oil is used in perfumery and pharmaceutical industries.

3.5.5.3 Soil and Climatic Requirements

Cool weather and cloudiness for about a week after sowing and occasional drizzling during active growth are conducive to successful cultivation of ajwain. Cool and dry climate favors good growth and yield of the crop. High atmospheric humidity after flowering stage invites diseases and insects (Shetty et al., 2012). Except sandy loam soil, it can be grown all types of soil. It grows best in loamy or clay loam soil rich in humus and well drained.

The important varieties of ajwain with their parentage, average yield (kg/ha) and the salient features are given in Table 3.22.

TABLE 3.22 Important Varieties of Ajwain

Sl. No.	Name	Parentage	Average Yield (kg/ha)	Special features
1.	Gujarat Ajwain	Selection from germplasm	2250	Non shattering type, mildly susceptible to powdery mildew and late maturity (about 180 days)
2.	RPA – 68	Selection from germplasm grown in Pratapgarh area	900	Medium maturity (150 days) and adopted for cultivation in Rajasthan.
3.	Pant Ruchika	Pure line selection from local collection	600–800	Suitable for growing both under rainfed and irrigated condition. Seed essential oil content is 3.4% and maturity is 160 days.
4.	Ajmer Ajwain-1	Selection from Pratapgarh local (NRCSS-AA-61)	1420	Susceptible to powdery mildew and suitable for growing both under rainfed and irrigated condition. Bold seeded with an essential oil content of 3.5%.
5	Ajmer Ajwain-2	Selection from Gujarat local (NRCSS-AA-19)	1280	Early maturity (145–150 days) moderately tolerant to drought and seed essential oil content is 3.0%.
6	Lamsel-1	Mass selection	1000	Early maturity (145–150 days) and it is adopted for cultivation in Andhra Pradesh.
7	Lamsel-2	Mass selection	1000	Spreading type and maturity is 160 days and it is adopted for cultivation in Andhra Pradesh.
8	Rajendra Mani	-	-	It is adopted for cultivation in Bihar.

3.5.5.4   Agrotechniques for Quality Production

Sowing is done in October–November in most part of the India, but in some parts it may be sown in May-June. Seed rate is 3–4 kg/ha. Seeds are sown in broadcast or line. Incase of line sowing a spacing of 30–45 cm × 20–30 cm may be given depending upon the locality and management condition especially irrigation and nutrient management. In most of the areas farmers grow local cultivar. Pant Ruchika, Gujarat Ajwain-1, Lam Selection-I and Lam Selection-II are some of the improved varieties released from different research institute for commercial cultivation. Application of farmyard manure @ 15 t/ha along with NPK @ 30:45:45 kg/ha is beneficial for higher growth and yield of the crop.

3.5.5.5   Harvest and Post Harvest Technology

The cop matures 160–180 days after sowing. Fruits become ready for harvesting when, the flower head turn brown. Average yield of ajwain varies from 0.6–0.8 t/ha. Collar rot, root rot and powdery mildew are the important diseases of this that affect the yield and quality of the crop.

3.5.5.5.1   Value Addition

Essential oil: Steam distillation of crushed ajwain seeds produces 2–4% essential oil. It is mainly used for preparations of different medicines. The main component of the essential oil is thymol (53–80%).

   Thymol: It is prepared by treatment of ajwain essential oil with aqueous alkaline solution and regenerating thymol from it by extracting with ether or steam distillation. It is also mainly used in pharmaceutical industries.

3.5.6   CUMIN, BLACK OR KALONJI OR NIGELLA (NIGELLA SATIVA L.)

3.5.6.1   Systematic Position

   Kingdom: Plantae
   Division: Magnoliophyta

Class: Magnoliopsida
Order: Ranunculales
Family: Ranunculaceae
Genus: *Nigella*
Species: *sativa*

3.5.6.2 About the Crop, National and International Scenario, Uses, and Composition

Black Cumin (*Nigella sativa* L.) belongs to the family ranunculace is an important promising underutilized seed spices. It is native to Mediterranean region. It is considered to be a highly aromatic spice cum medicinal plants. It is mainly cultivated in Punjab, Himachal Pradesh, West Bengal, Bihar, Orissa, and Assam. The commercial product is seed, which is used as spices and in pharmaceutical industry. Seed emits an aroma resembling strawberries when crushed. The drier tracts of different states are highly suitable for commercial cultivation and seed production of this particular spice. It is an important common spice in the Indian kitchen especially for the preparation fast food, oil fry food and for the preparation of the various delicious and tasty dishes.

Black cumin seeds contain 20.85% protein, 38.20% fat, 4.64% moisture, 4.37% ash, 7.94% crude fiber and 31.94% total carbohydrates. Potassium, phosphorus, sodium, and iron were the major elements present. Zinc, calcium, magnesium, manganese, and copper are some minor elements. Thirty-two fatty acids (99.9%) have been identified in the fixed oil. The major fatty acids were linoleic acid (50.2%), oleic acid (19.9%), margaric acid (10.3%), cis-11, 14-eicosadienoic acid (7.7%) and stearic acid (2.5%), myristic acid as well as β-sitosterol, cycloeucalenol, cycloartenol, sterol esters and sterol glucosides. The volatile oil (0.4–0.45%) contains saturated fatty acids, which includes: nigellone that is the only component of the carbonyl fraction of the oil, Thymoquinone (TQ), thymohydroquinone (THQ), dithymoquinone, thymol, carvacrol, α and β-pinene, d-limonene, d-citronellol, *p*-cymene volatile oil of the seed also contains: *p*-cymene, carvacrol, t-anethole, 4-terpineol and longifoline. Black cumin seed have two different forms of alkaloids: isoquinoline alkaloid that includes: nigellicimine, nigellicimine n-oxide and pyrazol alkaloid that includes: nigellidine and nigellicine.

3.5.6.3   Soil and Climatic Requirements

Cool and dry climate is ideal for its cultivation. Sunny weather in day and low temperature in night is highly congenial for its vegetative as well as reproductive growth. Rainfall during flowering is harmful whereas it is beneficial after seed set to get higher yield. Well-drained, high to medium land is suitable for its cultivation. Though it can be grown in a wide range of soil but it prefers loamy sand soils.

The important varieties of nigella with their maturity (days), average yield (kg/ha) and the special features are given in Table 3.23.

3.5.6.4   Agrotechniques for Quality Production

Datta et al. (2003) observed that variety, Rajendra Shyama performed best under West Bengal condition. Banafar et al. (2002) found that cultivar NS.- 44

TABLE 3.23   Important Varieties of Nigella

Sl. No.	Name	Maturity (days)	Average Yield (kg/ha)	Special features
1.	Azad Kalaunji	135–145	1000	Erect plant, foliage normal green, bold seeds, highly aromatic, no incidence of any diseases no serious pest problem.
2.	Pant Krishna	135	720	Medium statured, susceptible to damping off, bold seeded, suitable for cultivation in UP and Uttaranchal
3.	Ajmer Nigella-1	135	720	Seeds are dark black, bold, volatile oil content 0.7%, field resistant to root rot.
4.	Rajendra Shyama	-	-	They grow best under West Bengal condition.
5.	NS 32	140–150	-	It is a high yielding variety.
6.	NS 44	140–150	-	It is a high yielding variety.
7.	Kala Jeera	135–145	450	It is also a high yielding variety.

performed best under Jabbalpur condition of Madhya Pradesh. Sowing time varies from locality to locality. Under South Bengal condition the ideal time is first fortnight of November while in North Bengal ideal time is mid November. 7.5–10 kg of seeds are required for sowing depending upon the methods of sowing. Higher germination percentage can be achieved by soaking the seeds with GA_3 at 100 ppm or 200 ppm under dark condition (Mondal and Maity, 1993). Depending upon the edaphic condition seed is sown with a spacing of 20–30 cm × 5–10 cm. Datta et al. (2001) reported that 45:45:45 kg NPK/ha was beneficial for higher yield. A fertilizer dose of 30:30:40 kg NPK/ha produced maximum yield under Assam condition (Luchon and Sarat, 2003) whereas a fertilizer dose of 60 kg each of N &P were recommended for Kanpur (Singh et al., 1999). Nataraja et al. (2003) reported that application of 50:40:30 kg NPK/ha gave maximum yield under Bangalore condition. Irrigation at 35, 60 and 85 days after sowing is beneficial to get higher yield under Raipur condition. The important diseases of black cumin are blight and wilt. Generally pest is not a problem in black cumin cultivation. Sometimes cut worm and fruit borer may cause considerable damage.

3.5.6.5   Harvest and Post Harvest Technology

The crop should be harvested when plants start yellowing. Delay in harvesting causes shattering loss of seed at field. Generally plants are uprooted or cut the plant just above ground level. The seeds are separated by beating lightly with sticks. Seed are then cleaned, dried, and filled in gunny bag with polylyning for storing and it should be stored in cool and dry place. Black cumin produces 9.0–12.0 q/ha yield under irrigated condition and 5.0–7.0 q/ha under rain fed condition – 5.0–7.0 q/ha.

3.5.6.5.1   *Value Addition*

Black seed oil is the oil obtained from black cumin. The list of pharmacological and therapeutic properties of black seed oil is enormous.
Anti-rheumatic – it provides relief from rheumatoid arthritis.
Anti-inflammatory – black seed oil is strongly antiinflammatory.
Anti-allergic – oil can suppress food allergies.
Anti-diabetes – having blood glucose lowering effect.
Anti-cancer – the chief anticancer compound is thymoquinone.

Hypotensive – volatile oils aid in managing blood pressure.

Anti-tumor or antineoplastic

Radioprotective – safeguards our cells from harmful effects of damaging radiation.

Hepatoprotective – within safe limits, it protects the liver.

Renoprotective – exerts a protective effect on the kidneys.

Apoptosis induction – programs potentially harmful cells to die automatically.

Immunomodularity – keeps the immune system at a balance.

Analgesic – pain reliever.

Antipyretic – alleviates fever.

Antibacterial – kills many bacteria strains, even the most antibiotic resistant ones like MRSA.

Antiviral – kills virus

Antifungal – easily kills many fungi

Galactagogue – promotes lactation.

Adjuvant – modifies the effect of other therapeutic agents.

Bronchodilator – aids dilation of air passages.

Laxative – promotes timely bowel movement.

Antioxidant – protects from free radicals.

Lipid lowering – makes the lipid profile healthier.

Insulin-sensitizer – affects the secretion of insulin.

Interferon booster – stimulates the production of interferon, which boosts immunity against pathogens.

Antispasmodic – it relieves involuntary spasms of the respiratory tract, thus providing relief in asthma and other kinds of cough.

Anti-convulsive – calms seizures

Black cumin tea: In one teaspoon of crushed seeds 50 mL of boiling water is poured, and strained after 10 minutes. Usually a cup is taken twice a day as a remedy for colds and bronchitis. In folk medicine, tea from black cumin seed is used as a remedy for flatulence, diarrhea, and biliary colic, as a diuretic, cholagogue, anthelmintic, mild laxative, and stomachic. It is given for mothers with inadequate secretion of milk.

Black seed honey: Bees collect nectar from flowering fields of wild black cumin in a sunny day and a delicious black seed honey is naturally produced. In the Middle East, honey of black cumin is widely known and loved.

3.5.7   CARAWAY OR SIAH ZIRA (CARUM CARVI L.)

3.5.7.1   Systematic Position

Kingdom: Plantae
Division: Tracheophyta
Class: Magnoliopsida
Order: Apiales
Family: Apiaceae
Genus: *Carum*
Species: *carvi*

3.5.7.2   About the Crop, National and International Scenario, Uses, and Composition

Caraway or Persian cumin (*Carum carvi*) is a biennial plant in the family Apiaceae. It is originated from Europe and Western Asia. In India, the species is available in the wild in the high altitude regions of Jammu and Kashmir (Ladakh, Zanskar, and Lungna) Himachal Pradesh (Kinnaur, Lahaul, and Spiti and Pangi-Bharmour) and Uttaranchal. The species is now under large-scale cultivation in Europe and Morocco in Africa. In India, it is cultivated in Bihar, Orissa, Punjab, West Bengal, Andhra Pradesh and in the hills of Kumaon, Garhwal (Uttaranchal) and a few cool dry regions of Kashmir. The plant is similar in appearance to a carrot plant, with finely divided, feathery leaves with thread-like divisions, growing on 20–30 cm stems. Caraway fruits (frequently called seed) are crescent shaped achenes, around 2 mm long, with five pale ridges. The dried fruit seed is brown in color, has pleasant odor, aromatic flavor, is warm with somewhat sharp taste.

It is one of the world's oldest culinary spices. Seeds of it were found in the remains of food from the Mesolithic age. It is extensively used for flavoring rye-bread, biscuits, cakes, and cheese. It is also used in manufacturing of Kummel cordial and also become an ingredient for seasoning sausage and pickling spice, soups, and meat stews. It is a mild stomachic and carminative in nature. Occasionally used as flatulent and colic. Carvone (isolated form seed) is used as anthelmintic in Hookwarm disease. Caraway seed oil is used in oral preparations for overcoming unpleasant odor. It is also used for sending soaps.

There are two types:

i)   Normal Caraway (*Carum carvi*)

ii) Black caraway (*Carum bulbo castanum*). It is also called Kala zira'

The percentages of essential oil available in caraway seeds vary from 1–6%. Carvone and limonene are the main components available in their oil. It also contains trace amounts of other compounds including acetaldehyde, furfural, carveole, pinene, thujone, camphene, phellandrene, etc. limonene (43.5%), carvone (32.6%), and apiole (15.1%).

3.5.7.2.1   *Composition*

Moisture: 4.5%	Ca: 1.0%
Protein: 7.6%	P: 0.11%
Fat: 8.8%	Na: 0.02%
Fiber: 25.2%	K: 1.9%
Carbohydrates: 50.2%	Ash: 3.7%

3.5.7.3   Soil and Climatic Requirements

Caraway grows well in a variety of soils but moderately sandy loam to light clay soil that is rich in humus is ideal for its commercial cultivation. It can tolerate a pH range is 4.8 to 7.6. It is a sun loving plant and requires full sunlight although it can grow in semi-shade too. It can also tolerates light frost.

3.5.7.4   Agrotechniques for Quality Production

Caraway is propagated through seed. Seed viability ranges from 60–90% after harvesting and viability decreases after 6 months Seeds of the biennial varieties can be sown either late in summer or early in spring, but those of the annual type only in spring. The selected land is ploughed at least once after, which well decomposed farmyard manure (FYM) is mixed with soil. Seeds are sown as soon as they ripen in autumn, though they may be sown in up to February–March Sowings should be made at a spacing of 60 cm × 30 cm. The height of each bed is kept 6 inches above the ground. The seeds should be sown in light well-drained soil. Germination is slow as well as the growth of plants in the early part of the season; therefore, considerable care is necessary to keep down weeds.

Well decomposed FYM @ 15–20 t/ha should be applied at time of final land preparation. Among the inorganic fertilizers, application of 50–60 kg/ha nitrogen, 90–100 kg/ha phosphorous and 80–90 kg/ha of potassium is conducive for getting higher yield. The field is to be irrigated immediately after sowing if there is no rain or deficit in soil moisture. In the initial month irrigation should be done at an interval of 3–4 days after, which watering once in a week is sufficient. Irrigation is required during April-July in both the first and second year prior to monsoon rain. Hoeing for loosening soil must be done. Weeding should be done at least twice before the plants are full grown. The biennials starts flowering during early in the second season after planting and matures their seeds by midsummer. The seeds ripen in August and are collected when the oldest fruits turn brown. Normally, when the umbels turn brown they should be cut from the plant before shattering begins. The umbels should be dried thoroughly in the sun or shade, and the seeds separated from umbel and then cleaned and stored in a paper bag or closed container. Caraway seeds are normally stored in airtight containers for maintaining its flavor for periods of time.

3.5.7.5   Harvest and Post Harvest Technology

Harvesting time: When the seeds turn light brown (for commercial grade) and dark brown (for seed purpose). Threshing is done with sticks. Well-ripened fruits are to be harvested to maintain good quality. Otherwise, if fruits are collected before ripening, they will ultimately produce low quality seeds. There are four stages of post harvest management of caraway seed, as follows:

- Picking (threshing)
- Drying (cleaning)
- Packing (small bags)
- Storage

3.5.7.5.1   *Value Addition*

Volatile Oil: It is obtained from fresh seeds up on steam distillation yields a colorless or pale yellow oil. A terpene, formerly called carvone, has been recognized to be di-limonene and traces of carvancol are found. The official standard requires that the oil should contain not less than 53% & not more than 63% carvone.

Decarvonized Oil: It consists of limonene with trace of carvone, used to scent cheap soaps.

3.5.8 CELERY (APIUM GRAVEOLENS L. VAR CELERY)

Mostly cultivated as a salad crop, the dried fruits are used as spice. Fruits yield 2–3% of volatile oil, used as a fixative and ingredient of novel perfumes. Seeds have many medicinal values.

3.5.8.1 Systematic Position

Kingdom: Plantae
Phylum: Tracheophyta
Class: Magnoliopsida
Order: Apiales
Family: Umbelliferae
Genus: *Apium*
Species: *graveolens*

3.5.8.2 About the Crop, National and International Scenario, Uses, and Composition

Celery (*Apium graveolens* L.) is an umbelliferous aromatic herbaceous plant grown for its leaves, seeds, oleoresin, and essential oil. It is believed to be originally from the Mediterranean basin. Before 850 B.C. in ancient literature documents that celery or a similar plant form was cultivated for medicinal purposes. It is claimed that, all plant parts, but mostly seeds contains volatile oils, which probably used for medicinal purposes. Hence, during ancient time's ayurvedic physicians used celery seed to treat the colds, flu, water retention, poor digestion, various types of arthritis and liver and spleen ailments. Although celery is thought to be from the Mediterranean, indigenous "wild" relatives of celery are found in southern Sweden, the British Isles,

Egypt, Algeria, India, China, New Zealand, California, and southernmost portions of South America. India's annual (April–March) Celery export is around 4 thousand. USA and Vietnam are the largest importer of celery.

3.5.8.2.1 *Composition*

Moisture: 5.1%	Ca: 1.8%
Protein: 18.1%	P: 0.55%
Fat: 22.8%	Fe: 0.45%
Crude fiber: 2.9%	Na: 0.17%
CHO's: 40.9%	K: 1.4%
Total ash: 10.2%	

3.5.8.3 Common Types of Celery Worldwide

1. CELERY (*Apium graveolens L.var. dulce*): It is grown in North America and temperate Europe for its succulent petioles. It is also known as stalk celery.
2. CELERIAC (*Apium graveolens L.var. rapaceum*): It is known by other name such as celery root or knob celery.
3. SMALLAGE (*Apium graveolens L. var. secalinum*): It is grown for its leaves and seeds mostly in Asia and Mediterranean regions. It is commonly known as lead celery.

Celery is grown for its fibrous leaf stalks (petioles). It is edible cluster of long, green stalk grows upright from crown of the plant. At the top of the stalks are leaves that look similar to Italian parsley. Celery leaves are finely divided, yellow-green to dark green in color and ridges run the length of the celery stalk. They are edible and sometimes used as seasoning.

3.5.8.4 Cultivars

There are few green type celery varieties viz., Utah, Tall Utah 52–70 and selections (52–70R, Tendercrisp, Florida 683, Pascal, Giant Pascal, Pascal 259–19 (Summer Pascal), etc. Golden Self-blanching (Golden Detroit) is a self-blanching variety. Golden Plume, Cornell 19, etc., are used for flavoring and decoration and not eaten fresh. Few other varieties are Sac Yuquin, Wolf-249, FBL 5–2M, Shangnong Yuquin, Florida Slobolt M68, Floribelle M9, UC-T3 Somaclone, RRL-85–1, Advantage, Pilgrim, Companion, Sigfrido, and Golden Spartan.

3.5.8.5 Soil and Climatic Requirement

Celery thrives well in a relatively cool growing season with a monthly average temperature of 12°C to 15°C and a well-distributed rainfall during the growing season. High temperature during growing period promotes bolting. It can be grown in wide range of soils. It performs best when grown on friable soil rich in organic matter as they have high water-holding capacity and ensure uniform moisture supply. It grows well on soils having pH range of 5.5 to 6.7.

3.5.8.6 Agrotechniques for Quality Production

Celery is mainly propagated by seeds. Seeds of celery are small weighting 3000 seeds per one gram. They germinate slowly. It requires 15–20°C for good germination, while poor at 25°C and none at 30°C. A seed rate of 100 g/ha is recommended. Sowing dates should be adjusted according to climatic conditions. Thus, in Low hills, it is sown in September–October, in mid hills in the month of August-September, in high hill conditions in March–April and in plains by the end of August. Seedlings are ready for transplanted at 4–6 weeks after sowing by adjusting the distance 60 × 20 cm for winter crop and 25×25 cm for summer crop. Summer crops are self-blanched but in case of winter it required blanching.

The nutrient requirements of celery are quite high and it removes 140 kg nitrogen, 55 kg phosphorus and 220 kg potash with yield of 20 t/ha. It is a shallow-rooted crop. So the majority of the nutrients are drawn from the upper surface. Good quality depends on continuous steady growth so, application of FYM @ 20 t/ha with the supplements of NPK @150:60:100 kg/ha is suggested. High rates of mineral forms of N adversely affected flavor in celery by increasing the nitrate content.

Irrigation significantly influences the yield and quality. As it increases dry matter and total sugar contents but decreased vitamin C contents. Apart from these it also decreased α and β-carotene as well as thiamine contents of celery. Shoot growth is better with sprinkler irrigation than drip irrigation. But drip irrigation is helpful in applying fertilizers and pesticides effectively.

Timely cultural operations, improve the growth and quality of the crop as, celery responds more to the cultivation than many other crops. Soil mulch has beneficial effect. As it is a shallow rooted crop, light cultivation should be practiced. Since celery is a long duration crop, weeds controls are difficult on most soils. Soil should be hoed lightly but frequently and all lateral shoots removed as they appear. Herbicidal spray can also be followed.

Excluding light from the stalk while the plants are still growing makes them devoid of chlorophyll are known as blanched. Blanching is done either by wrapping paper around the leaf stalks or by earthing up soil as the plants grow. Blanching is now being discouraged as the nutritive value is reduced by this process. There are self-blanched and green varieties also developed.

3.5.8.7 Plant Protection

Celery is prone to many diseases of which *Fusarium* Yellows (Fusarium oxysporum f. Sp. apii), Bacterial blight, Early blight (*Cercospora*), *Septoria* leaf blight/Late blight, Leaf curl are the major diseases that cause economic losses. Leaf miner, Leaf hoppers, Carrot weevil, Celery caterpillar, Green celery worm, Celery looper, Thrips, Carrot rust fly, Aphid are major pest affecting the quality and yield losses of celery.

Damping off: Causal organisms of this disease are *Pythium*s pp, *Rhizoctonia solani.* Common symptom of this disease is soft rotting seeds, which fail to germinate and rapid death of seedling prior to emergence for soil. Sometime seedlings collapse after they have emerged from the soil caused by water soaked reddish lesions girdling the stem at the soil line. Avoid planting celery in the soil having poor drainage capacity and wet soil. Planting in raised beds treat seeds with fungicide prior to planting to eliminate fungal pathogens.

Downy mildew: It is casued by *Peronospora umbellifarum.* Yellow spots appear on upper surface of leaves, white fluffy growth noticed on underside of leaves and lesions become darker as the mature. This disease can be controlled by sowing disease free seed, adoption of crop rotation with non-umbelliferous crops.

Early blight: This fungal disease is caused by *Cercospora apii.* Small yellow flecks are noticed on upper and lower leaf surfaces, which enlarge to brown gray spots with no defined border and ultimately lesions develop into a papery texture leaves; with many lesions may turn necrotic and die. This disease can be managed by sowing pathogen free seed, crop rotation and avoiding overcrowding plants and by applying suitable fungicides.

Fusarium yellow: It is caused by *Fusarium oxysporum.* Yellowing plants that are severely stunted, roots brown and water soaked. The most effective method of controlling *Fusarium* yellows is to prevent its introduction by regularly sanitizing tools and equipment, planting in pathogen free soil and planting resistant or tolerant varieties of celery.

3.5.8.8 Physiological Disorder

Black heart: It is major disorder of celery due to Ca-deficiency. It shows tip burn symptoms on young leaves. It can be controlled by foliar application of 0.10 molar solution of Ca-nitrate or Ca-chloride.

Chlorosis: It is due to Mg-deficiency, which affects the yield and market quality. Spraying of Mg-sulphate @12 kg/ha will effectively prevent it.

Cracked stem: B-deficiency results in cracked stem with lesions on the inner and outer surface of the petioles and over the vascular bundles. The adjoining epidermis curls outwards followed by a dark brown color of the exposed tissue. To control, apply commercial borax @12 kg/ha solution near the base of the plant. More amount of borax would be required on a neutral or alkaline soil than on acidic soil.

3.5.8.9 Harvest and Post Harvest Technology

There is no definite stage of maturity at, which celery is harvested. When the plants attained 40 cm height, side suckers are removed and sent to market. In general, the crop is ready within 75–90 days after transplanting. Each plant is cut down just below the soil surface with a sharp knife, then trimmed and prepared for the market. The yield of celery varies from 330–560 q/ha depending on variety with an average yield of 450 q/ha.

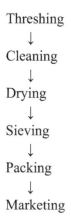

Threshing
↓
Cleaning
↓
Drying
↓
Sieving
↓
Packing
↓
Marketing

In storage, celery absorbs foreign flavors. So, it should be kept away from the odors of other products. Harvested produce are stored in trenches, cellars, and cold storage. In cold storage, they are kept at 0°C and 95–98% R.H. where it can be stored for 2–3 months in good condition.

3.5.8.9.1 Value Addition

Volatile oil: The seed contains 2–3% volatile oil. It is yellowish to dark greenish in color and main components are d-limonene (19.5–68%), B-selinene (8%), n-butyl lidene puthalide (8%). It is mostly used as fixative and as an ingredient of novel perfumes, pharmaceutical preparations and medicines.

Seed oleoresin: It is prepared by extraction of crushed dried celery seed seeds with suitable solvents like food grade hexane or ethylene dichloride. It is used as food flavorant and also used in pharmaceutical and perfumery industries.

Celery fatty oil: The fruit is also yield 7% of fatty oil. This oil is used as antispasmodic and nerve stimulant.

Celery Chaff oil: It has a harsher and coarser odor and flavor than that of celery seed oil.

Celery pepper: It substitutes ground black pepper for celery salt but should contain not more than 70% of pepper.

Celery salt: Prepared by mixing finely ground table salt with ground Celery seed or Celery seed oleoresin or ground dried Celery stems. It contains about 25% of celery seed powder or ground celery and 75% of common salt.

3.5.9 DILL/INDIAN DILL/SOWA SEED (ANETHUM SOWA ROXB. EX. FLEM.)

3.5.9.1 Systematic Position

Kingdom: Plantae
Division: Magnoliophyta
Class: Magnoliopsida
Order: Apiales
Family: Apiaceae
Genus: *Anethum*
Species: *sowa*

3.5.9.2 About the Crop, National and International Scenario, Uses, and Composition

Dill is a herbaceous annual with pinnately divided leaves. Seed is used as a condiment and foliage is used flavoring agent. The ripe, light brown seeds emit an aromatic odor. The leaves have pleasant aromatic odor and warm taste. Both seeds and leaves are used as spice. Dill is a herbaceous annual with pinnately divided leaves. The ripe, light brown seeds emit an aromatic odor. The leaves have pleasant aromatic odor and warm taste. Both seeds and leaves are valued as spice. There are two types dill is marketed in the world, namely, European and Indian dill. European Dill (*Anetheum graveolens*) is originated Europe and is commercially cultivated in England, Germany, Romania, Turkey, USA, and Russia. The Indian dill (*Anetheum sowa*) indigenous to of Northern India is bolder than the European dill. It is cultivated as a cold weather.

Dill seed is used both whole and ground as a spice in vegetable preparation, soups, salads, processed meats, sausages, and pickling. Its stems and blossom heads are used for dill pickles. The essential oil is used in the manufacture of soaps. Both seeds and oil are used in ayurvedic medicine preparations. Its common use in Ayurvedic medicine is in abdominal discomfort, colic, and for promoting digestion.

The main components of the volatile oil of dill seeds are monoterpenic hydrocarbons: α-phellandrene (30.26%) and limonene (33.22%), total monoterpenic hydrocarbons representing 65.21%. In flowers essential oil the content of α-phellandrene is 30.26% and 62.71% in the leaves. The fruit essential oil was rich in limonene 21.56% and cis-and trans-dihydrocarvone adding up to 3.06%. The amount of α-phellandrene in leaves oil and flowers oil was 62.71%, respectively 30.26%, and only 0.12% in fruits oil. The chemical composition of dill volatile oil varies depending on the plant parts. In the leaves oil monoterpenic hydrocarbons are predominant, amounting to 79.14% (62.71% α-phellandrene and 13.28% limonene). In the flowers oil the content of α-phellandrene and limonene is 32.26% and 33.22%, respectively. Anethofuran (dill ether) is present in leaves and flowers with 16.42% and 22%, respectively, but is missing in the fruit oil. The main compound in fruits essential oil is carvone (75.21%), while the content of α-phellandrene is only 0.12% and limonene is 21.56%.

3.5.9.3   Soil and Climatic Requirements

It prefers well-drained fertile sandy loam soil. It can tolerate a pH in the range 5.3 to 7.8. It prefers mild climate for growth, development, and seed production. It requires warm to hot summers with huge sunshine levels; even partial shade will reduce the yield substantially. The plant quickly runs into seeds in dry weather. It is susceptible to weather hazards like heavy hail, strong winds and heavy rains. Extreme heat during the critical periods of maturity is unfavorable for dill cultivation as it reduces fruit yield as well as oil yield considerably.

The important varieties of dill with their parentage, average yield (kg/ha) and the special features are given in Table 3.24.

3.5.9.4   Agrotechniques for Quality Production

The sowing of seed, land should be prepared well for better crop establishment. Seed maintains its viability for 3–10 years. Seed rate is 3–4 kg/ha. Seed germination takes a time of 10–12 days. Indian dill is normally sown during the month of October/November. European dill is cultivated in India as a winter crop in Jammu and as a summer crop in Kashmir. Normally a spacing of 30–45 cm × 15–30 cm is adopted for dill cultivation. Thinning is normally done at three weeks after sowing of seeds to remove the excess

TABLE 3.24   Important Varieties of Dill

Sl. No.	Name	Parentage	Average Yield (kg/ha)	Special features
1.	RSP-11	Selection from germplasm of Pratappur	600	Suitable for cultivation in Rajasthan
2.	Gujarat Dill-1	Selection from local germplasm.	1800	Non-shattering and non-splitting type. Seeds are medium in size with, pungent with good aroma. Seed essential oil content is 3.6%. Suitable for irrigated condition
3.	Gujarat Dill-2	Selection from early and dwarf plant Navagam Selection.	935	Early maturity type, bold seeded with non-splitting type. Susceptible to powdery mildew and suitable for growing under rainfed condition.
4.	Ajmer Dill-1	Single plant selection from Mammoth, an introduction from Europe	800 (rainfed) – 1460 (irrigated condition)	Susceptible to powdery mildew and suitable for growing both under rainfed and irrigated condition. Bold seeded with an essential oil content of 3.5%.
5	Ajmer Dill-2	Selection based on individual plant progeny performance of Nagpur Local	800 (rainfed) – 1460 (irrigated condition	Resistance to powdery mildew and moderately tolerance to drought. Suitable for growing both under rainfed and irrigated condition. Bold seeded with an essential oil content of 3.2%.

plants. Application of organic and inorganic is beneficial for getting higher yield and quality produce.

3.5.9.5   Plant Protection

Carrot motley dwarf (CMD): The symptoms of carrot redleaf virus (CRLV) and carrot mottle virus (CMoV) include yellow and red leaves with stunted plant growth. To manage this disease planting dill in close proximity of overwintered carrot fields is to be avoided.

Cercospora leaf blight: The causal organism is *Cercosporidium punctum.* The prominent visible symptoms are formation of small, necrotic flecks on leaves, which develop a chlorotic halo and expand into tan brown necrotic spots; lesions coalesce and cause leaves to wither, curl, and die. For effective management of this disease only pathogen-free seeds are to be planted, crop rotation to be followed; crop debris to be ploughed into soil ofter harvest and appropriate fungicides are to be sprayed.

Cutworms: Agrotis spp, *Peridromasaucia, Nephelodesminians.* In case of cut worm infestation, stems of young transplants or seedlings may be severed at soil line; if infection occurs later, irregular holes are eaten into the surface of fruits; larvae causing the damage are usually active at night and hide during the day in the soil at the base of the plants or in plant debris of toppled plant; larvae are 2.5–5.0 cm (1–2 in) in length; larvae may exhibit a variety of patterns and coloration but will usually curl up into a C-shape when disturbed. To manage this pest all plant residue are to be removed from soil after harvest or at least two weeks before planting. Alfalfa, beans or a leguminous cover crop is effective. Plastic or foil collars fitted around plant stems to cover the bottom 7–8 cm above the soil line and extending few centimeters into the soil can prevent larvae severing plants; all visible larvae are to be hand-picked after dark; appropriate insecticides are to be applied.

Root knot nematode (*Meloidogyne* spp.): Galls on roots are formed, which can be up to 3.3 cm in diameter but are usually smaller; plant vigor is reduced; plants start yellowing and generally wilt in hot weather. To manage nematodes, resistant varieties are to be planted; roots of plants are to be checked mid-season or sooner if symptoms indicate nematodes. Moreover, solarizing soil can reduce nematode populations in the soil and levels of inoculum of many other pathogens.

3.5.9.6 Harvest and Post Harvest Technology

The fruiting umbels are ready to harvest for seasoning when the fruit is fully developed but not yet brown in color. Crop is ready for harvest during the month of March–April. Normally, crop duration of 130–150 days is recommended for seed purposes. Plants are harvested by pulling the whole plants. The leaves are used only in the fresh state, but the fruiting tops may be used either fresh or dried form. The umbels may be dried on screens in the shade and stored in closed container. Seed yield normally varies from 800–900 kg/ha.

3.5.9.6.1   Value Addition

Ground dill: Ground dill is produced by grinding whole dill seed with no additions. The color is dependent on origin, sub-species, and crop conditions and varies from tan brown via green brown to dark brown.

Dehydrated soup mix: It is formulated with functional ingredients, modified potato flour (thickening agent), and dill leaf powder. They are prepared by blending dried ingredients with thickening agents or by spray drying the formulated slurry. It may increase health benefits since it contain higher levels of additional phenolic acids and have antioxidant properties.

Dill oil: It is an essential oil extracted by steam distillation from the seeds or leaves/stems of the plant. Dill oil is known for its grass-like smell and its pale yellow color, with a watery viscosity. It is used for the relief of flatulence, especially in babies. Dill essential oil has a characteristic fresh, herbaceous aroma, enhanced by minty and zesty notes. The oil has high concentrations of limonene, carvone, and a-Phellandrene.

The dill seed oil extraction is done by Supercritical carbon di oxide method and is obtained in good yield. The major constituents extracted by Supercritical carbon dioxide method are limonene, 27.93%; carvone, 9.76%; dihydrocarvone, 26.74%; and dillapiole, 34.05%. The yield of oil is 5 wt% (Nautiyal and Tiwary, 2011).

3.5.10   ANISEED OR ANISE (PIMPINELLA ANISUM L.)

3.5.10.1   Systematic Position

Kingdom: Plantae
Division: Magnoliophyta
Class: Magnoliopsida
Order: Apiales
Family: Apiaceae
Genus: *Pimpinella*
Species: *anisum*

3.5.10.2 About the Crop, National and International Scenario, Uses, and Composition

Anise or Aniseed (*Pimpineela anisum* L.) belongs to the family apiaceae is grown for its seed. It is an annual pubescent aromatic herb, which attains a height of 60–90 cm under cultivation. It is native to Mediterranean region and extensively cultivated in South and South East European countries and also in Egypt, Syria, Iran, and South America, Mexico, and India. In India It is grown to small extent as culinary herb or as garden plant. In India, commercial cultivation is confined to Bihar and Uttar Pradesh. To a small extent it is also cultivated in Gujarat and Rajasthan. Seed contains 2.1–2.4% essential oil. The main composition of the essential oil is anethole (84–87%) and methyl chavicol (12–15%). Fresh leaves are used as a garnishing agent and for flavoring salad. The seeds are used for flavoring food, confectionaries, beverages liquor industries and processed food. Essential oil is used in perfumery, soap, toilet article, dental, and mouth wash article preparation. Medicinally, the essential oil is used as an aromatic carminative and to relive flatulence.

3.5.10.3 Soil and Climatic Requirements

The crop is grown wide range of soil, ranging from sandy loam to clay loam, having a pH range of 6.0–8.5 and of moderate fertility. A fairly warm weather during sowing is desirable. Low temperature and higher humidity are conducive for vegetative growth, while the crop requires warm sunny weather during seed formation and development stage. The reported life zone for anise cultivation is 8 to 23°C with 1,000 to 1,200 mm annual precipitation, this produces excellent crop, however a rainfall of 2,000 mm is tolerated with a soil pH of the cultivated field ranging from 6.3 to 7.3 (Simon et al., 1984).

3.5.10.4 Agrotechniques for Quality Production

EC-22091 is the improved variety of anise, which has been developed at IARI, New Delhi. Seeds are sown at a spacing of 45 cm × 20 cm with a depth of 1–2 cm during the time of late October to early November. Like other umbelliferous crop, seed germination is rather problematic and uneven. To obtain uniform germination the seed should be soaked in water overnight and moist seed sowing is recommended. A light irrigation is to be given to get uniform and higher percentage of seed germination. Irrigation also should be provided at flowering and grain filling stages of the crop. Application of farmyard manure at the rate of 10 t/ha and inorganic fertilizer like NPK @ 80:60:60 kg/ha is beneficial for getting higher yield of the crop.

3.5.10.5 Plant Protection

Alternaria blight (Alternaria spp.): The symptom of this disease is formation of small round yellow, brown or black spots on leaves in a concentric ringed pattern with holes in leaves where lesion has dropped out. Treating seeds with hot water prior to planting, keeping plants well watered, removing all plant debris from soil to avoid survival of fungi on pieces of plant can effectively manage the disease.

Downy mildew (Peronospora umbellifarum, Plasmopara nivea): The most prominent symptoms are yellow spots on upper surface of leaves, white fluffy growth on underside of leaves and darker lesions with disease maturity. Using plant pathogen-free seed, avoidance of overcrowded plant stand; rotation of crops with non-umbelliferous varieties are some of the management options against this disease.

Powdery mildew (Erisyphe heraclei): Powdery growth appears on leaves, petioles, flowers stalks and bracts; leaves become chlorotic; severe infections can cause flowers to become distorted. To manage the disease, tolerant varieties are used, excess fertilization is avoided and adequate protection is provided with protective fungicide. Sulfur can also be applied if infection occurs early in the season.

Rust (Puccinia spp, *Uromyces* spp, *Nyssopsora* spp.): Symptoms are light-green discolored lesions on leaves, which become chlorotic; yellow-orange pustules on underside of leaves; stems bend and become swollen or distorted; plants may be stunted. Bordeaux mixture is effective to control rust.

Aphids (Cavariella aegopodii): These are small soft-bodied insects on underside of leaves and or stems of plant usually green or yellow in color. If infestation is heavy it may cause leaves to yellow and/or distorted, necrotic spots on leaves and/or stunted shoots. Aphids generally secrete a sticky, sugary substance called honeydew, which encourages the growth of sooty mold on the plants. Using tolerant varieties, using reflective mulches such as silver colored plastic and spray of insecticides are generally required to manage aphids. Insecticidal soaps or oils such as neem or canola oil are usually the best method of control.

Armyworm (*Pseudaletia unipuncta*)*:* Singular, or closely grouped circular to irregularly shaped holes are found in foliage. Heavy feeding by young larvae leads to skeletonized leaves. Fruits are characterized by shallow dry wounds. Organic methods of controlling armyworms include biological control by natural enemies, which parasitize the larvae and the application of *Bacillus thuringiensis.*

Root knot nematode (*Meloidogyne* spp.): Galls on roots are formed, plant vigor reduced and plants start yellowing and generally wilted in hot weather. Resistant varieties are to be planted periodic checking of roots are necessary and soil solarization is to be done to reduce nematode populations in the soil.

3.5.10.6   Harvest and Post Harvest Technology

The umbels come to maturity from March to April. The terminal umbels mature first and lateral ones 15–20 days later. When 80–90% fruits begin to turn grayish green in color, the top of the plants cut along with branches and tied in bundles. Under favorable climatic condition and good management practices anise yields 0.7–0.8 t/ha. Semilooper and cigarette beetle are the important paste attacking the crop. Among the different diseases, *Alternaria* leaf blight is an important disease that reduces the crop yield considerably.

3.5.10.6.1   Value Addition

Anise Essential Oil: The essential oil of anise is extracted by steam distillation of dried fruits of anise. It is a thin and clear oil of which Anethol is the prime constituent (about 90%) and is responsible for its characteristic aroma. The other constituents are alpha pinene, anisaldehyde, beta pinene, camphene, linalool, cis and trans-anethol, safrol, and acetoanisol. Anise essential oil has a narcotic and sedative effect; it can calm down epileptic and hysteric attacks by slowing down circulation, respiration, and nervous response, if administered in higher dosages.

3.5.11 POPPY SEED (PAPAVER SOMNIFERUM L.)

3.5.11.1 Systematic Position

Kingdom: Plantae
Division: Magnoliophyta
Class: Magnoliopsida
Order: Ranunculales
Family: Papaveraceae
Genus: Papaver
Species: *somniferum*

3.5.11.2 About the Crop, National and International Scenario, Uses, and Composition

Opium Poppy (*Papaver somniferum* L.) belongs to the family papaveraceae and is originated from the Western Mediterranean Region. The genus *Papaver* has 50 different species of which 6 species grown in India. It is mainly grown for its seed and latex. The seed is used as spice for preparation of culinary items and small portion of seed is exported mainly in the Asian and African countries. It is grown Near East and Asia, most part of Europe, North, and east Africa, Australia, Japan, and South America. In India its cultivation is mainly confined to states of Madhya Pradesh, Rajasthan, and Uttar Pradesh. Poppy seed is utilized as spice and as a source of protein and fatty oil. It is considered nutritive and used in breads, curries, sweet, and confectionaries. Because of its highly nutritive nature it is used in breads, cakes, cookies, pastries, curries, sweets, and confectionary. Its seeds are demulcent and are used against constipation. The capsules are used as a sedative against irritant coughing and sleeplessness in the form of syrup or extract. Poppy seed oil is useful in preparation of linoleic acid, soft soap, ointments, and preparation of skin care products. In Europe, used as a substitute and adulterant of olive oil, which it resembles very much.

Poppy seeds contain protein (21.1%), moisture (5.0%), ash (6.3%), crude fine (6.2%) and total carbohydrates (23.6%). Potassium and calcium are the predominant elements in the poppy seeds. Linoleic acid was the major unsaturated fatty acid (75.0% of total fatty acids) while palmitic acid was the main saturated fatty acid. The poppy seeds also contain amounts of α-, β- and δ-tocopherols.

3.5.11.3   Soil and Climatic Requirements

It is a crop of temperate climate but can also be grown successfully during winter season in subtropical regions. Cool climate favors higher yield, while higher day/ night temperature generally affect seed yield. The crop requires deep clay loam soil highly fertile in nature and well drained with an optimum pH around 7.0 and with adequate irrigation facility

3.5.11.4   Agrotechniques for Quality Production

The seed is either sown broadcast or lines. Seed should be treated with fungicide with mancozeb. Seed is usually mixed with sand before broadcasting to ensure uniformly spread in the bed. Optimum sowing time is best October to early November. Seed rate is 7–8 kg/ha and 4–5 kg/ha for line sowing. A spacing of 30 cm × 30 cm is adopted for line sowing. Germination takes 6–10 days depending upon the soil moisture content. Telia, Dhotia, and Ranjihatak are some of the local races recommended for commercial cultivation. MOP-3, MOP-16, MOP-572, MOP-1186 and UD-177-2 are some of lines of opium. Farmyard manure @25–30 tonnes/ha is generally applied at the time land preparation. Besides, NPK @ 80–90:90–100:40–50 kg/ha is recommended. Half of N and K and full of P should be applied as basal. Remaining N and K should be applied in two equal split at 30–40 days and 60 days after sowing. Kharwara et al. (1988) found that application of N @150 kg/ha produced significantly higher yield than lower dose of 75 kg/ha. The weeding cum hoeing and thinning should be done 15–20 days after sowing and irrigation. Second weeding should be given 50–60 days after sowing. A light irrigation is given immediately after sowing and another light irrigation should be applied after one week when the seeds start germinating. Chung (1987) reported that irrigation increases total dry matter production, LAI, and delayed senescence in dry season for getting maximum yield one irrigation of 5 cm should be applied 50% hooking stage, at 50 flowering stage and at the end of flowering stage. Tomar et al. (1990) reported that the greatest reduction in seed yield (294%) was obtained with stress imposed at bud and early capsule stage.

3.5.11.5   Plant Protection

Downy Mildew: Reddish lesions on poppies' upper leaf surfaces accompanied by fluffy gray, tan or purple areas on their undersides point to downy mildew (*Plasmopara*) infection. Cultural control includes watering from beneath, spacing poppies for adequate air circulation and removing plant debris. Serious outbreaks demand application of suitable fungicide like Mefenoxam and Mancozeb.

Gray mold (*Botrytis cinerea*): Fungus disfigures damaged or older poppies during cool, wet spring weather. Brownish lesions on newly infected plants decay and develop masses of gray spores. The crop is to be irrigated from beneath early in the day to speed evaporation. All the old or injured leaves and flowers are to be removed. Application of Captan, Chlorothanolil, and several other fungicides can limit gray mold's spread.

Powdery Mildew: White, powdery coating on the stems, leaves, and blooms are the typical symptom. Protecting the crop by giving them full sun and watering from above to wash spores from the leaves is beneficial.

Rhizoctonia Root Rot (*Rhizoctonias olani*): Rhizoctonia root rot fungus attacks the crop during midsummer, entering their root systems and sometimes progressing into their stems. Dry rot at the base of the stems is its unique feature. The disease blackens and shrivels roots, destroying their firm texture and white tips. Drenching soil with Quintozene fungicide before symptoms surface effectively manages the disease.

Sap-Sucking Insects: Aphids and thrips are sap-sucking insects that suck the juice of the plant. Aphids, tiny green or yellowish pests, are usually found on the undersides of the leaves and on the joints of stems. The leaves take on a deformed, curled appearance and the plant may eventually die in case of severe attack. Thrips are black, brown or yellow, oblong shaped pests similar in size to aphids. When the pests suck on the flowers and leaves, the foliage takes on a stippled appearance with deformed flowers or buds that fail to open. Application of Pyridine azomethines or Acetamiprid kills the sucking insects.

3.5.11.6   Harvest and Post Harvest Technology

The poppy seed harvest can be a by-product of opium poppy cultivation for opium, poppy straw, or both opium and poppy straw. Conversely, poppy straw can be a by-product of cultivation of poppy seeds. When the seeds are fully matured at this stage crop is harvested and seed yield varies from 3.5 to 6.0 q/ha. Powdery mildew and Downy mildew are important diseases of opium and aphid, thrips, cut worm and capsule borer are some of the pest.

3.5.11.6.1   Value Addition

In Europe, The sugared, milled mature seeds are eaten with pasta, or they are boiled with milk and used as filling or topping on various kinds of sweet pastry. Milling of mature seeds is carried out either industrially or at home, where it is generally done with a manual poppy seed mill.

In Maharashtra, poppy seeds are used to garnish anarsa (Pine apple), a special sweet prepared during the festival of Diwali. It is also added in boiling milk sometimes.

In Gujarat, poppy seeds are mostly used in sweets. The most common use is to garnish on a traditional Indian sweet-ladoo.

3.6   LEAFY SPICES OR HERBAL SPICES OR AROMATIC HERBS

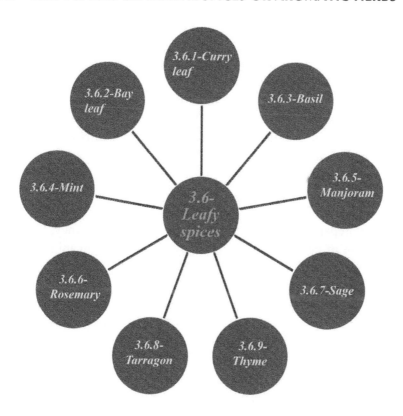

3.6.1   CURRY LEAF (MURRAYA KOENIGII L.)

3.6.1.1   Systematic Position

Kingdom: Plantae
Division: Tracheophyta
Class: Magnoliopsida
Order: Sapindales
Family: Rutaceae
Genus: *Murraya*
Species: *koenigii*

3.6.1.2   About the Crop, National and International Scenario, Uses, and Composition

Curry leaf (*Murraya koenigii* Sprengel) belongs to the family rutaceae, popularly used for flavoring food for its typical flavor and taste. It is mainly cultivated for its aromatic leaves. It is found throughout the India. Its wild form is reported to occur in Western Ghats, West Bengal, Assam, Orissa, Tamil Nadu, Kerala, Gujarat along with foot hills and plains of the Himalayas from Kumaon to Sikkim. It is cultivated in Tamil Nadu and Andhra Pradesh. Curry leaf is rich Vitamin A (12600 mg/100 g dry leaves), Calcium (830 mg/100 g dry leaves) and iron (730 mg/100 g dry leaves). Curry leaf is traditionally and extensively used in South Indian cooking for preparation of curries, rassam, sambar, and chutneys. Leaves, bark, and root of curry leaf are used in ayurvedic and yunani system of medicine.

Dried curry leaf contains 12% protein, 5.4% fat, 64.3% carbohydrate and cellulose, 12 mg/100 g iron and 5292 µg/100 g β-carotene. major compounds identified in the oil are Linalool (32.83%), elemol (7.44%), geranyl acetate (6.18%), myrcene (6.12%), allo-ocimene (5.02), α-terpinene (4.9%), (E)-β-ocimene (3.68%), and neryl acetate (3.45%). These compounds are classified into four groups that are oxygenated monoterpenes (72.15%), monoterpene hydrocarbons (11.81%), oxygenated sesquiterpenes (10.48%), and sesquiterpenes hydrocarbons (03.12%).

3.6.1.3   Soil and Climatic Requirements

It does not require exact specific climate and can also be grown in dry climate too. But the growth of the shoot will be affected if the temperature goes below 13°C. It can be cultivated in most of the soil types, but comes up well in light textured red soil. DWD-1 and DWD-2 are the most common varieties of the curry leaf.

The Table 3.25 will describe the following varieties of curry leaf with regard to their parentage, average yield (kg/ha) and the special features.

3.6.1.4   Agrotechniques for Quality Production

Good and healthy seeds selected from vigorously growing plants should be using for raising nursery. Young seedling of 3 month age is planted in pit of 30 cm × 30 cm × 3°C size. Normally a spacing of 1.0–1.20 m × 1.0–1.20 spacing is adopted for cultivation of curry leaf. Better grow and yield can be obtained by applying farmyard manure @ 15–20 kg/plant along with NPK @ 150:25:50 g/plant. During the seedling stage, the crop should be irrigated at an interval of 5–7 days and there after 15 days interval. Citrus psylla (*Diaphornia citri*) is considered a major pant and from this pest aphid also causes a considerable damage to the crop. Leaf spot and root rot are the diseases of the crop. Despite pest being major constraints in curry

TABLE 3.25   Important Varieties of Curry Leaf

Sl. No.	Name	Parentage	Average Yield (kg/ha)	Special features
1.	DWD – 1	Clonal selection from root suckers.	-	Leaves are dark green in color, shiny, and aromatic in nature. Sensitive to low temperature and leaf essential oil content is 5.22%.
2.	DWD – 2	An op seed progeny.	-	Leaves are pale green in color and less aromatic in nature. Leaf essential oil content is 4.09%. It is not very sensitive to low temperature.

leaf production, they are easily manageable when plant protection measures are into time, of course with close monitoring. Preferably, the non-residual chemicals like dimetoate, malathion, and monocrotophos could be sprayed to bring down pest population well below the threshold level. Even for these chemicals, the farmer must allow a safe period not less than 2–3 weeks for harvesting after spraying pesticides.

3.6.1.5   Plant Protection

Scale: Scales are tiny insects that appear as small, flat bumps on the surface of a leaf. They damage plants by sucking out their fluids, causing the leaves to yellow and wilt. Scales can be controlled by introducing natural predators such as ants or applying natural horticultural oil. Close monitoring can also help to catch scale problems before they become infestations.

Aphids: Aphids are yellow, green, brown or white small pear-shaped insects. They suck the juices from a plant, causing the leaves to mottle and curl, and can also introduce mold fungus. This insect pest can best be controlled by introducing natural predators such as ladybugs.

Citrus Mealybugs: These are soft pinkish-white insects with a waxy appearance the nymphs of which feed on plant sap. In addition to causing leaves to shrivel, large infestations of citrus mealybugs can cause a tree's fruit to drop prematurely. Mealybugs can best be controlled by introducing natural predators such as ladybugs or mealybug destroyers.

Asian Citrus Psyllids: The mottled brown insects feed directly on the leaf of the curry tree. This causes damage to the leaves and stems, and can also introduce bacteria to the tree. Infestations of this pest spread quickly and cannot be effectively controlled chemically, rather can only be managed through detection and destruction of infected trees.

3.6.1.6   Harvest and Post Harvest Technology

The plant should be maintaining a bush of 1 m height. After each periodical harvest thoroughly weeding and irrigation is practiced. At the end of the first year the crop comes to first harvest and the yield is about 400 kg/ha. During 2nd and 3rd year the crop is harvested at an interval four months and yield varies from 2000–2200 kg/ha. At the 4th year the crop is harvested at an interval of three month and yield is about 2500 kg/ha. From the 5th year and onwards the crop is harvested at interval two and half month to three months interval and yield varies 3500–5000 kg/ha. Plants can be maintained upto 20–25 years depending upon the management practices.

3.6.1.6.1 Value Addition

Curry Leaf oil: Fresh leaves are steam distilled under high pressure to yield curry leaf oil. The yield is poor under normal pressure. The oil is pale yellow in color with a strong aroma and a blend of sweet bitter tone. The average yield is 0.12% to 0.20%. The composition of essential oil is alpha and beta pinene, and caryophyllene.

3.6.2 BAYLEAF (LAURUS NOBILIS L.)

3.6.2.1 Systematic Position

Kingdom: Plantae
Division: Tracheophyta
Class: Magnoliopsida
Order: Laurales
Family: Lauraceae
Genus: *Laurus*
Species: *nobilis*

3.6.2.2 About the Crop, National and International Scenario, Uses, and Composition

L. nobilis is a native of the Mediterranean and grows spontaneously in scrubland and woods in Europe and in California. In biblical times, the bay was symbolic of wealth and wickedness, and in the classical world heroes and victors were decorated with a laurel wreath. In addition to being a very well known culinary herb, the leaves and fruits of *L. nobilis* are used medicinally throughout the world.

L. nobilis is an evergreen shrub, or more rarely a tree attaining a height of 15–20 m. The chemical composition of the flower essential oil is quite different from other parts of the plant, namely leaves, stem bark and stem wood. The presence of 1,8 cineole in appreciable amounts makes the oil of bay leaves an important perfumery item. Bay leaves are used as flavoring in soups, stews, meat, fish, sauces, and in confectionaries. Both leaves and fruits possess aromatic, stimulant, and narcotic properties. Leaf essential oil is also used as spice and food flavoring agent and widely used in traditional medicines of different countries. The major functional properties are antimicrobial, antifungal, hypoglycaemic, antiulcerogenic, etc.

According to the ASTA, the chemical composition of the dried bay leaves is as follows-moisture: 4.5%; protein: 7.6%; fat: 8.8%; fiber: 25.2%; carbohydrates: 50.2%; total ash: 3.7%; calcium: 1.0%; phosphorus: 0.11%; sodium: 0.02%; potassium: 0.6%; iron: 0.53%; vitamins (mg/100 g)-vit. B1 (thiamine): 0.10%; vit. B2 (riboflavin): 0.42; niacin: 2.0; vit. C (ascorbic acid): 46.6 and vit. A: 545 international units (IU), calorific value (food energy): 410 calories/100 g.

Bay leaves contain approximately 1.5–2.5% essential oil, the principal component of which is cineole. Bay oleoresin contains about 4–8% volatile oil. Essential oil of bay is also available. Oil of bay and bay oleoresin is used in soluble pickling spices.

Bay leaves are delicately fragrant but have a bitter taste. In classic and contemporary cuisines, bay leaves are regarded as a popular culinary flavoring leaves greatly stimulating appetite. This popular spice is used in sauces, pickling, marinating, flavoring stews, and stuffings. The smooth and lustrous dried bay leaves are usually used whole and then removed from the dish after cooking; they are sometimes marketed in powdered form. The crushed form is a major component in pickling spices in processed meats and pickle industry. Ground bay is utilized in many seasoning blends and products.

Bay leaf has legendary medicinal properties. It has astringent, diuretic, and digestive qualities and is a good appetite stimulant. When pulped, these leaves can be applied as an astringent to burns and bruises. Oil from ripe berries is used in liqueurs, perfume, and in veterinary field. The acid from the leaves discourages moths.

3.6.2.3   Soil and Climatic Requirements

It is usually found in Northern areas of India. The trees are grown best in a moist and warm climate, which is like that of the Mediterranean area. The trees of the Bay leaves require plenty of sunlight for better growth so proper care should be implemented if the tree is grown inside and the hot spots should be avoided near its plantation. The plant grows best in the soil, which is rich in humus and is moist.

3.6.2.4   Agrotechniques for Quality Production

The rooted cuttings are placed in small pots containing sandy loam soil with good drainage. *L. nobilis* stem cuttings produce roots better in July/August, under Mediterranean conditions, than in other seasons. These trees grow in rich, well-drained soil in full sun. Young plants are to be protected from cold winds and frost. They are generally planted in early autumn or spring and trimmed to shape in summer. Propagation is mainly done by layering shoots or from cuttings of side roots. The leaves are harvested by hand, dried in shallow layers in shade and lightly pressed flat. Ripe berries are pressed for oil.

3.6.2.5   Plant Protection

Canker disease: This is caused by fungus. On affected plants leaves sore on with dark sunken nectars. Infected branches should be pruned. Otherwise no effective controls are there.

Wood decay disease: These are fungal diseases. The woods are rotten and generally decay may contained in the trees internal heartwood. The plants are to be kept well-watered, mulched, and fertilized.

Borers: They burrow into the wood of trees and make their way into branches and twigs to lay eggs and feed on it. The infested parts require pruning.

3.6.2.6   Harvest and Post Harvest Technology

The leaves of *L. nobilis* are plucked from September to March and dried under shade for use as a flavoring material in a variety of culinary preparations, especially in French cuisine. The leaves contain an essential oil of aromatic, spicy odor and flavor, which can be isolated by steam distillation. The oil is a valuable adjunct in the flavoring of all kinds of food products, particularly meats, sausages, canned soups, baked goods, confectionery, etc.

The leaves on steam distillation yield 1–3% essential oil. Fresh leaves and terminal branch-lets yield 0.5% oil, while dried leaves yield about 0.8%.

3.6.2.6.1   Value Addition

Bay oil: A sweet and spicy oil is obtained from bay. Its principle constituent (up to 50%) is cineol, a colorless liquid with a strong aromatic, camphoraceous odor, and a cooling taste.

Bay fat: Commercial fat is obtained from whole berry by pressing or by boiling with water and skimming off the separated fat. The mixed fatty acids contain lauric acid: 30.35%; palmitic acid: 10–11%; oleic acid: 33–40%; and linoleic acid: 18–32%. It is used in pharmacy, veterinary practice and perfumery.

3.6.3   BASIL OR SWEET BASIL OR TULSI (OCIMUM BASILLICUM L.)

3.6.3.1   Systematic Position

Kingdom: Plantae
Division: Tracheophyta
Class: Magnoliopsida
Order: Lamiales
Family: Lamiaceae
Genus: *Ocimum*
Species: *basillicum*

3.6.3.2   About the Crop, National and International Scenario, Uses, and Composition

Basil, French Basil or sweet basil, tulsi is an annual herb and native of North-Western India and Persia. Hindus consider it sacred and also good for health when its fresh leaves are taken raw and also their decoction. There are four types of basil namely American, French, Egyptian, and Indian of which the later is also called French basil or sweet basil. The flavors of basil have a special affinity and this is the basis of today's increased usage of this herb in the Western World, especially in the pizza and other fast foods. there is a considerable commercial-scale production of basil in the USA, France, Hungary, Egypt, Bulgaria, West Germany, Poland, Yugoslavia, Belgium, Turkey, Italy, and the Netherlands. There are numerous varieties of *O. basilicum*, of which 4 are identified in India. These are (i) var. *album* Benth. (lettuce-leaf basil); (ii) var. *differme* Benth. (curly-leafed basil); (iii) var. *purfurascans* Benth. (violet-red basil); and (iv) var. *thyrsiflorum* Benth, (common white basil). Curly-leafed basil is considered most suitable for cultivation. It is grown in France and is reported to give good yields of high quality oil.

According to the analysis report of the American Spice Trade Association (ASTA), USA, the composition of basil or sweet basil herb is as follows: moisture: 6.1%; protein: 11.9%; fat (ether extract): 3.6%, fiber: 20.5%; carbohydrates: 41.2%; total ash: 16.7%; calcium: 2.1%; phosphorus: 0.47%; sodium: 0.04%; potassium: 3.7%; iron: 0.04%; vitamins (mg/100 g) – vit. B1 (thiamine): 0.15; niacin: 6.90; vit. B2 (riboflavin): 0.32; vit. C (ascorbic acid): 61.3 and vit. A: 290 international units/100 g; calorific value (food energy): 325 calories per 100 g of dried herb.

Basil is mainly used as food flavorant, in preparing perfumery and cosmetics, in medicine as well as having insecticide, insect-repellent, and bactericidal properties. The plant is considered stomachic, anthelmintic, alexipharmic antipyretic, diaphoretic, expectorant, carminative, and stimulant. The juice of the leaves is considered useful in the treatment of croup and is a common remedy for coughs. An infusion of seed is used in gonorrhea, dysentery, and chronic diarrhea.

3.6.3.3 Soil and Climatic Requirements

Basil can be grown under a wide range of climates and soils. It is very versatile.

3.6.3.4 Agrotechniques for Quality Production

Sweet basil (*O. basilicum*) is propagated by seeds and is commonly grown in gardens as an aromatic herb. The best season for sowing it in the plains of India is October–November and in the hills, it is March–April. Seedlings are raised in the nursery beds and transplanted 30 cm apart in rows spaced 40 cm apart. The crop is ready for harvesting in 2–3 months after planting. Wilt is reported to be an important disease of basil.

3.6.3.5 Plant Protection

Fusarium wilt (*Fusarium oxysporum* f.sp. *basilicum*): Plants become stunted and wilted with yellowish leaves, which may develop when the plants are 6–12 inches tall. Afterwards, brown streaks on the stems, discoloration of stem tissue, severely twisted stems and eventually sudden leaf drop are common symptoms. Sweet basil is more severely affected than other basil varieties.

Mycostop is a biological fungicide that will safely protect crops.

Bacterial leaf spot (*Pseudomonas cichorii*): Typical symptoms are the water-soaked brown and black spots on leaves and streaking on the stems. The leaf spots are angular or irregular or delineated by the small veins. Weekly application of sulfur or any other copper-based fungicide as spray at first sign of disease prevents its spread.

Downy mildew (*Plasmopora* spp, *Peronospora* spp.): An initial symptom is yellowing of leaves, typically from around the middle vein and then spreading to most of the leaf surface. At later stages, the more characteristic symptom of fuzzy grayish-purple sporangia appears. Application of Calcium chloride (20 mM) + *S. cerevisiae* 10 x 1010 cfu/mL (10 mL/L) + Chitosan (0.05 mM) is effective. Potassium bicarbonate (20 mM) + Thyme oil (5 mL/L) is also equally effective.

Root-knot nematode: Nematodes are soil-borne microscopic roundworms that damage the roots. Once established in soil, it is hard to eradicate this pathogen from contaminated soil unless there is no more suitable hosts. The symptoms are similar to nutrient deficiency including wilting, discoloring, and low yield. Crop rotating is not very effective because these nematodes have a wide host range. Applying organic matter to the soil and soil solarization may be practiced.

3.6.3.6 Harvest and Post Harvest Technology

Several (3–5) cuttings of leaves and flowering tops may be made during the season. Plants are cut close to the ground, bunched, and dried. The dried leaves and flowering tops are stripped from stems and packed in the closed containers. While harvesting the crop, precaution should be taken that the root-system of the plant is not injured, otherwise yield of subsequent harvests will be affected adversely. The crop is to be harvested at the stages of maximum bloom to seed set. A yield of 6,800 kg of leaves and flowers per hectare in 2 cuttings is reported from trial cultivation in Kanpur. In the USA, a yield of 20–25 tonnes of fresh herb/hectare has been reported, and 3–5 harvests are made per annum. The dried leaves are generally packed in tin containers, which are kept closed. The enzymatic browning of basil has been attributed to oxidation of phenolic compounds by polyphenol oxidase, naturally present therein.

3.6.3.6.1 Value Addition

The main product manufactured from basil leaves and a flower top is the essential oil known as 'Oil of Basil.' Oil of sweet basil is produced by the hydro-distillation of the herb. The flowers, on an average, yield 0.4% oil while the whole plant (Indian basil) contains 0.10–0.25% oil. The oil obtained from flowers is better than the oil from whole herb in quality. Methyl chavicol, the main constituent of sweet basil oil, is oxidized on ageing and on exposure of oil to light and air; thus older oils usually show a higher specific gravity and higher refractive index. Sweet basil oil must, therefore, be stored carefully.

3.6.4 MINT (MENTHA PIPERITA L.)

3.6.4.1 Systematic Position

Kingdom: Plantae
Division: Tracheophyta
Class: Magnoliopsida
Order: Lamiales
Family: Lamiaceae
Genus: Mentha
Species: *piperita, arvensis, spicata*

3.6.4.2   About the Crop, National and International Scenario, Uses, and Composition

Mentha is a commercially important spice, medicinal, and aromatic perennial herb belongs to the family of labiateae. It is commonly known as Pudina. Mints are extensively cultivated in USA, South Eastern European countries and Latin American countries. India too has the potentially to growing good quality mint. Its commercial cultivation is restricted in Uttar Pradesh, Punjab, Jammu-Kashmir, and some parts of Tamil Nadu. Attempts have been made to cultivate in the states of Himachal Pradesh, Madhya Pradesh, Andhra Pradesh, West Bengal, Assam, and Meghalaya. Today mint oil is an important item if trade. Among the 40 species of menthe only four species Viz. *Mentha arvense* (Japanese mint), *M. piperita* (Pepper mint), *M. spicata* (Spear mint) and *M. citrata* (Bergamot mint), are commercially cultivated in India. The chemical constituent of mint oil varies from species to species.

Dried herbs of pudina (except Bergamot mint) are commonly used as spices. The fresh leaf tops of all types of mint are used in beverages, apple sauces, meat preparations also to flavor vegetable curries. The main use of mint is extraction of Volatile oil. The Japanese mint and peppermint oil is used as flavoring agent in mouth washes, tooth paste, different pharmaceutical preparation and also for cigarette industry. Spear mint oil is mostly used in pharmaceutical and flavoring industry, while use of Bergamot mint oil is restricted in cosmetic industry.

3.6.4.3   Soil and Climate

Mints prefer well drain, deep, fertile, rich in organic matte and with good moisture retention capacity. Though it can be grown successfully in a wide range of soil from sandy loam to clay loam, but sandy loam soil with a pH range of 6.5 to 7.5 is ideal for its quality and yield. Mint is a crop of temperature and sub temperature climate but in India. It is also cultivated in humid tropics. Long sunshine hour is determining factor for higher yield and quality. It can be grown in a wide range of temperature of 5°C to 40°C but ideal temperature is 20–30°C and with annual rainfall of 50–150 cm.

3.6.4.4   Agrotechniques for Quality Production

3.6.4.4.1   Improved Variety

Japanese mint :   Hybrid-77, RRL-118/3, EC-41911
Pepper mint :   EC- 41911
Spear mint :   MSS-1, MSS-5, Arka, Neera
Bergamot mint :   Kiran

3.6.4.4.2   Land Preparation

Mint is a deep-rooted crop. So deep ploughing is preferable for its normal growth and production. Depending upon soil types three-five ploughing followed by planking is required. Then it should be properly leveled and laid out into beds of 10 m × 10 m in size.

3.6.4.4.3   Propagation

Mentha is propagated by suckers, stolons, runners, although in sometimes young plants are used. Suckers are usually obtained from the field that has been planted during preceding years, whereas the stolon is an underground stem that is formed during winter to overcome dormancy period.

3.6.4.4.4   Planting

In case of Japanese mint suckers should be planted in furrows at 30–45 cm apart. The sucker should be cut into pieces of 10–12 cm length and planted at a distance of 10–15 cm in furrows. In pepper mint a spacing of 45 cm × 30 cm is normally adopted. Generally 500 kg of sucker is sufficient for planting one-hectare land. January to early February is the best season for stolen propagation.

3.6.4.4.5   Manures and Fertilizers

Depending upon edaphic factor and cultivable species FYM @ 15–25 tonnes/hectare and 40–50 kg N, 60–75 kg P_2O_5 and 60–80 kg K_2O/ha should be well incorporated during final land preparation. Apart from basal application 70–80 kg N/ha should be applied in two-three equal split doses. First, top dressing is when the plants are about 15 cm high and remaining dose should be applied after each harvest.

3.6.4.4.6   Irrigation

Mentha crop requires considerable moisture distributed the growing season. Being a shallow rooted crop, it is better to irrigate at frequent intervals. Frequency of irrigation depends upon soil texture and weather condition. During hot summer months of April-June irrigation is essential after every 10–15 days interval. Both lack and excess of water reduce essential oil content of the herb. On an average 8–12 irrigation is required for commercial cultivation.

3.6.4.4.7   Mulching

Mulching is an important cultural aspect for obtaining higher herbage yield and ultimately oil yield. Mulching is done during winter. Among the different mulching materials like paddy and wheat straw, dried leaves, sugarcane trash, dried grass, paddy straw is considered more suitable.

3.6.4.4.8   Interculture Operation

Slow sprouting and initial slow growth rate of the crop is favorable for germination and establishment of weeds. Generally manual weeding is not very effective, so chemical weed control is recommended. The most critical period of wee control lies in between 30–90 days after planting. Regarding chemicals, pre-emergence application of Diuron (2 kg ai/ha), Pendimethalin (1 kg ai/ha) and Oxyflurofen (0.25 kg ai/ha) is effective to keep the weed flora at subsistence level.

3.6.4.4.9   Crop Rotation

In India, mint is generally grown as an annual crop. The ratoon crop facilitates occurrence of high incidence of soil born diseases like stolon rot. Crop rotation is found to overcome these problems to a great extent. A number of crop rotation is practiced in mint cropping. The cultivators in different mint growing areas of our country follow the following crop rotations.

Mint–Maize–Potato
Mint–Paddy/Potato
Mint–Maize–Mustard
Mint–Red gram

3.6.4.5   Harvest and Post Harvest Technology

The common practice is to maintain the field for two years is temperate climate of Jammu and Kashmir. But in northern India the field should be maintained maximum for 7 to 8 months. The crop is harvested when it comes to flowering, which indicates maturity with a view to obtain the maximum oil yield. The oil content decreases rapidly after full bloom as the leaves begin to fall. In the plains of North India where menthe does not flower crop is harvested at the time when lower leaves start falling.

The oil is extracted through steam distillation. Except Bergamot mint, the harvested materials are staked and kept as such for one day and then allow for distillation. It helps to achieve more oil yield rather than fresh herbage distillation. In case of Bergamot mint freshly harvested materials are more suited for oil extraction.

The herbage and oil yield of different Mentha species are given in Table 3.26.

The major diseases of mint are stolon rot (*Macrophomina phaseoli*), root rot (*Thielavia basicola, Rhizoctonia*), leaf spot (*Corynespora cassicola, Curvularia lunata, Alternaria sp.*), fusarium wilt (*Fusarium oxysporum*), stolon rot (*Meloidogyne incognita*), rust (*Puccinia menthae*), powdery mildew (*Erysiphe cichoracearum*), etc., and the major insects are mint leaf roller (*Syngamia abrupatalis*) and root knot nematode.

TABLE 3.26   Herbage and Oil Yield of Different Mentha Species

Species	Herbage yield (t/ha)	Oil yield (kg/ha)
Japanese mint	30	150
Bergamot mint	20–25	100–120
Pepper mint	12–15	80–100
Spear mint	15–20	60–100

3.6.4.5.1   Value Addition

Peppermint oil: Spearmint and peppermint oil is derived from the respective plants. They are commonly used as flavoring in foods and beverages and as a fragrance in soaps and cosmetics. They are also used for a variety of health conditions and can be taken orally in dietary supplements or topically as a skin cream or ointment.

3.6.5   MARJORAM (MARJORANA HORTENSIS L.)

3.6.5.1   Systematic Position

Kingdom: Plantae
Division: Tracheophyta
Class: Magnoliopsida
Order: Lamiales
Family: Lamiaceae
Genus: *Marjorana*
Species: *hortensis*

3.6.5.2   About the Crop, National and International Scenario, Uses, and Composition

Sweet Marjoram (*Majorana hortensis* Monech) is an aromatic perennial herbaceous shrub belongs to the family lamiaceae and originated from Southern Europe. Marjoram or Sweet Marjoram is characterized by strong spicy and pleasant odor. The typical odor of marjoram is due to the presence of essential oil. The oil is colorless or pale yellow-to-yellow green, with a tenacious odor reminiscent of nutmeg and mace. Sweet marjoram is characterized by a strong spicy, slightly sharp, bitterish, and camphoraceous. The Greeks called this plant "joy of the mountain." They believed it was precious to Aphrodite, goddess of love, and they used it to crown newlyweds on their wedding day. Dried leaves with or without flowering tops in small proportion constitute the spice of commerce. Marjoram grows as an upright compact bush under 45–60 cm in height with a main stem and many softer branches. It grows indifferent part of Asia, Europe, Africa, and America. France and Egypt are the major exporter of marjoram oil in the world. In India, it is being grown in the home gardens of Karnataka, Andhra Pradesh and Tamil Nadu. It is highly esteemed as foodstuff. Fresh leaves are used garnishing agent in salads and different food preparations. Marjoram also used to flavor soups, stews, and sauces. Seeds are used in confectionary industry. Essential oil is used for preparation high-grade perfume, soap, and liquors industries. Medicinally, it has carminative, expectorant, and tonic properties.

3.6.5.3   Soil and Climatic Requirements

It is primary a warm climate plant and as such rather sensitive to cold. It thrives well in 22°C day temperature and 15°C night temperature. It requires sunny situation and though mid day shade is preferred. Well-drained nutrient rich soil is ideal successful cultivation of the crop. It has strong flavor when grown in nutrient rich soil. Though ideal pH is 6.9 for its cultivation but it can also successfully grown in saline and alkaline soil. The plant can resist drought condition but prolonged drought and drought in early few months of growth is harmful for the crop.

3.6.5.4   Agrotechniques for Quality Production

It is commonly propagated stem cutting. It can also be propagated by seeds. Seeds of marjoram are small. In the indoors, during midspring seeds start germinating after 14 days. It can also be propagated from cuttings, layerings or root divisions made in the late spring. Seedlings are transplanted in the main field when they are about 50–60 days old at a spacing of 30 cm × 20–25 cm. Adequate amount of organic manure and NPK@ 240:40:80 is to be applied for getting higher yield. It grows quickly and pinching should be done to make them bushy.

3.6.5.5   Harvest and Post Harvest Technology

The leaves should be harvested just before they begin to flower and if harvested when the flowers have set seed, the taste becomes bitterer. While harvesting, about 10 cm of the shoot above the ground should be left for further growth. The subsequent harvest carried out at an interval of 45 days from the previous harvest. After each harvest top dressing with urea is beneficial for getting subsequent herbage yield. In the advance stage of growth pruning is practiced at the end of winter because the herb has a tendency to become hardy. On an average it may yield about 16–18 t/ha fresh herbage per year, which in turn may yield about 35–40 kg marjoram oil by steam distillation. Herbage can be dried properly and stored in an airtight container as whole or crumbled leaves for three months or more.

3.6.5.5.1 Value Addition

Marjoram oil: This essential oil is extracted by steam distillation of both fresh and dried leaves of the marjoram plant. The main components of marjoram oil are sabinene, alpha terpinene, gamma terpinene, cymene, terpinolene, linalool, sabinene hydrate, linalyl acetate, terpineol, and gamma terpineol.

3.6.6 ROSEMARY (ROSEMARINUS OFFICINALIS L.)

3.6.6.1 Systematic Position

Kingdom: Plantae
Division: Tracheophyta
Class: Magnoliopsida
Order: Lamiales
Family: Lamiaceae
Genus: *Rosemarinus*
Species: *officinalis*

3.6.6.2 About the Crop, National and International Scenario, Uses, and Composition

Rosemary (*Rosamarinus officinalis* L.) belong to family lamiaceace is an under utilized herbal spices grown for its aromatic leaf and flowering tops. It is a native of the Mediterranean regions of Europe, Asia Minor and North Africa. It is cultivated in Yugoslavia, Spain, Portugal, and certain parts of Europe and USA. In India, it is cultivated in temperate Himalayan region of

North India and Nilgiris in South India. It is an evergreen, perennial dicot shrub. Rosemary is considered as a symbol of faithfulness. In ancient time, people of Greece associated with memory. During the Medieval time, rosemary was rewarded as an evil protector (Patil and Madhusoodan, 2002). The leaves and flowering tops, on steam-distillation, yield the essential oil. The oil has 1,8 cineole (20–50%), borneol (20%), camphor, linalool, α-pinene, camphene, β-pinene. The oil is valued for its use in culinary, medicine, perfumery, and cosmetic industries. It is also used in flavoring meat, sauces, candy, baked products, perfumery, cosmetics including soaps, cream, lotions, deodorants, and hair tonics. The leaves are used in cooking.

3.6.6.3 Soil and Climatic Requirements

It prefers a Mediterranean type of climate with low humidity, warm winters and mild summers for its successful growth. Climatic condition of Himalayan region, Nilgiri hills and surroundings Bangalore is highly suited for cultivation in India. It cannot tolerate frost. It is very hardy plant and is found growing on rocky terrains in the temperate parts of the World. But it prefers well-drained calcareous soil. In India it comes up well in loamy and sandy loamy soil with a pH range of 6.5–7.0.

3.6.6.4 Agrotechniques for Quality Production

There are two types of rosemary under cultivation. They are the 'French Rosemary' and the 'Italian rosemary.' Rosemary can be propagated by seed or stem cutting. In case of seed propagation seed should be sown in nursery during September to November. When the seedlings are about 8 to 10 weeks old, they are ready for transplanting into the main field. In case of stem cutting, cuttings are ready planting in the main field after 6–8 weeks after cutting. At time of land preparation apply farmyard manure at the rate of 20 t/ha and during time of final land preparation apply NPK @ 20:40:40 kg/ha. After each harvest, 80 kg/ha of N is applied in four equal split does as a side dressing to promote vegetative growth. Initially the crop should be irrigated once in a week and thereafter, one irrigation in a week is sufficient for normal growth and development of the crop.

Rosemary is vulnerable to insects like spider mites, mealy bugs, whiteflies, and thrips. Spidermites feed preferentially on the lower stem and then move on to the upper section. Mealy bug females feed on plant sap. Whiteflies suck sap from the leaves and also excrete honeydew, which serves as a growth medium for sooty mould. Thripos feed on leaves and damage the plants. They can also be the vectors of other diseases. Crop rotation and application of insecticidal soap is helpful to control the insects. Regarding disease, fungal problems viz., powdery mildew and root rot may arise when the plants are over irrigated. Early detection and management of disease along with regular scouting is essential.

3.6.6.5 Harvest and Post Harvest Technology

Depending upon the exposure of plantation, the plants start flowering earlier in warmer and low altitude areas and later on the high altitude. In some areas harvesting commences from the second year in August after the full flowering, which commences in May-June. Harvesting should begin at the time of 50% blossoming and continue till 75–90% inflorescence emerges and must end when the flowers have finished blossoming.

3.6.6.5.1 Value Addition

Rosemary essential oil: Essential oil is obtained by steam distillation of freshly harvested twigs or leaves. The leaves can also be shade dried, stored, and distilled at convenience without any loss of oil. It is always advisable to do continuous distillation for 120 minutes for maximum recovery of rosemary oil.

3.6.7 SAGE (SALVIA OFFICINALIS L.)

3.6.7.1 Systematic Position

Kingdom: Plantae
Division: Tracheophyta
Class: Magnoliopsida
Order: Lamiales
Family: Lamiaceae
Genus: *Salvia*
Species: *officinalis*

3.6.7.2 About the Crop, National and International Scenario, Uses, and Composition

Sage (*Salvia officinalis* L.) belongs to the family lamiaceae is an under-utilized herbal spice. Dried leaf of sage is the commercial part utilized as spice. It is native Southern Europe. It is grows and cultivated in Yugoslavia, Portugal, Spain, England, and Canada. Now it is also cultivated in India mainly in Jammu and Kashmir. The characteristics aroma and flavor of the sage is due the presence of essential oil. The main composition of the oil is thujone. It is a hardy sub shrub with a height of 15–30 cm tall and aromatic leaves. It has been extensively used in the food industry as a standard spice in making stuffing for meat and sausages. Dried and powdered leaves can be utilized as condiment for preparation of cheese dishes, meat, and other preparation. Fresh sage leaves are used in salads and sandwiches. Sage oil finds use in perfumes, deodorant, and insecticidal preparation. It is also used for adulterating rosemary and lavender oils.

Three important subspecies of sage are under cultivation. These are as follows:

a. Major (Dalmatian): This is the term for the product of Yugoslavia, derived from Dalmatian coast. It is highly aromatic, with a pleasing, particularly mellow flavor. Its mellowness is being its hallmark among sages.
b. Lavandulifola (Spanish).
c. Triloba (Greek).

Sage is the most important herb in every kitchen for flavoring fish and meat dishes and in making poultry stuffings. Dried and powdered leaves are mixed with cooked vegetables and sprinkled on cheese dishes, cooked meats and other similar preparations. Fresh sage leaves are used in salads and sandwiches. The young leaves are pickled and used for making tea. Dried leaves are used in tooth and mouthwashes, gargles, poultices, tooth-powders, hair-tonics, hair-dressings. It is considered to be a mild tonic, astringent, carminative, diaphoretic, and antipyretic. Sage oil is used in perfumes as a deodorant, in insecticidal preparations, for the treatment of thrush and gingivitis. Sage and sage oil also exhibit antioxidant properties. After steam-distillation, the residual plant material still contains constituents of considerable flavor value.

3.6.7.3   Soil and Climatic Requirements

It is a cool season crop and hence its performance in hot and dry climate is not up to the mark. A hot and dry climate is not suitable for its cultivation, since it produces an inferior crop. The crop can be grown as a perennial or annual crop. Sage is said to thrive on any soil but better quality product is obtained from clayey and loamy soil. Sage grows better in well-drained calcarious soils under full sunlight.

3.6.7.4   Agrotechniques for Quality Production

Sage plantations are preferably raised from rooted stem cuttings as the crop raised from seed is likely to be a mixture of different types. The cuttings are obtained preferably from broad-leaved plants that have least tendency to flower. The rooted cuttings are planted 30–45 cm apart in rows, 0.9 m apart. One kg of seeds provides about 20,000–30,000 seedlings, enough to plant a hectare. Seeds are drilled directly in the field or seedlings are raised in the nursery for transplantation later. In case of seed propagation, seed rate is 1 kg/ha. Sage plantations are preferably raised from rooted stem-cuttings, as the crop raised from seeds is likely to be a mixture of different types. The cuttings are obtained preferably from broad-leaved plants that have the least tendency to flower. In case of stem cutting, basal cutting rooted better than apical cutting. It has been observed that April–May cuttings rooted better than October- November (Raviv et al., 1984). Cuttings are usually planted in the minefields at spacing of 60–90 cm × 30–45 cm under Indian condition. Bezzi (1987) reported that application nitrogen (at the rate of 0–150 kg/ha) increased dry matter production, but potassium and phosphorus had no effect on dry matter production in sage.

3.6.7.5   Harvest and Post Harvest Technology

The first year crop is usually light, but after the plants is established and two or more cuttings in a year may be possible. From the second year onwards a crop yield of 1.75–2.25 t/ha may be obtained. Sage plantation should be replanted every 4–5 years, by that time they become logy and show sign of die-back. The first year crop is usually light, but after the plants are established, 2 or more cuttings a year is possible. In the second and sub-sequent years, a crop of 1,750–2,250 kg of herbage per hectare may be obtained.

3.6.7.5.1 Drying

Sage is dried in the shade in order to retain as much of the natural color and flavor as possible. The leaves and small tops are tied into small bundles or spread on screens and dried in a well-ventilated warm room, away from direct sunlight. If the leaves are dusty or gritty, they are washed in cold water before drying. The dried bunches can be sold without further treatment or the leaves may be cut, rubbed or pulverized for the packeted herb trade. Other products such as Spanish sage or Greek sage, derived from other species of *Salvia*, are used as adulterants of genuine product.

3.6.7.5.2 Value Addition

Sage leaf is available in the market in following forms.

i.   whole
ii. cut
iii. rubbed
iv. ground

Whole leaf sage: Cleaned and dried whole leaves.

Cut sage: It is also called as cracked, sliced, chopped, and butchers chop, etc. it refers to leaves that have been cut into smaller pieces. Particle size ranges from 0.3–0.6 cm.

Rubbed sage: This sage has under gone minimum grinding and a coarse sieve.

Rubbed sage is a fluffy, almost cotton like product, unique among ground herbs. Many sausage makers prefer this, as they believe that it preserves flavor longer and blends into the product very easily.

Ground sage: Ground form of sage leaves.

3.6.8   TARRAGON (ARTEMISIA DRACUNCULUS L.)

3.6.8.1   Systematic Position

Kingdom: Plantae
Division: Tracheophyta
Class: Magnoliopsida
Order: Asterales
Family: Asteraceae
Genus: *Artemisia*
Species: *dracunculus*

3.6.8.2   About the Crop, National and International Scenario, Uses, and Composition

Tarragon (*Artemisia dracunculus* L.) belongs to the family composite is a well known spices for its unusual flavor. Sometime, it is also called as French Tarragon. *Artemisia* is a large genus and about 280 species found in the Northern hemisphere. About 34 species reported in the temperate region of North Western Himalayas. Its aroma is warm, aromatic, and reminiscent of anise. It is cultivated in France, Spain, temperate zone of USA and colder zone of England. The herb on steam distillation, yields about 0.3% essential oil, which is responsible for the aromatic aniseed like odor of the herb. Metyl chavicol is the chief constituent of the essential oil (Werker, et al., 1994). Phellandrene and ocimene are additional constituents of essential oil. Both leaves and essential oil are used for flavoring vinegar, pickles, prepared mustard and to a limited extent for flavoring soups, salad, meat dishes, certain cheeses and vegetables. The aromatic leaves are credited with stomachic, stimulant, and febrifuge properties.

3.6.8.3   Soil and Climatic Requirements

Tarragon is a xerophytic plant. In central Asia, they grown in semi desert areas where extremes of temperature, both high and low prevails. It prefers a little shade during the warmest part of the day. Long days promoted shoot growth where as plant grown under shot day produce rosette plant. Under ground buds removed from roots or crowns of tarragon plants in August rooted well in sand in the glass house, and plantlets are ready for filled planting after 3 weeks.

It prefers a saline sandy soil. Tarragon will grow in a pH range between 6.5 (neutral) and 7.5 (mildly alkaline) with a preferred pH of 6.5. It grows well in a rich loamy soil that holds moisture, but drains well. Mulching is beneficial to this end.

3.6.8.4 Agrotechniques for Quality Production

It may be grown outdoors, in containers, and hydroponics. Tarragon grown outdoors prefers full sun but can tolerate some shade. Russian tarragon seeds are sown indoors in sunny location or under plant grow lights six weeks before last frost. Soilless potting mixes (Pro-Mix, Sunshine Mix, etc.), perlite, vermiculite, rockwool, coco peat, Oasis Rootcubes. French tarragon only propagates via division, stem cuttings, or layering. By using bud crop establishment is much more rapid than the rooted tip cuttings and survival rate was much higher. Tarragon propagates best through root division, planting the divisions at least 18 inches apart. Since tarragon has a shallow root system, care must be taken not to damage the roots when weeding, and special care must be shown during the winter after transplanting, as the root systems will not have developed fully. Rapid propagation of tarragon can be done by using *in vitro* technique. Russian tarragon seeds will germinate in soil in approximately 10 to 14 days, but can germinate in as few as 7 to 10 days in dedicated propagation media such as Oasis Rootcubes, Rapid Rooters, or Grodan Stonewool. Tarragon plants should be spaced 45–6°C apart. There are approximately 6, 000 tarragon seeds per gram. Tarragon will grow indoors satisfactorily under standard fluorescent lamps, Standard fluorescent lamps are kept between 2 and 4 inches from the tops of the plants, high output and compact fluorescents approximately one foot above the plants, and HID lights between 2 and 4 feet above the plants, depending on wattage. Average water needs, water on a regular schedule. Allow soil to go almost dry between watering, then soak thoroughly. Do not overwater. Tarragon can be susceptible to whitefly and spider mites but has minimal disease issues.

3.6.8.5   Harvest and Post Harvest Technology

The crop is picked from May until September. Shoot tips are generally removed with a pair of secateurs and leaves are striped with fingers. Leaves are best used fresh, but can be dried and stored in airtight bags or containers.

3.6.8.5.1   Value Addition

The light to dark green leaves possess a bittersweet and herbaceous taste. Highly aromatic with a licorice-like flavor, tarragon is essential in French cuisine. It is used in roast chicken, eggs, herb butter, vinaigrettes, and the classic Béarnaise sauce.

3.6.9   THYME (THYMUS VULGARIS L.)

3.6.9.1   Systematic Position

Kingdom: Plantae
Division: Tracheophyta
Class: Magnoliopsida
Order: Lamiales
Family: Lamiaceae
Genus: *Thymus*
Species: *vulgaris*

3.6.9.2   About the Crop, National and International Scenario, Uses, and Composition

Thyme (*Thymus vulgarisi* L.) belongs to the family lamiaceae commonly known as thyme or golden thyme. It is an important herbal spice used by man since ancient times. It is low evergreen perennial under shrub reaching a height of 20–30 cm. It is one of the very frequently used spices in fresh or dried form for several European delicacies. It is common garden plant, which survives for many years under good management condition. The dried leaves and floral tops constituting the thyme of commerce is known as "Thymi Herba." The dried leaves are curled, are of brownish green color, usually not longer than 6–7 mm and 2–3 mm broad and marketed in whole or ground form. On steam distillation thyme produce essential oil and the main composition of essential oil is thymol. Thyme is grown in Europe, Australia, and North Africa. It is originated from Mediterranean region. It is cultivated in France, Germany, Spain, Italy, Greece, North Africa, Canada, and USA. In India, it is found in the western Himalayan from Kashmir to Kumayan between altitudes of 1525 to 4000 m. The leaves and flowers are find use as food flavorants and seasoning various food items, especially fish and meat preparations. Thyme oil is used in soap, perfumes, and flavoring food products such as sausages, sauces, meat, and canned products.

3.6.9.3 Soil and Climatic Requirements

The plant prefers a light but fertile and calcareous soil for good and oil content. It can be grown both hill and plains but hilly region is most suitable for its growth. It prefers mild climate for its better growth and yield.

3.6.9.4 Agrotechniques for Quality Production

Late summer planting is the ideal time for transplanting of seedlings or planting of rooted cuttings. Thyme can be seeds and vegetatively by division of old plants or by cutting or by layering of side shoots in March–April. The seeds are sown directly in rows or the seeds are sown in well-prepared nursery beds in good soils. A spacing of 90 cm × 30–45 cm may be maintained in case direct sown crop. While planting the seedlings or rooted cuttings or layer, etc. they are planted in a spacing of 60 cm × 30 cm. alight irrigation is provided after planting. Application of farmyard manure and nitrogen is reported to promote the formation of numerous leafy shoots. Frequent irrigation is required during the dry periods. In the hills, in order to avoid frost injury to plants during winter mulching is done.

3.6.9.5   Plant Protection

The plant is not attacked by any pests of serious nature, but wilt disease is a major problem in this crop. The disease can be controlled by improving the phyto sanitation and by the use of suitable fungicides like blitox or Dithane M-45 @ 0.3% concentration for soil drenching.

3.6.9.6   Harvest and Post Harvest Technology

The leaves and flowers, which are used for culinary or medicinal purposes are harvested five months after sowing or planting. The leaves and flowers are plucked from plants are dried immediately in shade or in mechanical drier and the dried product should be kept in air tight container to prevent the loss of flavor. Under favorable condition and good management condition, yield (in dry form) of thyme varied from 1.6–2.2 t/ha. The yield is comparatively low during the first year. The plants become woody and replanting become necessary after 3–4 years. The plant is not damaged by any pest of serious nature but grey mould, root rot and wilt diseases are major problems of this crop.

3.6.9.6.1   Value Addition

Thyme essential oil: The oil has been recognized for thousands of years in Mediterranean countries. This substance is also a common agent in Ayurvedic practice. Today, among the many producers of thyme oil, France, Morocco, and Spain emerge as the primary countries. The oil contains thymol, linalool, carvacrol, thujanol, alpha-terpneol, geraniol, 1,8-cineol, p-cymene, phenol, etc. Thyme oil can be used as a preservative against spoilage and several foodborne germs that can contribute to health problems. It is effective against other forms of bacteria such as Salmonella, Enterococcus, Escherichia, and Pseudomonas species.

3.7   LESSER KNOWN SPICES

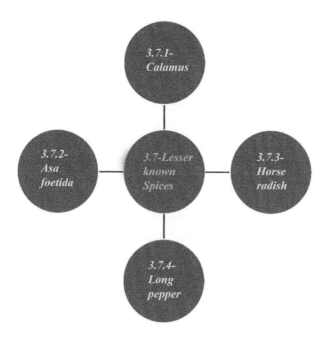

3.7.1   CALAMUS OR SWEET FLAG (ACORUS CALAMUS L.)

3.7.1.1   Systematic Position

Kingdom: Plantae
Division: Tracheophyta
Class: Magnoliopsida
Order: Acorales
Family: Acoraceae
Genus: *Acorus*
Species: *calamus*

3.7.1.2   About the Crop, National and International Scenario, Uses, and Composition

Sweet flag (*Acorus calamus* L.) commonly known as 'Bach or Gorbach in Hindi; as Vacha, Ugragandha or Bhadra in Sanskrit' is an important minor spice cum medicinal and aromatic plant belongs to the family araceae. It is a semi-aquatic perennial herb with long, creeping, much branched, aromatic, rhizomes, and fibrous root that occurs widely all over India especially in hilly tracts (Selvi et al., 2003). It is mainly cultivated in the Netherlands, Persia, United Kingdom, India, and Sri Lanka. In India, it is common in Kashmir and Kumayun region of the Himalaya. However, it is mainly cultivated in Karnataka, Kashmir, Manipur, and Nagaland. The fresh rhizomes are used in confectionary and also used as substitute for ginger (Farooqi et al., 2000). In West Bengal, the ground dried rhizome and rhizome powder is used in baits for fishing. Root is used for treatment of Kwashiorkor disease of children. The rhizomes are used as carminative, stimulant, and tonic (Jain, 2001). In Ayurveda it is highly valued as a rejuvenator for the brain and nervous system and as a remedy for digestive disorders. It is used internally in the treatment of digestive complaints, bronchitis, sinusitis, etc. It is said to have wonderfully tonic powers of stimulating and normalizing the appetite.

The essential oil composition of Acorns calamus (sweet flag) leaves at different growing phases was examined by GC and GC/MS. The content of the oil in dried sweet flag rhizomes was $1.20 \pm 0.12\%$ and in the leaves, depending on the vegetation phase, was from 0.56 to 1.01%. BetaAsarone [(Z)-asarone] was the major constituent in the leaves (27.4 to 45.5%), whereas acorenone was dominant in the rhizomes (20.86%) followed by isocalamendiol (12.75%) (Venskutonsis et al., 2003). Besides Monoterpene hydrocarbons, sequestrine ketones, (Trans- or Alpha) Asarone (2, 4, 5-trimethoxy-1- propenylbenzene), and β-asarone (cis- isomer) and eugenol were also identified (Kindscher Kelly, 1992).

3.7.1.3   Soil and Climatic Requirements

It is a hardy plant found growing from tropical to sub-tropical climates. Plenty of sunshine should be available to the plant during its growth and after harvesting for drying the rhizomes. Temperature ranging from 10°C to 38°C and annual rainfall between 70 and 250 cm are best suited. Sweet flag comes up well in clayey loams, sandy loams and light alluvial soils of riverbanks.

3.7.1.4   Agrotechniques for Quality Production

The land should be ploughed twice or thrice prior to the onset of rains. The land should be prepared like paddy fields. It is propagated through rhizomes. Rhizomes obtained from earlier planting are kept preserved in the soil and constantly kept moist. After emergence the rhizomes are cut into small pieces and planted. Sprouted rhizome pieces are planted at a spacing of 30 × 30 cm and depth of 4 cm in the month of June–July. FYM @15–20 t/ha along with nitrogen and phosphorus are applied. One third of N along with 50 kg of P and 25 kg of K is the basal requirement. The second dose of N should be given after one month of planting as broadcast and a third dose should be applied after two months of planting. Datta et al. (2009) reported that the maximum fresh and dry rhizome yield (3013.23 kg/ha, and 1389.15 kg/ha, respectively) was recorded with 50 t/ha FYM along with 100 ppm GA_3 followed by application of 50 t/ha FYM (1342.65 kg/ha in dry form). Combined application of N @ 90 kg/ha and K 275 kg/ha recorded the maximum dry rhizome yield (1600 kg/ha) and the highest B:C ratio of 2.06 under terai (foot hills) zone of west Bengal. The initial level of water standing in the field should be 5 cm and later increased to 10 cm. Timely weeding and hoeing to control the spread of weeds and to obtain good yield is essential.

3.7.1.5   Plant Protection

Major disease is leaf spot and a spray of Captan 10 g with Chloropyriphos 20 mL/10 L controls leaf spot as well as mealy bugs and caterpillar.

Mealy bugs and caterpillar are the pests occurring on this crop. Control measures: Spraying the shoots and drenching the roots of plants with 10 mL methyl parathion or 20 mL Quinolphos in 10 L of water can be effective in controlling the shoot and root mealy bugs.

3.7.1.6   Harvest and Post Harvest Technology

Harvesting is done after 6–8 months after planting, usually in December. The lower leaves turn yellowish and dry indicating their maturity. The field should be partially dried only leaving sufficient moisture for uprooting the plant. In case of large scale cultivation rhizomes may be removed by passing the plough. The uprooted rhizome is cleaned after washing with water and cut into size of 5–7.5 cm length and fibrous roots removed Mealy bugs and caterpillar are the pests and leaf spot is the common diseases occurring on this crop. The yield is expected to be 6–10 t/ha of fresh rhizomes per hectare.

3.7.1.6.1 Value Addition

Essential oil: Sweet flag contains 1.5 to 3.5% essential oil. The essential oil is extracted from the rhizome, which is utilized in perfumery. Due to presence of acorin in its essential oil it is commonly as a remedy for asthma and chronic diarrhea. Its aroma makes its essential oil valued in the perfume industry.

3.7.2 ASAFOETIDA OR HING (FERULA ASAFETIDA L.)

3.7.2.1 Systematic Position

Kingdom: Plantae
Division: Tracheophyta
Class: Spermatopsida
Order: Apiales
Family: Apiaceae
Genus: *Ferula*
Species: *asafetida*

3.7.2.2 About the Crop, National and International Scenario, Uses, and Composition

Asafoetida (*Ferula foetida* L.) belongs to the family apiaceae is commonly known as hing or hingra. It is the gum oleoresin of the genus *Ferula* that includes 60 different species found mostly in Central Asiatic Countries, Europe, and Tropical North Africa. Some major varieties are also grown in Pakistan, Afghanistan, and Persia. Most of the asafetida obtained from the variety of the species *Ferula foetida*. The other related species are *Ferula alliaceae* and *Ferula galbani*. *Ferula foetida* is a perennial plant, reaches a height of 1 to 1.5 m and have massive carrot shaped roots. The rhizome produce two types plants, viz. male and female plant. Male plant produces inflorescence whereas female plants do not produce inflorescence. The female plant produces an exudation in the form of very thick and sticky paste sap from the underground rhizome. This exudation is called gum resin, which is the commercial part of asafetida. It is extensively used for flavoring curries, sauces, pickles, and with other spices. Asafoetida is mainly used to impart taste. The gum resin possesses antispasmodic, carminative, expectorants, and stimulant properties. Fried asafetida powder is used in many ayurvedic preparations to cure convulsions and nervous breakdowns. According to the different workers, huge variations in physiochemical characteristics of different samples of asafetida have been reported. Such variation might be occurred due to certain factors, such as part of the plant from, which prepared, the season of collection, method of preparation, the degree of adulteration and nature of adulteration (Pruthi, 1976).

3.7.2.3   Soil and Climatic Requirements

Horseradish needs a temperature between 5 and 19°C with an annual precipitation of 50–170 cm and a soil pH of 5.0–7.5. The hardy horseradish thrives in moist, semi-shaded environments of the north-temperate regions of North America. Although the plant will grow on any soil type, best growth is in deep, rich loam soil, high in organic matter. It prefers a rocky dry soil and is usually found above 1350 m.

3.7.2.4   Agrotechniques for Quality Production

The plant is a perennial and may grow as high as 3.6 m. After four years, when it is ready to yield asafetida, the stems are cut down close to the root, and a milky juice flows out that quickly sets into a solid resinous mass. A freshly exposed surface of asafoetida has a translucent, pearly white appearance, but it soon darkens in the air, becoming first pink and finally reddish brown.

Dried asafetida consists mostly of a resin (25 to 60% of the total mass, 60% or, which are esters of ferula acid) and a complex carbohydrate part (25 to 30%).

3.7.2.5 Harvest and Post Harvest Technology

The plant possesses a perennial rootstock through, which they multiply vegetatively. During the onset of spring sprouts comes out from the rhizome of the female plants and put 4th foliage. It takes nearly one month for green foliage to turn yellow. It is the right stage for tapping (by incision) the rhizomes for asafoetida. Initially, the yellow foliage is removed and then top of the rhizome and brush like mass is covered with loose earth and left for five days. Then brushy mass on the top of the rhizome pulled out and the root and gravel around the rhizome are cleaned, exposing the top of the rhizome. After that incision is made at the base. A thick white fluid is discharged from the place of incision and pours slowly downwards. After a few days, the end of the root is again cut off and the accumulated fluid sponged out. This is continued till the plant stop discharging the gum resin. The whole process may continue up to three months. However, the get dried out and to be mixed with some chemicals to increase the amount and also giving proper shape.

3.7.2.5.1 Value Addition

Asafoetida is marketed in three forms tears, mass, and paste. *Tears* are the purest form, they are round or flat, about 15–30 mm diameter and have greyish or dull yellow color. *Mass* is agglutinated tears mixed with extraneous matters. *Paste* is semi-solid and contains extraneous matter.

Asafoetida is often adulterated with gum arabic, other gum resins, barley, and wheat flour, red clay, gypsum, chalk, etc. The so-called "powdered asafetida" is the resin mixed with rice flour and therefore much less strong in taste, but easier in application.

3.7.3 HORSERADISH (COCHLEARIA ARMORACIA L.)

3.7.3.1 Systematic Position

Kingdom: Plantae
Division: Tracheophyta
Class: Magnoliopsida
Order: Brassicales

Family: Brassicaceae
Genus: *Cochlearia*
Species: *armoracia*

3.7.3.2 About the Crop, National and International Scenario, Uses, and Composition

Horseradish (*Cochlearia armoracia* L./*Armoracia rusticana* Garten) belongs to the brassicaceace, a near relative of turnip, cabbage, and mustard and originated from middle Europe. It is on of the oldest condiments in the world and its white fleshy and tasty roots are used as commercial part. It is a large leaved hardy perennial herb has been used as culinary herb in the different European countries more than 3000 years. It is very popular condiment for foods in Germany and surrounding areas. It is grown in in Eastern Europe, UK, USA, India, and Sri Lanka. In India it is grown in gardens both in north India and hill station of South India. The root contains a pungent, acrid, and vesicating volatile oil. The pungency is due to the presence of an allyl isothiocyanate and butylthiocyanate similar to mustard oil, occurring in combination with the glucoside 'sinigrin.' Organoleptocally active substances of horseradish are mainly due to the presence of 2-propenyl isothiocyanate and 2-phenyl ethyl isothiocyanate. Horseradish is used as an appetizing condiment. It is highly prized as condiments especially with oyster and cold meats. A freshly grated horseradish root, when mixed with vinegar and salt, is much appreciated with as an appetizing condiment to enhance the flavor of roasted beef. Inhaling of horseradish essential oil or taking horseradish by mouth promotes blood flow in nasal and sinus tissues and helps possibly relieving respiratory congestion.

3.7.3.3 Soil and Climatic Requirements

Horseradish thrives well in temperate climate and in the cool season and high altitude of tropical countries. It grows best in deep, rich, moist loamy soil and sunny situation.

3.7.3.4   Agrotechniques for Quality Production

Horseradish plant is highly sterile in nature and hence the plant is propagated asexually mainly by root cuttings. Usually cuttings are made from side roots, which are trimmed off in preparing roots for market. Before planting the cutting main field should be well prepared and cutting are planted at a spacing of 30 cm × 30 cm. Regarding fertilizer requirement for 30,000 plants/ha, 140–170 kg K_2O/ha and 80–200 kg mineral Nitrogen was sufficient depending on the amount of farmyard manure used (Kraxner et al., 1986). Horseradish is generally two types viz. Common and Bohemain type. Among the two types common types produce better quality roots. Edelkafener, Jugoslavischer, Humberger, and Steirischer are some of the ecotypes of horseradish (Nebel et al., 1988). Cabbage worms, the larvae of *Pieris rapae*, the Small White butterfly, are a common caterpillar pest in horseradish.

3.7.3.5   Plant Protection

White rust: It is caused by *Albugo candida*. White rust, sometimes called white blister, is easily recognized by the chalk-white, cheesy, raised spore masses (sori), which occur mostly on the under leaf surfaces. The first and often overlooked symptom of white rust is the appearance of small, irregular yellow areas (chlorotic lesions) on the upper leaf surface, which range in size from minute dots to one-half inch or more in diameter. The leaves of systemically infected horseradish plants are usually smaller than normal and may curl inward.

Planting materials are to be chosen only from the terminal ends of healthy primary All systemically infected horseradish are carefully removed and destroyed Field inspection is to be carried out at least every 10 to 14 days during spring and early summer. A 3-years crop rotation between susceptible crops in the same area or field is to be adopted.

Bacterial leaf spot: Small dark translucent spots appear initially and later enlarge turning black. It is usually scattered over the entire leaf surface. Disease free planting materials are to be used. Application of Serenade is found effective.

Turnip mosaic virus: Ring spots and mosaic or mottling symptoms appear on leaf and black streaks on leaf stalks. Virus free plants are to be used.

3.7.3.6 Harvest and Post Harvest Technology

To grow high quality horseradish, lift, and strip the roots two times, first when the biggest length is 20 to 25 cm long and again 6 weeks later. The main root is harvested and one or more large offshoots of the main root are replanted to produce next year's crop. Yield of horseradish is about 2 t/ha.

3.7.3.6.1 Value Addition

Horseradish oil: It is obtained through steam-distillation of the oil-rich roots. It is pale yellow in color with a reminiscent smell of hot mustard seed oil. The pungency of horseradish is found in the outer part of the roots, is quickly dispelled when the root is grated, and is not formed at all if the root is cooked. The main ingredient is sinigrin, a glycoside which, combined with water, yields the isothiocyanates.

Horseradish sauce: Freshly grated horseradish, along with thick cream (or Greek yogurt), freshly chopped parsley, white wine vinegar and salt are used as the ingredients. After placing the cream and horseradish in a bowl and mixing gently, parsley is added along with other ingredients, mixed well and kept at room temperature.

3.7.4 LONG PEPER OR PIPLI (*PIPER LONGUM* L.)

3.7.4.1 Systematic Position

Kingdom: Plantae
Division: Tracheophyta
Class: Magnoliopsida
Order: Piperales
Family: Piperaceae
Genus: *Piper*
Species: *longum*

3.7.4.2 About the Crop, National and International Scenario, Uses, and Composition

Long pepper or Pepper long or Pipli is the dried fruit of *Piper longum* under the family Piperaceae. It is an important medicinal cum spice crop. It is slender, aromatic plant with creeping jointed stems and perennial woody roots. The taste of the long pepper is similar like black pepper but long pepper is somewhat hotter than the black pepper. *Piper longum* is of South Asian origin (Deccan Peninsular). It is found growing wild in the tropical rain forests of India, Nepal, Indonesia, Malaysia, Sri Lanka and Philippines. Indian long pepper is mostly derived from the wild plants. In India, it is distributed from Central Himalayas to Assam, Lower hills of West Bengal, evergreen forests of Western Ghats, Nicobar Islands, Uttar Pradesh and Nepal.

It is used as a spice and also in pickles and preserves. The fruits and roots are used as medicine for respiratory disease and as counter irritant and analgesic for muscular pains and inflammation. In ayurveda system of medicine, the roots are used as a carminative, tonic to the liver, stomachic, emmenagogue, abortifacient, aphrodisiac. Whereas the fruits possess haematinic, diuretic, digestive, general tonic properties, besides being useful in inflammation of the liver, pains in the joints, lumbago, and night blindness.

The fruit of long pepper contains a large number of alkaloids and related compounds, the most abundant of which is piperine, methyl piperine, iperonaline, piperettine, asarinine, pellitorine, piperundecalidine, piperlonguminine, piperlonguminine, refractomide A, pregumidiene, brachystamide, brachystamide-A, brachystine, pipercide, piperderidine, longamide, and tetrahydropiperine, terahydro piperlongumine, dehydropipernonaline piperidine, piperine, terahydropiperlongumine, and trimethoxy cinnamoylpiperidine and piperlongumine. The essential oil of the long pepper fruit is a mixture of three major components of which are (excluding the volatile piperine) caryophyllene and pentadecane (both about 17.8%) and bisaboline (11%). Some other components are thujine, terpinoline, zingiberine, p-cymene, p-methoxy acetophenone and dihydrocarveol. Long pepper essential oil consists of sesquiterpene hydrocarbons and ethers (bisabolene, β-caryophyllene, β-caryophyllene oxide, each 10 to 20%; α-zingiberene, 5%), and saturated aliphatic hydrocarbons such as 18% pentadecane, 7% tridecane, 6% heptadecane.

3.7.4.3 Soil and Climatic Requirements

It performs well-drained fertile loamy soil with organic matter content. Well drained forest soil and laterite soils with high organic matter content and moisture-holding capacity are also suitable for long pepper cultivation.

Long pepper prefers hot moist climate for growth and development. It can be grown from 0–1000 m elevation from mean sea level. However, optimum elevation for its cultivation is between 100 to 1000 m. Higher elevations are not favorable for higher yields. It requires provision of a partial shade for its ideal growth. Partial shade of 20–25% intensity is found to be the optimum.

3.7.4.4 Agrotechniques for Quality Production

It is usually propagated by rooted vine cuttings and suckers during the onset of monsoon. in beginning of rainy season. The establishment percentage of 3–5 node-rooted cuttings is very high. The suckers are suitable for planting in heavy rainfall areas. Vine cuttings can be rooted in poly bags filled with the common pot mixture during March–April and planted in main field from end of May to middle of June. Excess moisture in the nursery invites rotting and wilt diseases. Before planting the cutting or sucker land should be prepared with repeated ploughing and harrowing. Pits are prepared at spacing of 60 cm × 60 cm. Raised bed should be in case of high rainfall receiving areas. Before planting pits are filled with top soil and organic manures. It can also be planted as an inter crop in coconut, areca nut, and other plantation crops. Apply 20 t/ha of farmyard manure or compost at the time of final land preparation. Among the inorganic fertilizers long pepper requires 50 kg N, 20 kg P_2O_5 and 70 kg K_2O per ha for its optimum growth and yield. Apply ½ amount of N and entire dose of P_2O_5 and K_2O as basal doses and the remaining N should be applied as top dressing in two equal splits. The crop usually grown as rainfed crops but providing irrigation during dry season will induce continuous spike formation in off-season. Sprinkler system of irrigation is found more effective. Hand weeding is done as and when necessary. Mulching during summer season with dry leaves or straw will help for conservation soil moisture around base of plant.

3.7.4.5   Plant Protection

Wilt and pollu: Symptoms: The wilt is characterized by the death and decay of the roots, yellowing, and shedding of the leaves and ultimately drying of the plant. It not only causes hollowness of the fruits but also leads to their complete destruction. Control measures: Spraying Bordeaux mixture (0.1%) in the month of May and 2–3 times in rainy season found effective in reducing the extent of damage.

Leaf and vine rotting: during warmer months is common in long pepper. It can be controlled by spraying 1% Bordeaux mixture during May and 2–3 sprays subsequently during rainy season.

A virus like disease showing yellowing and crinkling of leaves produces spikes of reduced size and inferior quality. The affected plants should be uprooted and destroyed.

Mealy bugs: Mealy bugs infest the roots and suck its sap resulting stunted growth. The severity is more in summer. It can be controlled by spraying Rogor.

Helopeltis thivora, a pest, which cause destruction of leaves. To control this pest decoction prepared out of neem (0.25%) has to be sprayed.

3.7.4.6   Harvest and Post Harvest Technology

The first harvest is normally done 6 months after planting. The spikes are ready for harvest 2 months after fruit setting. The yield of dry spike during first year is around 400 kg/ha., it increases up to 1000 kg/ha in the third year. Spikes are picked when they are blackish green and most pungent. Delayed in picking loses its pungency to a great extent. The harvested spikes are dried in the sun for 4–5 days until they are perfectly dry. The green to dry ratio of long pepper is around 10:1.5. The dried spikes are then stored in the moisture proof containers.

3.7.4.6.1   Value Addition

Pippali powder: Pure finely ground ayurvedic herb, which stimulates digestion by encouraging and supplementing the production of digestive enzymes, eases liver function, provides chemical heat to the GI tract to burn up toxins and expel them from the body, assists in the break down of food, and cleans out the digestive tract, Improves blood flow and circulation, detoxifies the liver.

The *fruits* are used in pickles and preserves. They have a pungent pepper-like taste.

The *roots* and *thicker parts of stem* are cut and dried and used as an important drug (Piplamool or Pippali) in the Ayurvedic and Unani systems.

3.8   OTHER SPICES

3.8.1   SAFFRON (CROCUS SATIVUS L.)

3.8.1.1   Systematic Position

Kingdom: Plantae
Division: Tracheophyta
Class: Magnoliopsida
Order: Asparagales
Family: Iridaceae
Genus: *Crocus*
Species: *sativus*

3.8.1.2   About the Crop, National and International Scenario, Uses, and Composition

Saffron (*Crocus sativus*) is an important high value, low volume spice in the World. It is one of oldest and costliest spices. The yellow style is deeply divided into three branches and the stigmata are bright red. Flowers are arising directly from the corms. Flowers have tri-lobed stigma, which along with the style tops yield the saffron of commerce. Saffron has several names-Zafran, Kesar, Kang, and Kang Posh, etc. Kang posh, the flowers of Saffron is a symbol of freshness and purity. It is small winter active, perennial bulbous plant. It is originated from Southern Europe. It is mostly cultivated in Mediterranean countries, particularly in Spain, Austria, France, Greece, England, Turkey, Iran, and India. More than 50% World saffron is produced from the La Mancha areas of the Central Spain. In India, it is cultivated in Jammu and Kashmir. Efforts are made to popularize the crop in dry and temperate hills of Himachal Pradesh and Arunachal Pradesh. In India, it is cultivated in an area of about 5700 ha.

The saffron contains safranal (for ordor), picrocrocin (for taste) and crocin (for pigment) are localized in the red stigmatic lobes of the flower. Four major bioactive compounds viz., crocin (Mono-glycosyl polytene esters), crocetin (a natural carotenoid dicarboxylic acid precursor of crocin), picrocrocin monoterpene glycoside precursor of safranal and product of xeaxanthin degradation) and safranal are present in the saffron. The uses of saffron (kesar) are multiple. It is famous for its medicinal, coloring, and flavoring properties. On account of its coloring and aromatic properties it is used as food additives in culinary preparation, bakery industries and confectionary preparation. It is used for preparation exotic dishes especially Spanish rice and French fish preparation. It is commonly used as a flavoring agent in Indian sweet dishes especially those made from milk, it is also used to flavor curries and rice based dishes such as vegetable pulao, etc. Saffron is also used as a perfume in cosmetics. It has a lot of medicinal properties and is widely used to prepare medical formulations in several countries. It is used in the making of tonic and digestive liqueurs to, which it adds its golden color, In medicine saffron is used in fevers, melancholia, and enlargement of liver and spleen. In Ayurvedic medicine it is used to heal arthritis, impotence, and infertility.

3.8.1.3   Soil and Climatic Requirements

Sandy loam and loamy soils rich in well-decomposed organic matter are ideal for saffron cultivation. Similarly well drained loamy soil or light soil neutral to slight alkaline soils is also suitable for its cultivation. The organic matter content of be soil should be 22%. It prefers very well drained, clay loam soils of *Karewas* of Kashmir. The soils should be deep and free from stones.

Water logging and poor drainage condition of soil causes corm rot disease and therefore, the saffron field requires proper drainage of excess water.

It is a crop of sub temperate to dry temperate region. It is cultivated in Greece at an altitude of 650–750 meter but in Kashmir it is grown at an altitude ranges from 1500 meter to 2400 meter. It thrives best in these cold areas, which receives good snow during winter and no rains and snow during October–November. It requires cool and sunny situation for better growth. An optimum of 12 hours light duration is essential for growth and flowering. The night temperature of 6–8°C and 15–20°C during the month of October–November provides a congenial climate for better blooming. Photoperiod exerts a considerable influence in the flowering of saffron. An optimum period of 11 hours illumination is desirable. Unusually low A good amount of rainfall during August-September facilities flowering and increased yield. Dry weather condition during flowering period is essential for realizing higher yields. In general locations, which receive 30–40 cm rainfall and are covered with snow during winter are good for its cultivation. Spring rains are favorable for promoting corm multiplication whereas, a second spell of rains at the beginning of autumn encourages profuse flowering.

3.8.1.4 Agrotechniques for Quality Production

3.8.1.4.1 Propagation

It is propagated through corms. Size of the corm influences the flower production. It has been observed that a corm diameter of 3.5 cm or more are the best and produced maximum number of shoots and flowers as compared to corms of other size.

Land preparation starts during the month of April–May. The field is ploughed four- to five times to a depth of 30–35 cm. it is planted in Kashmir in the month of August though the best time of planting is from Mid-May to Mid-June. The corm should be planted at a depth of 282–18 cm and with a spacing of 20 cm × 10 cm. Shallow planting is undesirable as the corm gets exposed to freezing cold in winter months and high temperature in summer, which markedly effect on growth and development of the crop.

3.8.1.4.2   Manures and Fertilizers Application

Apply 15–20 t/ha of well-decomposed farmyard manure or compost at the time of final land preparation. Application of vermicompost @350 500 kg/ha is beneficial for getting higher yield of saffron. A fertilizer dose of 15:10:15 kg/ha NPK under Ranikhet condition and a fertilizer dose of 20:80:30 kg/ha NPK under rainfed condition of Kashmir.

Irrigation management:

In Spain, saffron is cultivated under irrigation condition while in Kashmir it is totally grown as rainfed crop. Rains during the month of August–September is considered beneficial for flowering.

3.8.1.4.3   Intercropping and Crop Rotation

Saffron cultivation is mainly confined in *Karewas* land of Kashmir, but new cultivation is extended to the different areas as intercrop in the orchard. It has been noticed that after few years of cultivation there is a decline in yield is noticed apart from disease pest incidence.

Corm rot is a problem of saffron cultivation. To manage the disease crop is is practiced in case saffron cultivation. In Kashmir, fields are kept fallow for one year and thereafter crops like wheat or mustard is grown to the incidence of the corm rot diseases.

3.8.1.5   Plant Protection

Corm rots: This fungal disease is caused by *Fusarium moniliforme var. intermedium, Fusarium oxysporium.* The first symptom of the disease is yellowing of leaves, which soon dry. In severe cases, discoloration extends to the surface of the corm at the nodes and surface lesions are readily visible. The external symptoms appear as sunken dark spots that are hard and rough. Severely infected corms rot before they sprout when planted in soil.

For control, the corms to be planted should be put in a fungicidal solution containing Mancozeb 75WP (0.03%), Carbendazim 50WP (0.01%) for 5–10 min and dry in shade for another 10–15 minutes. Treatment with Carbendazim 50WP(0.2%) or Myclobutanil 10WP(0.2%) proved most effective in reducing the corm rot severity.

Bulbrots: It is caused by *Bacillus sp.* Symptoms appear as brown to dark brown, sunken, irregular patches below corm scales. Severely infected corms have foliage that dries from the tip downward. White fungal mycelia appeared on the bulbs that rot at later stages of disease development. Sclerotia formation may also be observed.

Bacterial rot: Causal organism of this disease is *Sclerotium rolfsii.* The characteristic symptoms of the disease on saffron plants are emergence of yellowing of leaves and death of plant. In the field, the disease is destructive and reduced flowering by about 80%. Infected corms are round, wrinkled, and produce yellowish pigment; while in some case the pigmentation is absent.

Saffron thrips: Most species of plant feeding thrips have piercing and rasping mouthparts. The surface of the leaf develops a crinkled silvery appearance as a result of damage to cells below the surface. Lightly-infested plants show silvery feeding scars on the under surface of leaves, especially alongside the mid rib and veins. Heavily-infested plants show silvering and browning of leaves, stunting of young leaves and terminal growth, with fruit scarred and deformed.

Mites: Damage of mites on the saffron leaves is yellow to white spots. Severely infected leaves turn yellow. Chemical control is not usually effective. Natural enemies of mite are predators like predatory mites, *Orius* spp. (pirate bug), hover fly, mirid bug, etc.

3.8.1.6 Harvest and Post Harvest Technology

The flowering period is confined to three weeks from middle of October to first week of November. The flowers are picked daily in the morning before sun gets too hot. The duration of picking depends upon the time of blooming. The flowers are trimmed during the day time and style and stigma are separated from the perianth. This operation requires skilled laborer. Separation of stigma from perianth has to be carried out everyday and stigmas and styles are trimmed immediately. About 1,60,000 flowers are hand picked to produce one kg of good quality dried saffron. From one hectare of land about 160 kg of flower is obtained and gives a yield of 5 kg of dried saffron.

Usually drying of saffron is done in in sun but color fading (discoloration) is a common features of sun drying. Properly dried saffron contains 8–10% moisture. In Kashmir, stigmas are picked from the flowers and dried, which constitute the first grade saffron. Shahi saffron and other grades are Lachha and Mongra grade. In Spain, the drying procedure of saffron is known as toasting. In toasting method, stigmas are spread out in sieves in layer of 2–3 cm thickness and then sievies are placed in fire for drying of saffron.

Because of high cost of saffron, it is frequently adultered with style and other parts of the crop. It is also adultered by mixing with the floral parts of compositae like *Calendula sp.* and *Carthamus tinctorious.*

3.8.1.6.1   Value Addition

Saffron extract: It is a concentrated liquid containing saffron flavor or essence. Saffron extract is a natural supplement, made from the saffron crocus. It also has a longer shelf life than dried saffron threads. The best reason to use saffron extract is that it is quick and easy to use.

Saffron essence: It is an essence that aids in skin repair with saffron extract, reduce skin inflammation while it offer high moisturizing and antioxidant properties.

3.8.2   VANILLA (VANILLA PLANIFOLIA ANDREWS)

3.8.2.1   Systematic Position

Kingdom: Plantae
Division: Tracheophyta
Class: Magnoliopsida
Order: Asparagales
Family: Orchidaceae
Genus: *Vanilla*
Species: *planifolia*

3.8.2.2   About the Crop, National and International Scenario, Uses, and Composition

The vanilla pods of commerce are the cured fruits or beans of of climbing orchid *Vanilla planifolia*. It is a herbaceous perennial vine, climbing up trees or other supports. It is originated from Mexico and introduce in India in 1835. The cultivation of vanilla started in Mauritius in 1827 and in Madagascar in the year 1848. Vanilla flavor was first introduced to France and England during the early part of the seventeenth century. The important vanilla growing countries are Madagascar, Indonesia, Java, Mauritius, Tahiti, Seychelles, Brazil, Jamaica, Guatemala, West Indies, Tahiti, and India. At present, Madagascar, and Indonesia are the major vanilla producing countries in the World. Vanilla is the second most expensive spice next to saffron traded in the world market. The climatic condition prevailing in the South India particularly in Kerala is highly favorable for vanilla cultivation. The spices board observes considerable export potential in vanilla and has taken the initiative to popularize vanilla cultivation in India. It has a programme for subsidized supply of planting material to small growers.

Vanillin is mainly responsible for the fragrance, flavor, and aroma of vanilla essence. Vanilla constitutes the World's most popular flavoring agent for numerous sweeten foods. It is used in the preparation of ice creams, chocolates, cakes, pastries, puddings, soft drinks, pharmaceuticals, liquors, perfumery, and in nutraceuticals industries. At present, synthetic products such as ethyl vanillin and synthetic vanillin are used for the above-mentioned purpose. But such synthetic products are being rejected by discerning customers worldwide, and there is an increasing demand for natural vanillin.

About 85% of the volatiles of vanilla are vanillin ($C_8H_8O_3$) and around 130 different chemical compounds that contribute to the greater part of the vanilla flavor have been detected in the fermented fruit like as phenols, phenol ether, alcohols, carbonyl compounds, acids, ester, lactones, aliphatic, and aromatic carbon hydrates and heterocyclic compounds. Important aroma components are p-hydroxybenzaldehyde (up to 9%), p-hydroxybenzyl methyl ether (1%), phenols, alcohols, lactones, etc. Vanilla additionally contains 25% sugar, 15% fat, 15 to 30% cellulose and 6% minerals. Water content is unusually high (35%). The quite different fragrance of Tahiti vanilla is due to additional presence of heliotropin, (3, 4-dioxymethyl benzaldehyde) and diacetyl butandione. A minimum vanillin content of 1.18–2% and a moisture content ranging between 20 to 22% are accepted by the users.

3.8.2.3   Soil and Climatic Requirements

It can be successfully cultivated in a wide range of soil types with rich in humus and having good drainage capacity. It prefers land with gentle slope, light porous soil having good drainage facility. Forest soil rich in humus is ideal for its cultivation. Optimum soil pH range for vanilla is 6.0–6.5. It can also be grown as intercrop in coconut and arecanut garden.

It prefers warm and humid climate conditions with well distributed annual rainfall of 150–300 cm and a temperature with range of 25–32° C. It grows well at an elevation of 1000–1500 m altitude from the mean sea level. The crop requires a dry spell of three months for uniform flowering. However, very high temperature, strong wind and dry weather are not suited, which affects growth of vanilla.

3.8.2.4   Agrotechniques for Quality Production

Vanilla is generally propagated by stem cutting. Vines of 60–120 cm long are selected as planting material. To minimize the fungal infection stem cutting are treated with 1% Bordeaux mixture of 0.3% copper oxychloride. Plant raised from the long cutting commences flowering early and superior in terms of yield as compared to plants raised from shorter cutting. Vanilla is a climbing vine, needs support for its growth. It also flourishes well in partial shade. Trees like *Plumeria alba, Casuarina equisetifolia, Erythrina spp.* and *Glyricidia spp.* can be used as shade trees. The shade trees/standards are generally planted at spacing of 2.5 m to 3.0 between rows and 2.0 m within a row making a population of 1600–2000. Seedling of standards may be established 6 months before planting of vanilla.

Vanilla cutting are planted during the onset of Southwest Monsoon. The cuttings are planted with two nodes bellow the soil and generally two cuttings are planted against the each standard. Basal portion of the cutting should be kept just above the soil to prevent rotting. Provision of shade is necessary to protect the newly planted cuttings. Regular mulching during dry months and combined with providing irrigation once in 4–7 days during summer increases the growth and yield. Application of 40–60 g N, 20–30 g P_2O_5 and 60–100 g K_2O per plant per hectare is recommended for vanilla. The above quantity of inorganic fertilizer should be applied in 2–3 splits in between June to September. Decomposed mulch materials are the main source of organic matter in vanilla cultivation. The height of the vine should restrict with 2.0 m to facilitate hand pollination and easy harvesting of the pods. Hence, vines are allowed to grow upto 1.50 m and then trained horizontally on the branch of the supporting trees and later coiled round them. The top 7.5–10 cm is pinched off 6–8 months before the flowering season to encourage the production of inflorescence in the vines.

Vanilla commences flowering in the second or third year depending upon the length of cutting used. The usual flowering season is from December to March and sometimes in October–November. Unlike most of the orchids, its flower last for a day only from early morning to late evening. Artificial pollination is necessary for fruit set in vanilla. Usually about 10–12 inflorescences in a vine are pollinated by manually. Plant growth is affected due to unfavorable conditions like poor soil nutrient availability, excessive shade or the lack of it, damage to roots and stems, over crowding of vines, excessive use of manures and inorganic nutrients, excessive moisture, water stagnation and poor drainage that often lead to pest and disease infestation in field.

3.8.2.5 Plant Protection

Bean rot: Two kinds of rot caused by two different species of fungi are recorded. *Phytopthora* induced rot develops at the tips of beans slowly extends towards the pedicel and affected beans show water soaked lesion, which become dark green leading to rotting of the beans. The rotting extends to whole bunch of beans exhibiting abundant external growth of fungal mycelium. In later stages of infection the rotting advances to the stem, leaves, aerial roots and extend to the entire vine. *Sclerotium* induced rot is characterized by rotting of bean tips. Affected portion shows white thick mats of fungal mycelium forming a mantle around the bunch of beans and leaves. Excess shade, continuous heavy rains, overcrowding of vines, water logged conditions and presence of pathogen inoculum in the field are the predisposing factors for bean rot.

Bean rot of vanilla can be effectively managed by removal and destruction of infected plant parts and mulching during rainy season. Regulation of shade during monsoon period in order to prevent excess shade is effective. Spraying of Bordeaux mixture (1.0%) and drenching of soil with 0.25% Copper oxy-chloride 2–3 times depending on the severity of infection effectively controls the rot. If rotting is due to scloretium, Carbendazim-Mancozeb mixture @ 0.25% can be sprayed twice at 15 day interval.

Stem rot: Causal organism is *Fusarium oxysporum f.sp. vanilla*. The disease usually appears during the post monsoon period of November–February. The disease appears as yellowing and shriveling of the inter- nodal area extending to both sides of the stem. When the basal or middle portion of the vine decay and shrivel, the remaining distal portions of the vines show wilting symptoms.

Root rot/ wilt: It is a fungal disease and caused by *Fusarium batatis Wollen var. vanilla*. Initially the disease appears in the form of browning and death of underground and aerial shoots. Aerial shoots die before entering the soil resulting in flaccidity and shriveling of the stem and finally the vine drops.

It can be effectively controlled by removing and destroying the infected plant parts (Phytosanitation). Foliar spray with carbendazim 0.2%, soil drenching with carbendazim (0.2%), Copper oxy-chloride (0.25%) or a mixture of Carbendazim-Mancozeb (0.25%). Application of biocontrol agents such as *Tricoderma harzianum* and *Pseudomonas fluorescence* having a cfu of 10 g@50 g/vine is also effective.

Mosaic disease: Various kinds of mosaic such as mild mottle, mild mosaic and mild chlorotic streak are observed. In a few cases, such mosaics are also associated with leaf distortion with wavy margin. The size of the leaves also gets reduced and in advance stages, leaves become brittle and show severe crinkling.

Leaf feeding Caterpillars and Beetles: A few species of leaf feeding caterpillars and beetles feed on leaves and tender stems. They can be controlled by spraying quanilphos 0.05%.

Sucking bug: Adults and nymphs of the sucking bug *Halyomorpha* sp. infest tender shoot tips and emerging inflorescences resulting in their drying and rotting. Spraying quinalphos (0.05%) each on tender shot tips and emerging inflorescences is effective for management of the pest.

Snails: Snails and slugs feed and damage tender shoot tips and leaves especially in moist and shaded areas in the plantations. Handpicking and poison baiting helps in preventing the pest.

3.8.2.6   Harvest and Post Harvest Technology

After about 6–8 months after flowering, the beans are ready for harvest. Immature vanilla bean is dark green in color. The maturity of the beans can be judged by color changing from green to pale yellow. Ripe yellowish coloration appears from its distal, which the optimum time is for harvesting is. Delay in harvesting after results splitting of beans. Daily harvesting is essential operation in vanilla cultivation. The bean can be harvested by cutting with knife. The harvested beans at this stage do not have any aroma as vanillin is not present in them. A good vanilla crops gives a yield of 300–60 kg of cured beans/ha. Vanilla is developed as a result of enzyme action on the glucosides contained in the beans during the process of curing.

3.8.2.6.1   Different Stages of Curing

1. Killing the vegetative life of the beans to allow the onset of enzymatic reaction.
2. Raising temperature to promote this action and to achieve rapid drying to prevent harmful fermentation.
3. Slower drying for the development of different fragment substances.
4. Conditioning the produce by storing for a few months.

Curing and Drying: The commercial value of vanilla depends mainly on the care imparted on curing. Eventually they become dry, brittle, and finally become scentless. Therefore, different artificial methods are employed to cure vanilla.

i. Peruvian process (wet process): Curing is done by hot water. Pods are dipped in boiling water for killing of beans. The ends of harvested beans are tied and hanged in the open for drying for upto 20 days for slow drying of vanilla beans. Then it is coated with castor oil followed by tied up in bundles.

ii. Guiana process (dry process): Pods are collected and dried in the sun till they shrivel. They are subjected to wiping and rubbing with olive oil. The ends are tied up to prevent splitting and finally bundled

iii. Mexican process: Harvested pods are kept under shade till they shrivel. Then they are subjected to sweating (for two days) in warm weather. After then a blanket is spread over those and exposed to the sun, during midday, blanket is uncovered and bundled and left in the open for rest of the day. They should be wrapped in blankets in the night to maintain continuous fermentation and sweating. This process is repeated for 7–12 days till they become dark brown in color, soft, and flexible. After then they are packed in tins and sealed containers.

Vanillin is the principal compound responsible for the flavor and aroma of vanilla beans. The Yield of vanillin is about 4.15–4.40%. Divanillin is the compound formed from vanillin during the enzymatic curing process that adds a creamy, fatty mouth feel to the flavor of vanilla beans.

3.8.2.6.2   Value Addition

- Vanilla whole beans (organic/non-organic)
- Vanilla extract (Indian/Madagascar)
- Organic vanilla top note
- Vanilla oleoresin
- Vanilla crystal
- Vanilla emulsion
- Vanilla bean specks
- Mint vanilla flavor
- Cardamom vanilla flavor
- Cinnamon vanilla flavor
- Star anise vanilla flavor
- Coffee vanilla flavor
- Almond vanilla flavor
- Creamy vanilla flavor

Pure vanilla extract: It is made from whole vanilla beans extracted using 35% or more alcohol. It is dark brown in color. Imitation and clear vanilla utilizes artificial flavors and harmful chemicals.

Vanilla essence: It is an artificial essence prepared by using chemicals to recreate the flavor of vanilla. It is extracted from vanilla beans and is used to flavor several desserts and dishes. It is the extract that is made from vanilla beans, which are soaked in alcohol. It is widely used as a flavoring and vanilla ice cream is the most common flavor. Natural vanilla essence contains 2% to 3% of alcohol. Coumarin is a common adulterant of vanilla essence, which can prove to be harmful for the liver if in taken in a large amount.

3.8.3 LARGE CARDAMOM (*ELETTARIA CARDAMOMUM* L.)

3.8.3.1 Systematic Position

Kingdom: Plantae
Division: Tracheophyta
Class: Magnoliopsida
Order: Zingiberales
Family: Zingiberaceae
Genus: *Elettaria*
Species: *cardamomum*

3.8.3.2 About the Crop, National and International Scenario, Uses, and Composition

There has been controversy over the grouping of cardamom. But after complete discussions, the ISO (International Standards Organization) has officially documented nine species of cardamom under three main groups (Pruthi, 1977):

1. Group I: *Elettaria cardamomum*
2. Group II: 4 species of *Aframomum*
   a. *A. augustifolium* (Sonn) K. Schum: Madagascar cardamom
   b. *A. hanburyi* K. Schum: Cameroon cardamom
   c. *A. korarima* (pereira) Engler: Korarima cardamom
   d. *A. melegueta* (Roscol) K. Schum: Grains of paradise or Guinea grains.
3. Group III: 4 species of *Amomum*
   a. *A. aromaticum* Roxburgh: Bengal cardamom
   b. *A. kepulaga* Spraque et: Round cardamom Burkill, Syn. A. cardamom Roxburgh or Chester cardamom or Siam cardamom.
   c. *A. krervanh* pierre et Gagnipain: Cambodian cardamom

d. *A. subulatum* Roxburgh: Greater Indian cardamom, Nepal cardamom or large cardamom

The *Amomum* species are known in the Northeast Indian and South East
Asian countries, while the *Aframomum* species are known in the African
regions of Sierra Leone, Guinea Coast, Madagascar, and Tanzania. The fruits
of the *Amomum* and *Aframomum* are much larger in size in comparison with
Elettaria cardamomum and it is easy to distinguish them, but the seed size
and anatomy are similar in all the three genera.

Amomum subulatum Roxburgh is a perennial herb having subterranean
rhizomes, which give rise to leafy shoots and spikes. Plant height ranges
from 1.5 to 3.0 m. Leafy shoots are formed by long sheath-like stalks encir-
cling one another. The leaves are green or dark green, glabrous on both surfaces with acuminate apex. Inflorescence is a dense spike on a short peduncle
bearing 40 to 50 flower buds in an acropetal sequence. The fruit are on an
average 25 mm long, trilocular, oval to globose; greyish brown to dark red
brown capsules, which contains 40–50 seeds, held together by a viscous
sugary pulp.

It has a significant position in the global trade and it is being cultivated
in a larger extent, hence considers as large cardamom. This species is cultivated in marshy places along the sides of mountain streams in Nepal, West
Bengal, Sikkim, and Nagaland (eastern Himalayas) and forms one of the
cash crops of eastern India.

The seeds of large cardamom contain volatile oil, which is the principal
constituents responsible for providing the typical odor. The highest volatile
oil content was recorded as 3.32% in variety Golsey Dwarf, whereas the
lowest was 1.95% in variety White Ramna (Gupta 1986). The major constituent of essential oil is 1,8-cineole (65–80%). The monoterpene hydrocarbon
content is in the range of 5–17% of which lamonene, sabeinene, terpinenes,
and pinenes are significant components. The high cineole and low terpenyl
acetate probably account for the very harsh aroma of this spice in comparison with that of true cardamom (Pruthi 1993).

3.8.3.3 Soil and Climatic Requirements

Deep and well-drained loamy soils with plenty of leaf mould are the best for
sustainable large cardamom production. Large cardamom growing soils are
generally rich in organic matter and nitrogen, medium in available phosphorus

content and medium to high in potassium content. Soils are acidic in nature with a soil pH ranges in between 4.5 to 6.0. But the ideal soil pH ranges in between 5.8 to 6.5.

The crop grows under the of the forest trees in Sub-Himalayan mountain at an attitude in between 600 to 2000 m. It thrives well at an altitude range of 765- 1675 m above MSL. It prefers maximum temperature of 14–33°C and minimum temperature of 4–22°C but thrives even 6 to 30°C with a well distributed annual rainfall of 200–250 cm throughout the year.

3.8.3.4 Agrotechniques for Quality Production

As like small cardamom, large cardamom is also propagated through seed, division of rhizome, sucker, tissue culture plantlets, etc. Seedling and tissue culture plantlets are free from viral diseases like Chirkey and Foorkey.

During the onset of monsoon pits with a size of 30 cm × 30 cm × 30 cm are prepared. After 15 days pits are filled top soil along with organic manures like farmyard manures or compost or leaf mould. Thereafter, seedling or division of rhizome or suckers or tissue culture plant let are planted in the main field, in case of robust varieties like Ramsey, Golsey, and Swaney are planted at a spacing of 1.5 m × 1.5 m and as the case non robust varieties like Dzongu Golsey are planted at a spacing of 1.25 m × 1.25 m. June is the ideal time of planting for large cardamom. Generally 18–20 days are required for establishment in the main field.

There are three popular varieties of large cardamom in Sikkim, viz., Ramsey, Golsey, and Sawney. The varietal characters of large cardamom were described by Rao et al. (1993) as follows:

Ramsey: This cultivar is suitable for cultivation in high altitude. More than 50% area of large cardamom is covered by this particular cultivar. Plant is tall in nature with vigorous clump growth. Stem color is maroonish with dense foliage. Flowers are yellowish and small, corolla tip with pink tinge at base. Capsule is smaller in size (with 16- 30 seeds per capsule) and low in essential oil content (1.0–1.8%). This cultivar is susceptible to Chirkey and Foorkey at lower altitudes.

Golsey: This cultivar is suitable for cultivation in low to medium altitude. Near about 30% area of large cardamom is covered by this particular cultivar. Plant is less vigorous with erect leafy stem bearing stout upright leaves. Stem is greenish to maroonish in color. Flowers are yellowish orange in color. Capsule is bold in size (with 40 50 seeds per capsule) and roundish nature with high in essential oil content (2.3–5.0%). It is tolerant to Chirkey and Foorkey but susceptible to leaf spots.

Swaney: This cultivar is suitable for cultivation in medium altitude. Plant is tall and vigorous in nature. Stem is pinkish in color with dark green foliage. Flowers are Yellowish in color with pink tinge at base of corolla. Capsule is medium bold (with 30–40 seeds per capsule) and roundish nature with medium high in essential oil content (1.8–2.5%). It is susceptible to viral diseases.

In addition to these, there are several other varieties such as Ramla, Chivey Ramsey, Garday Seto Ramsey, Ramnag, Madhusay, Seto Golsey, Slant Golsey, Red Sawney, Green Sawney and Mingney (Gupta, 1986). Rao et al. (1993) reported a promising variety Barlanga from higher altitudes.

It is seen to grow under a shade of 60–70% of full day light interception to light shade, i.e., 26% full day light interception. For optimum growth of large cardamom, the light interception should be 40–50%. In open areas planting of shade tree is equally important. The most common shade trees in high altitude is Utis (*Alnus nepalensis*). It has been observed that yield large cardamom increased by 2.2 times. Shade trees can be planted at spacing of 5 m × 5 m.

The crop is grown under the canopy of forest trees and generally application of organic manure is no common in case of large cardamom cultivation. However, ensuring good yield, application of farmyard manure @ 30 t/ha along with inorganic fertilizer is beneficial. A fertilizer dose of 20:30:40 kg/ha of $N:P_2O_5:K_2O$ is beneficial. At the time of final land preparation, apply 1/3rd dose of nitrogen and full of P_2O_5 and K_2O in the month of April and remaining dose of nitrogen should be divided into two equal split once in the month and June and remaining in the month of September.

It is also noticed that yield performance is better in plantation where perennial; water source is available. Providing irrigation through surface channel, hose sprinkler once in 15 days from the end of October to February is essential for better growth and production.

3.8.3.5 Plant Protection

Seedling rot: Causal organism of this fungal disease is *Fusarium oxysporum.* Leaves turn pale and their tips become yellow. Gradually these symptoms spread over the entire leaf extending to leaf sheath resulting in wilting of seedlings. The collar portion decays and the entire seedlings die. In grown up seedlings rotting extends from the collar region to the rhizomes resulting in their decay and ultimate death of the plant.

Chirke disease: The disease is characterized by mosaic with pale streak on leaves. The symptom is more prominent on young emerged leaves where discrete pale green to yellow longitudinal strips running parallel to each other can be seen. The disease is readily transmitted by mechanical sap inoculation and in field it is spread by aphids, *Rhopalosiphum maidis* Fitch within a short acquisition feeding period of 5 minutes. Primary spread of diseases from one area to another area is through infected rhizomes and further spread within the field is by aphids.

Foorkey disease: The affected plants produce profuse stunted shoots, which fail to produce flowers. The leaves become small, lightly curled and pale green in color. The inflorescences become stunted, thereby producing no flowers and fruits. The diseased plants remain unproductive and gradually degenerate. Foorkey symptom appears both on seedlings and grown up plants.

Unlike Chirke, Foorkey virus is not transmitted through sap but by the aphids *Pentalonia nigronervosa* Cog and *Micromyzus kalimpongensis* Basu. The primary spread of disease from one area to another is through infected rhizomes and further spread within the plantation by aphids. The diseased plants are uprooted and destroyed as and when they are traced. Uprooted plants are taken to an isolated place, chopped into small pieces buried in deep pits for their quick decomposition.

Leaf streak disease: It is caused by causal organism: *Pestalotiopsis royenae (D. Sacc) Steyaert*. It is a fungal disease and is a serious disease among foliar diseases and is prevalent round the year. The disease symptom is the formation of numerous translucent streaks on young leaves along the veins. The infection starts from emerging folded leaves; infected leaves eventually dry up causing loss of green part and reduce the yielding capacity of the plant. Three rounds of 0.2% spray of copper oxy-chloride at 15 days interval, two schedules in a year, i.e., February–March and September–October can control this disease.

Leaf thrips: This pest feeds primarily on the foliage of ornamental plants. It attacks the lower surface first and, as feeding progresses and the population increases, the thrips move to the upper surface. The leaves become discolored and develop distortion between the lateral vines. Severely damaged leaves turn yellow. Natural enemies of leaf thrips are Parasitoid (*Megaphragma mumaripenne*) and Predators: Predatory thrips (*Franklinothrips orizabensis, F. vespiformis, Leptothrips mali*)

Aphid: Symptoms: Panicles become stunted. Shedding of flowers and immature capsules thus reducing the total number of capsules formed. Natural enemies are Parasitoids: *Aphidius colemani, Aphelinus* spp., and Predators: Lacewing, ladybird beetle, spider.

White grub: Affected plants show yellowing of the foliage, scorching of leaves, defoliation, and dieback. Inspection of the root system will reveal that the roots have been chewed off leaving calloused stumps. White grub also feeds on the at soil level, causing of the stem followed by death of the plants.

Rhizome weevil: Grubs tunnel and feed on the rhizome causing death of entire clumps of cardamom.

3.8.3.6   Harvest and Post Harvest Technology

A cardamom plant has average life of 20 years. It starts bearing fruits from second year of planting and gives steady yield throughout the year upto 6–10 years. The flowering starts in May and continues up to August. It takes about four months for the fruits to mature. Peak period of flowering and fruiting is April to July. Harvesting starts from September and continues up to the month of January with peak period during late October to mid-December. Harvesting is done by collecting panicles containing ripe fruits with the help of a special chisel-shaped narrow knife, which is specially made for this purpose. Harvesting is done once a year; hence there will be some immature fruits in the harvested lot, which give wrinkled appearance on curing.

After harvesting, the individual capsules are separated from spikes by hand for drying and curing. They are dried on a mud-plastered threshing floor for seven to ten days and sold in markets. This contains about 50% moisture thus, dried again by traders to avoid fungal contamination. Mainly three types of curing systems are available:

a. *Traditional bhatti system*: In this system, about 200–250 kg capsules are heaped per m² on a 25–70 cm thick bed and heated directly over a fire by firewood. The temperature during drying is 100°C and the drying operation may continue from two to three days. The capsules dried in this system are dark and have a smoky flavor because of direct exposure to heat and smoke. Thus, the original color of the capsules is lost and cannot be stored for a long time (Roy, 1988; Rao et al., 1993a). The volatile losses are as high as 35%.

b. *Flue pipe curing houses*: In this method flue pipes are laid inside a room (curing house) and connected to a furnace installed outside. Fresh cardamom is spread over wire meshes fixed above the flue pipes. This is an indirect system of drying and smoke does not come into contact with the produce at any stage. This type of drier resulted in early drying and gave better quality capsules, including a better color.

c. *CFTRI system*: The Central Food Technological Research Institute (CFTRI), Mysore, has designed and developed a low cost natural convection dryer. In this system the flue ducts are arranged in double-deck fashion and connected in series to the furnace. The convection current passes upward through the bed of capsules. Thermal efficiency is much better, the cost of drying cheaper, the quality of the product superior and the annual product output higher, than in the case of a curing house or any other existing system.

3.8.3.6.1 Value Addition

The Indian grading system for cardamom capsules separates them into different types:

- Alleppey green cardamom
- Coorg green cardamom
- Bleached or half-bleached cardamom
- Bleached white cardamom
- Mixed cardamom

Empty and malformed capsules: Capsules, which have no seeds or are scanty filled with seeds. To measure this, 100 capsules are selected at random from the sample, opened, and the number of empty and malformed capsules is counted.

Immature and shriveled capsules: Capsules, which are immature and shriveled capsules are not fully developed.

Black and splits: The former includes capsules that have a visible blackish color and the latter include those, which are open at the corners for more than half the length.

KEYWORDS

- agrotechniques
- climate
- plant protection
- post harvest value addition
- spice crops
- uses

REFERENCES

Achut, S. G., & Bandyopadhyaya, C., (1984). Characterization of Mango-Like Aroma in Curcuma Amada Roxb. *Journal of Agriculture and Food Chemistry, 32*, 57–59.

Ahmed, J., Pawan, P., Shivhare, U. S., & Kumar, S., (2003). Effect of processing temperature and storage on color of garlic paste. *J Food Sci Technol, 39*(3), 266–267.

Bailer, J., Aichinger, T., Hackl, G., Hueber, K. D., & Dachler, M., (2001). Essential oil content and composition in commercially available dill cultivars in comparison to caraway. *Indus Crops Prods., 14*, 229–239.

Banafar, R. N. S., Gupta, N. K., & Pathak, A. C., (2002). Suitability of black cumin varieties (*Nigella sativa* L.) for Madhya Pradesh. *Advances in Plant Sciences, 15*(1), 165–166.

Bharadwaj, S., Mahorkar, V. K., Panchbhai, D. M., & Jogdande, N. D., (2007). Preparation of Tamarind Kernel Powder for Value Addition. *Agricultural Science Digest, 27*(3), 194–197.

Chatterjee, R., & Chattopadhyay, P. K., (2009). *Curcuma amada*: an underutilized spice crop. *Indian Agriculturist, 53*(3–4), 107–109.

Chempekam, B., & Peter, K. V., (2000). Processing and Chemistry of Tamarind (*Tamarindus indicus* Linn), *Spice India 13*(4), 9–12.

Chung, B., (1987). The effect of irrigation on growth and yield components of poppies (*Papaver somniferum* L.). *Journal of Agricultural Science, UK, 108*, 389–394.

Crassina, K., & Sudha, M. L., (2015). Evaluation of Rheological, Bioactivities, and Baking Characteristics of Mango Ginger (*Curcuma amada*) Enriched Soup Sticks. *J Food Sci Technol., 52*(9), 5922–5929.

Datta, S., Chatterjee, R., & Ghosh, S. K., (2003). Evaluation of some black cumin accession for yield and quality. *The Orissa J. Hort., 31*(1), 34–36.

Datta, S., Dey, A. N., & Maitra, S., (2009). Effect of FYM and GA3 on growth and yield of sweet flag (*Acorus calamus* L.) under terai zone of West Bengal. *J. Hortl. Sci., 4*(1), 59–62.

Datta, S., Poduval, M., Basak, J., & Chattergee, R., (2001). Fertilizer Trial on Cumin Black (*Nigella stiva* Linn.). *Env., & Eco., 19*(4), 920–22.

Dey, P., & Khaled, K. L., (2013). An Extensive Review on *Allium ampeloprasum*: a magical herb. *International Journal of Science and Research, 4*(7), 371–377.

Farooqui, A. A., Sreerammu, B. S., & Srivasapa, K. N., (2000). Cultivation Practices of Sweet Flag. *Spice India, 13*, 18–21.

George, C. K., (1996). Ginger: quality improvement at farm level. In.: *Quality Improvement of Ginger* (Eds., Sivadasan, C. R., & Kurupu, P. M.), Spices Board (Govt. of India) Kochi, 7–15.

Gupta, P. N., (1986). Studies on Capsule Morphology of Large Cardamom Cultivars (*Amomumsubulatum Roxb*), *J. Plantation Crops, 16*, 371–5.

Hort Net Kerala, (2016). Pepper (*Piper nigrum*). http://hortnet.kerala.nic.in/pepper.htm.

Jain, S. K., (2001). Calamus. In: *Medicinal Plants*. National Book Trust Publishers, India, pp. 11–12.

Joshi, D., (1962). White pepper. *Indian Patent, 70*, 349.

Khare, C. P., (2004). Rational Western Therapy, ayurvedic, and other traditional usages, botany. In: *Indian Herbal Remedies*, Springer, Berlin, New York, pp. 60–61.

Kharwara, P. C., Awasthi, O. P., & Sing, C. M., (1988). Effect of sowing dates, nitrogen, and phosphorus levels on yield and quality of opium poppy. *Indian J. Agron, 33*, 159–163.

Kindscher, K., (1992). *Medicinal Wild Plants of the Prairie*. University Press of Kansas, Lawrence, Kansas, pp. xi+340.

Korikanthimath, V. S., (1993). Cardamom: An ecofriendly plantation spices crop. In: *Proc. Workshop Farm Forestry Management*, Indian Institute of Forest Management, Bhopal, pp. 36–38.

Kraxner, U., Weichmann, J., & Fritz, D., (1986). How should Horseradish be Manured? *Gemuse, 22*(9), 363–364.

Lakshmanachar, M. S., (1993). Spices oil and oleoresin. *Spice India, 6*(8), 8–10.

Luchon, S., & Sarat, S., (2003). Black Cumin (*Nigella sativa* L.): a new aromatic spicy medicinal plants. Abstract of paper presented in *Nat. Sem. On "New perspectives in spices, Medicinal, and Aromatic Plants,"* At ICAR Research Complex, Goa, p. 153.

Mandal, A. R., & Maity, R. G., (1993). Effect of some chemicals and microclimate factors on germination of black cumin (*Nigella sativa* L.) seed. *Horticultural Journal, 6*(2), 115–120.

Mani, B., Paikada, J., & Varma, P., (2000). Different Drying Methods of Ginger. *Spice India, 13*(6), 13–15.

Mathew, A. G., (1994). Blackening of pepper. *International Pepper News Bull., 18*(1), 9–12.

Matthew, M. D., Gance-Cleveland, B., Hassink, S., Johnson, R., Paradis, G., & Resnicow, K., (2007). Recommendations for prevention of childhood obesity. *Pediatrics, 120*(4), S229–S253.

Narayanan, C. S., Sreekumar, M. M., & Sankarikutty, B., (2000). Industrial processing. In: *Black Pepper (Piper nigrum)*, Ravindran, P. N. (ed.). Harwood Academic Publishers, Amsterdam, pp. 367–379.

Natarajan, C. P., Padma Bai, R., Krishnamurthy, M. N., Raghavan, B., Shankaracharya, N. B., Kuppuswamy, S., Govindarajan, V. S., & Lewis, Y. S., (1972). Chemical composition of ginger varieties and dehydration studies on ginger, *J. Food Sci., & Technol.*, *9*, 120–124

Nautiyal, O. P., & Tiwari, K. K., (2011). Extraction of dill seed oil (*Anethum sowa*) using supercritical carbon dioxide and comparison with hydrodistillation. *Ind. Eng. Chem. Res.*, *50*(9), 5723–5726.

Nebel, N., Weichmann, J., & Fritz, D., (1986). Yield of different Horseradish Ecotypes. *Gemuse*, *24*(6), 272–273.

Nybe, E. V., Sivaraman Nair, P. C. S., & Mjhankumaran, N., (1980). Assessment of yield and quality components in ginger. *National Seminar on Genger and Turmeric*, Calicut, pp. 24–29.

Pruthi, J. S., (1976). Cumin, garlic, ginger, mustard, red chili and turmeric. In: *Spices and Condiments*. National Book Trust, India, New Delhi.

Pruthi, J. S., (1977). Cardamom, greater cardamom and lesser cardamom. In: *Spices and Condiments*. National Book Trust of India, New Delhi, pp. 53–63.

Pruthi, J. S., (1980). *Spices and Condiments: Chemistry, Microbiology, and Technology.* Academic Press Inc.: New York, p. 450.

Pruthi, J. S., (1993). *Major Spices of India–Crop Management and Post Harvest Technology.* ICAR Publications, New Delhi, pp. 114–179.

Pruthi, J. S., (1997). Diversification in pepper utilization. *International Pepper News Bull.*, *15*(8), 5–9.

Quer, F., (1981). *Plantas Medicinales, El Dioscorides Renovado.* Barcelona: Editorial Labor, SA, p. 500.

Rao, S.A., Bandaru, R., & Ramachandran, R., (1989). Volatile Aroma Components of *Curcuma amada* Roxb. *Journal of Agriculture and Food Chemistry*, *37*, 740–743

Rao, Y. S., Gupta, U., Kumar, A., & Naidu, R., (1993). A Note on Large Cardamom (*Amomum subulatum Roxb*) Germplasm Collection, *Jour. of Spices and Aromatic Crops*, *2*(1–2), 77–80.

Ravindra, V., Sreenivas, K. N., & Shankarappa, T. H., (2012). Value Addition and Technology Development: Shatavari, aloe, and mango ginger blended nectar beverage. *Environment and Ecology*, *30*(3), 495–500.

Ravindran, P., & Balachandran, I. (2005). Under utilized medicinal spices II, *Spice India*, *17*, 32–6.

Sadanandan, A. K., Reddy, B. N., & Hamza, S., (1991). Effect of long term NPK fertilization in a laterite soil on availability and utilization of major and micronutrients for black pepper (Abstract). *National Seminar on Recent Advances in Soil Research, Dapoli.*, 79.

Sagani, V. P., Patel, N. C., & Golakia, B. A., (2005). Studies on Extraction of Essential oil of Cumin. *J. Food Sci. Technol*, *42*(1), 92–95

Sankar, V., Mahajan, V., & Lawande, K. E., (2005). Prospect of Garlic Processing Industry. *Spice India*, *18*(10), 23–26.

Selvi, B. S., Selvaraj, N., & Raghu, R., (2003). Sweetflag, *Spice India*, *14*(8), 4.

Simon, J. E., Chadwick, A. F., & Craker, L. E., (1984). *Herbs: An Indexed Bibliography. 1971–1980.* The scientific literature on selected herbs, and aromatic and medicinal plants of the temperate zone. Archon Books, Hamden, C.T.P., 770.

Singh, S. K., & Singh, S., (1999). Response of nigella (*Nigella sativa* L.) to nitrogen and phosphorus. *Crop Research*, *18*(3), 478–479

Singh, S., Kumar, J. K., Saikia, D., Shanker, K., Thakur, J. P., Negi, A. S., & Bannerjee, S., (2010). A bioactive labdane diterpenoid from *Curcuma amada* and its semisynthetic analogues as antitubercular agents. *European Journal of Medicinal Chemistry, 45*(9), 4379–4382.

Sreekumar, V., Indrasenan, G., & Mammen, M. K., (1980). Studies on Quantitative and Qualitative Attributes of Ginger Cultivars, *In: Proc. Nat. Sem. Ginger and Turmeric,* Calicut, pp. 47–49.

Srivastav, R. P., & Kumar, S., (2002). *Fruits and Vegetable Preservation* (Third and revised and enlarged). International Book Distributing Co., India, 169–172, 228–233, 267–270, 273, 290, 305–306.

Sudarshan, M. R., (2000). White pepper, A simple value addition. *Spice India, 13*(6), 5–7.

Syamkumar, S., & Sasikumar, B., (2007). Molecular marker based genetic diversity analysis of Curcuma species from India. *Scientia Horticulturae, 112,* 235–241.

Thankamony, V., Menon, N. A., Amna, O. A., Sreedaharan, V. P., & Narayanan, C. S., (1999). Bacterial removal of skin from white pepper. *Spice India, 12*(9), 10–11.

Tomar, S. S., Nigam, K. B.; Pachori, R. S., & Kahar, L. S., (1990). Critical irrigation stages in unlaced poppy for seed yield. *Current Research,* Bangalore, *19*(5), 77–78.

Varghese, (1991). Add a touch of green. *Pepper News, 15*(4), 9–11.

Vasantakumr, K., (2006). Processing and product development of spices subsidiary and minor products of black pepper, *Spice India, 19*(11), 6–10.

Venskutonis, P. R., & Dagilyte, A., (2003). Composition of essential oil of sweet flag (*Acorus calamus* L.) leaves at different growing phases. *Journal of Essential Oil Research, 15*(5), 313–318.

Zachariah, T. Z., (2005). Value added products from black pepper, *Spice India, 14*(5), 16–21.

CHAPTER 4

RECENT APPROACHES TO IMPROVED PRODUCTION TECHNOLOGY OF SPICES AROUND THE WORLD

CONTENTS

ABSTRACT

In the present scenario, one of the biggest concerns of the world is the burgeoning human population coupled with decreasing per capita land day-by-day, which is more acute in the developing countries like India. As a mitigation measure, the primary focus of the growers, researchers as well as the policy makers is, therefore, to concentrate on more and more food grain production taking into account the inevitable shift of primary agriculture towards subsistence farming, diminishing efficiency in optimal utilization of available natural resources, vagaries of climate change, etc. The scenario of spice crop production, in the renewed context of human health consciousness, and changes in dietary habits of the people around the world, too, has a significant advancement and the research works carried on throughout the world during the last few decades has been substantial. The following paragraphs will describe glimpses of those outstanding efforts as reviewed hereunder.

4.1   USAGE PATTERN OF COMMON SPICES

4.1.1   BLACK PEPPER

Black pepper, the "king of spices," has an incredible popularity and has been used traditionally as a utility spice crop since time immemorial. Among all the spices, the black pepper (*Piper nigrum*) is one of the most widely used spices. It is valued for its distinct biting quality attributed to the alkaloid, piperine. Black pepper is used not only in human dietaries but also for a variety of other purposes such as medicinal, as a preservative, and in perfumery. Many physiological effects of black pepper, its extracts, or its major active principle, piperine, have been reported in recent decades. Dietary piperine, by favorably stimulating the digestive enzymes of pancreas, enhances the digestive capacity and significantly reduces the gastrointestinal food transit time. Piperine has been demonstrated in *in vitro* studies to protect against oxidative damage by inhibiting or quenching free radicals and reactive oxygen species. Black pepper or piperine treatment has also been evidenced to lower lipid peroxidation *in vivo* and beneficially influence cellular thiol status, antioxidant molecules, and antioxidant enzymes in a number of experimental situations of oxidative stress. The most far-reaching attribute of piperine has been its inhibitory influence on enzymatic drug biotransforming reactions in the liver. It strongly inhibits hepatic and intestinal aryl hydrocarbon hydroxylase and UDP-glucuronyl transferase. Piperine, while it is non-genotoxic, has in fact been found to possess antimutagenic and antitumor influences (Srinivasan, 2007).

Piperine and piperic acid could be used as natural antioxidant and anti-bacterial agents in both food preservation and human health (Zarai et al., 2013). Kapoor et al. (2009) reported that essential oil and oleoresins (ethanol and ethyl acetate) of *Piper nigrum* were extracted by using Clevenger and Soxhlet apparatus, respectively. GC-MS analysis of pepper essential oil showed the presence of 54 components representing about 96.6% of the total weight. β-caryophylline (29.9%) was found as the major component along with limonene (13.2%), β-pinene (7.9%), sabinene (5.9%), and several other minor components. The major component of both ethanol and ethyl acetate oleoresins was found to contain piperine (63.9 and 39.0%), with many other components in lesser amounts. The antioxidant activities of essential oil and oleoresins were evaluated against mustard oil by peroxide, p-anisidine, and thiobarbituric acid. Both the oil and oleoresins showed strong antioxidant activity in comparison with butylated hydroxyanisole (BHA) and butylated hydroxytoluene (BHT) but lower than that of propyl gallate (PG). In addition, their inhibitory actions by FTC method, scavenging capacity by DPPH (2, 2'-diphenyl-1-picrylhydrazyl radical), and reducing power were also determined, proving the strong antioxidant capacity of both the essential oil and oleoresins of pepper.

4.1.2 SMALL CARDAMOM

Majdalawieh and Carr (2010) reported that black pepper and cardamom extracts significantly enhance the cytotoxic activity of natural killer cells, indicating their potential anticancer effects. Our findings strongly suggest that black pepper and cardamom exert immunomodulatory roles and anti-tumor activities, and hence they manifest themselves as natural agents that can promote the maintenance of a healthy immune system. Black pepper and cardamom constituents can be used as potential therapeutic tools to regulate inflammatory responses and prevent/attenuate carcinogenesis.

4.1.3 GINGER

Ginger is traditionally used in culinary for its flavor and pungency. It is also used as carminative, stimulant, and for its antiemetic properties due to gingerols and shogaols. A ready to drink appetizer ginger beverage has been developed by Wadikar and Premavalli (2012) using response surface methodology. The sensory score, acidity and degrees Brix were the responses in the central composite designs of experiments with three independent variables. The ingredients ginger and lemon were suitably processed. The optimized composition of ingredients was processed further through blending and in-pack pasteurization. The appetizer had 1.2 mg/100 mL vitamin C and 8 mg/100 mL total pungency. The beverage packed in polypropylene bottles had a shelf life of 6 months both at ambient conditions (18–33°C) as well as at 37°C storage.

Appetite loss is one of the problems faced at high altitudes and the appetizers based on ginger may be useful for appetite stimulation. The fruit munch and ginger munches based on fresh and powdered gingers respectively were developed using response surface methodology (RSM) (Wadikar et al., 2010). The sensory score, acidity and total sugars were the responses in the central composite designs of experiments with three independent variables. The ingredients raisins, dates, almonds were pre-processed by frying in stable fat while juice was extracted from pseudolemon and lemon. The optimized composition of ingredients was processed further through concentration. The carbohydrate rich munches had vitamin C content in the range 37–43 mg/100 g and calorific value of about 90 kCal per munch. The munches packed in metalized polyester pouches had a shelf life of 8 months at ambient conditions (18–33°C) as well as at a fixed temperature of 37°C storage.

Ali et al. (2008) reported that the main pharmacological actions of ginger and compounds isolated there from include immunomodulatory, antitumorigenic, antiinflammatory, antiapoptotic, antihyperglycemic, antilipidemic and antiemetic actions. Ginger is a strong antioxidant substance and may either mitigate or prevent generation of free radicals. It is considered a safe herbal medicine with only few and insignificant adverse/side effects. More studies are required in animals and humans on the kinetics of ginger and its constituents and on the effects of their consumption over a long period of time.

4.1.4   TURMERIC

Turmeric, a spice derived from the rhizome of the plant Curcuma longa, contains the chemical curcumin, which is responsible for turmeric's taste, color, and biologic properties. Curcumin is used as a spice in foods, as a treatment in traditional medicine, as a dye for fleece, and as a component in nutritional supplements. A few cases of allergic contact dermatitis from curcumin have been reported. Though turmeric is common it is an uncommonly beautiful plant with orange red lily-like flowers and deep green long slender leaves that smell like mangos. Most of the ayurvedic doctors that I meet in India consider turmeric, to be one of the best herbs of India, and many go as far as saying that it is the best (Tiwari and Agrawal, 2012).

Turmeric (*Curcuma longa*) contains biologically active coloring con-stituents, curcuminoids, which are isolated from the turmeric rhizome by solvent extraction. Nampoothiri et al. (2012) reported that curcuminoids enriched fraction (CEF) can reduce the risk of hypertension and cardiovas-cular diseases.

4.1.5   FENUGREEK

Kassaian et al. (2009) reported that fenugreek seeds can be used as an adju-vant in the control of type 2 diabetes mellitus in the form of soaked in hot water.

Devasena (2009) suggested that fenugreek act as antitumorigenic agent by influencing 1, 2-dimethylhydrazine (DMH) induced colon tumor inci-dence and oxidative stress through its constituents flavonoids, saponin, pro-tease inhibitors and dietary fiber.

The hexane and ethanol extracts (5, 50 and 100 mg/mL) of the seeds of *Trigonella foenum-graecum* had prominent antibacterial activity against *Pseudomonas aeruginosa* and *Escherichia coli* compared with the chloro-form extract. Two fractions of ethanol extract isolated by vacuum liquid chromatography showed similar antibacterial activity against *P. aeruginosa* and *E. coli* (Saleem et al., 2008).

4.1.6   BLACK CUMIN

Al Mofleh et al. (2008) reported that an aqueous suspension of black seed (*Nigella sativa*) significantly prevented gastric ulcer formation induced by necrotizing agents. It also significantly ameliorated the ulcer severity and basal gastric acid secretion in pylorus-ligated Shay rats. Moreover, the sus-pension significantly replenished the ethanol-induced depleted gastric wall mucus content levels and gastric mucosal non-protein sulfhydryl concentration. The antiulcer effect was further confirmed histopathologically.

Aydin et al. (2008) reported that black cumin (*Nigella sativa* L.) supplementation into the diet of the laying hen positively influences egg yield parameters, shell quality, and decreases egg cholesterol. Black cumin in the food supplement at the level of 2 or 3% would positively influence egg production, egg weight, and shell quality and decrease the concentration of cholesterol in the egg yolk.

4.1.7 FENNEL

The essence of fennel is effective as much as grip water syrup in relief of infantile colic, so it can be used as a safe, effective and inexpensive herbal drug for infantile colic (Attarha et al., 2008).

In an experiment Muhammad et al. (2008) studied on the composition and antimicrobial properties of essential oil of *Foeniculum vulgare*. According to them the major component was trans-anethole (70.1%). The analysis of ethanolic and methanolic seed extracts showed the presence of nine components, including linoleic acid (56%), palmitic acid (5.6%), and oleic acid (5.2%). Fennel oil showed inhibition against *Bacillus cereus, B. magaterium, B. pumilus, B. subtilis, Eschericha coli, Klebsiella pneumoniae, Micrococcus luteus, Pseudomonas putida, P. syringae,* and *Candida albicans* compared to methanolic and ethanolic seed extracts. The lowest MIC values of fennel oil were recorded against *C. albicans* (0.4% v/v), *P. putida* (0.6% v/v) and *E. coli* (0.8% v/v). It was observed that essential oil and seed extracts of *F. vulgare* exhibit different degree of antimicrobial activities depending on the doses applied. Therefore, fennel oil could be a source of pharmaceutical materials required for the preparation of new therapeutic and antimicrobial agents.

4.2 PROPAGATION, NURSERY MANAGEMENT, AND PLANTING

4.2.1 BLACK PEPPER

Sharangi and Kumar (2011) carried out an experiment to study the influence of different organic substitution of nitrogenous fertilizers in nutrient schedule on different growth parameters of the black pepper cultivar Panniyur-1.

Recommended dose of P and K in the form of single super phosphate and muriate of potash was given to the vines. The plants were allowed to grow for upto 36 months of age for taking observations on different growth parameters. Considering the realization of highest response for four important growth parameters viz., plant height (269.37 cm), plant fresh weight (533.80 g), plant dry weight (178.01 g), and relative growth rate, RGR (5.10 g g^{-1} day^{-1}) after 36 months of planting, it may be concluded that the organic matter supplementation by 25% farm yard manure along with (75% urea) may be the best nutrient schedule under this agro-climatic condition.

Sharangi et al. (2010) also conducted an experiment on survivability of black pepper (*Piper nigrum* L.) cuttings from different portions of vine and growing media. In their experiment, cuttings were taken from the upper, middle or lower portions of the vine with 1, 2 or 3 nodes in each portion. The cuttings were both dipped quickly in 100 ppm IBA or water and then planted in polythene packets (both sides open) filled with different growing media (consisting of different combinations of sand, soil, farmyard manure and coconut husk). The polythene packets were kept in partial shade and irrigated frequently for proper rooting and growth. Other management practices were done regularly following standard recommendations. Survivability percentage of cutting (assessed by discarding the rotting/dried/dead ones) was recorded at 15, 30, 45 and 60 days after cutting. The best survivability was recorded for the middle portion of the vine with at least 2 nodes dipped quickly in 100 ppm IBA and grown on a medium consisting of sand:soil:farmyard manure:coconut husk in a 1:1:1:1 ratio.

Sangeeth et al. (2008) evaluated the inoculation effects of indigenous *Azospirillum* spp. selectively isolated from various black pepper growing locations of Kerala and Karnataka for enhancing the growth and nutrient uptake of black pepper cuttings. *Azospirillum* isolates BPaz4 and BPaz9 recorded 67% more plant height in rooted cuttings than untreated cuttings. The nitrogen, phosphorus and potassium uptake and total dry weight was significantly superior in treated plants. Increase in the uptake of iron, manganese, zinc and copper was found in BPaz9 treatment. Colonization of rhizosphere soil and non-rhizosphere by these isolates was found to be high. The N_2 fixation capacity of the isolates BPaz4 and BPaz9 showed 8.68 and 8.52 mg g^{-1} malate. Thus, the application of these inoculants viz. BPaz4 and BPaz9 is suggested for the ecofriendly production of rooted cuttings of black pepper.

Thankamani et al. (2008) reported that black pepper plants raised in solarized potting mixture had better growth than plants raised in nonsolarized potting mixture (soil, sand, and farm yard manure 2:1:1 proportion). Among the various treatments, plants raised in solarized potting mixture with recommended nutrients (urea, superphosphate, potash and magnesium sulphate 4:3:2:1) showed significant increase in number of leaves (5.3), length of roots (20 cm), leaf area (177 cm^2), nutrient contents and biomass (3.7 g pl^{-1}). The results indicated the superiority of solarized potting mixture for reducing the incidence of diseases besides yielding vigorous planting material. Cost of production of rooted cuttings with biocontrol agents was found to be cheaper in the case of rooted black pepper cuttings raised in solarized potting mixture. Biocontrol agents or biofertilizers can be mixed with solarized potting mixture.

Thankamani et al. (2007) also carried out another experiment at Peruvannamuzhi (Kerala) to study the feasibility of using soil-less medium containing coir pith compost and granite powder for raising black pepper (*Piper nigrum*) cuttings in the nursery. Plant height, leaf production, leaf area and total dry matter production were significantly higher in the medium consisting of coir pith compost and granite powder in 1:1 proportion along with *Azospirillum* sp. and phosphobacteria as nutrient sources whereas, the cost of production of rooted cuttings was cheaper in the medium consisting of coir pith compost, granite powder, and farmyard manure in 2:1:1 proportion compared to conventional potting mixture (soil:sand:farmyard manure in 2:1:1 proportion).

Black pepper is a perennial crop, which takes a minimum of five years for yield stabilization. Hence, any seedling character, which reflects its productivity, may be of use in identifying the probable high yielding types at an early stage. Krishnamurthy and Chempakam (2009) carried out an experiment to study the influence of some physiological and biochemical parameters such as leaf and stem carbohydrate status, nitrate reductase and sucrose phosphate synthase activities and leaf gas exchange parameters during pre-bearing and bearing stages on productivity in black pepper. Field grown plants (1–2 years) were used for the study. Results revealed that leaf starch and sucrose phosphate synthase activity did not show significant correlation with productivity. Leaf photosynthesis, nitrate reductase activity and stomatal conductance showed significant positive correlation while leaf temperature had significant negative correlation with productivity. It was concluded that in black pepper, physiological traits such as high nitrate reductase activity, photosynthetic rate and stomatal conductance and low leaf temperature may be useful in identifying high yielding types during juvenile stage itself.

Vijayaraghavan and Abraham (2007) carried out an experiment to select efficient antagonists from black pepper nurseries and use them alone or in combination with soil solarization/fungicides in the integrated management of the Phytophthora diseases in black pepper nursery. Among the different native fungal isolates obtained, two were found to be effective and were identified as *Trichoderma viride* and *T. longibrachiatum*. Observations on the incidence and severity of Phytophthora rot in black pepper nursery showed that in general soil solarization, application of native fungal antagonists and spraying of Ridomil MZ had a favorable effect in checking the disease. It was also noticed that solarization of potting mixture and application of *Trichoderma* spp. had a positive effect in reducing the mortality of cuttings.

To study the effect of different spacings on yield and yield attributes of black pepper var. Panniyur-I an experiment was conducted by Thangaselvabai et al. (2010) at Horticultural Research Station, Pechiparai, Kanyakumari the high rainfall zone, during 1997–2005 to standardize the spacing requirement for pepper var. Panniyur-1 under mono cropping system. Six spacing levels viz., 2 x 2, 2.5 x 2.5, 2.5 x 2, 3 x 3, 3 x 2, and 3 x 2.5 m were tried. Among the treatments 3 x 3 m spacing recorded the highest per plant yield (1.72 kg dry grains/vine) whereas, the 2 x 2 m spacing registered the highest total yield of 3.3 t/ha as it accommodated the maximum number of 2500 plants/ha.

4.2.2 SMALL CARDAMOM

A study was undertaken by Ankegowda (2008) to find out the optimum stage for transplanting cardamom seedlings from primary nursery to polybag nursery for better establishment and growth. Healthy seedlings at the two-leaf, three-leaf, four-leaf and five-leaf stage were selected and dipped in carbendazim (0.1%) solution for 30 minutes. Mortality of seedlings due to transplanting shock was significantly low in the four- and five-leaf stage and higher in the two and three-leaf stage. Mortality was reduced when transplanting was done with grown up seedlings of the four- to five-leaf stage. Heights of seedlings at the four- and five-leaf stage were at par with each other at the different growth stages and were significantly higher than those of two-leaf stage seedlings. At 180 days of transplanting, plant height was maximum in the four- and five-leaf stage seedlings but was at par with three-leaf stage seedlings. The four- and five-leaf seedlings recorded higher number of leaves at the initial stages; and at 170 days after transplanting, two-, three- and four-leaf seedlings were *at par* with each other for the number of leaves. The five-leaf stage seedlings maintained more number of leaves even at 180 days after transplanting. The number of roots after 180 days after transplanting recorded significant variation. The number of roots was the maximum in the five-leaf stage seedlings, and other leaf stages were at par with each other. Four- and five-leaf stage seedlings recorded higher fresh weight of stems, leaves, and roots. Total fresh weight was maximum in the five-leaf stage and minimum in the two-leaf stage seedlings. Four- to five-leaf stage seedlings maintained better growth and accumulated higher biomass in the polybag nursery and attained transplanting stage in six months of transplanting. The results clearly indicated that transplantable seedlings can be produced in 6 months versus 10 months in conventional nursery and 18 months in raised bed nursery.

4.2.3   GINGER

Rahman et al. (2010) conducted an experiment on evaluation of micro propagated ginger plantlets in different soil composition of pot culture. According to them, rooting performance was good (11.1 cm root length) in MS basal liquid rooting media with 82% root. Field establishment of plant was better in soil combination of garden soil: sand (1:1) with 11.8 cm plant height.

Jagadev et al. (28008) reported that the major limitation in increasing production and productivity of ginger is lack of adequate disease-free planting materials of high yielding varieties. As the major diseases are spread through contaminated seed-rhizomes, the possibility of producing pathogen-free planting materials using tissue culture is attractive. Therefore, they conducted an experiment to standardize a rapid, efficient, and reliable regeneration protocol for *in vitro* propagation of a high yielding ginger, cv. Suravi, collected from the high altitude research station at Pottangi (Koraput, a tribal district of Orissa), India. The axillary bud (0.2–0.5 mm size) from the sprouted rhizome was taken as the explant. The most ideal surface sterilant was found to be 0.1 $HgCl_2$ for 13 min, which reduced the total infection (fungal + bacterial) significantly to 3.3% and took shortest time for bud emergence (9.3 days) in standard Murashige and Skoog (MS) medium. The extent of survival (96.7%) and production of buds per explant (2.7) were the maximum with this sterilant. MS medium supplemented with 3.0 mg/L benzyl amino purine (BAP) and 0.4 mg/L naphthalene acetic acid (NAA) was ideal for shoot proliferation and resulted in maximum number of total shoots from a single explant (36.0), maximum shoot length (6.1 cm) with 4.7 leaves after a second sub-culturing. For rooting, MS supplemented with NAA (0.5 mg/L) was found to be more effective and produced the maximum number of roots per shoot (13.3) and the maximum root length (2.0 cm) plus taking the least time for root initiation (10.3 days). The *in vitro* plantlets were prehardened in 1/2 MS liquid medium. The hardening and acclimatization media mixture of soil: sand: farmyard manure (1:1:1) was found to be best for survival of the plantlets in ginger.

4.2.4 TURMERIC

Adelberg and Cousins (2007) conducted an experiment on development of micro and minirhizomes of turmeric, *in vitro*. Microrhizomes are *in vitro* storage organs employed in micropropagation of turmeric and a variety of other geophytes. Microrhizomes develop as engorged stem tissues in response to high concentrations of sucrose in liquid medium. Active storage organ function is indicated since microrhizome tissues have higher dry/fresh weight ratios than leaves or roots. In a system optimized for turmeric in 180 mL vessels, 115 and 180 mg microrhizome dry weight per explant was accumulated during 4 and 6 weeks of culture, respectively. These microrhizomes were ovoid like primary rhizomes of young field-grown turmeric plants. In this chapter, we introduce the concept of "minirhizomes." In contrast with microrhizomes, minirhizomes produce cylindrical, lateral, secondary rhizomes comprising a large fraction the plant at harvest. Minirhizomes occurred during 15 to 24 weeks of *in vitro* culture in larger vessels (approx. 2.5 L) with periodic media supplementation to maintain high sucrose concentrations. Approximately 65% of explants possessed secondary rhizomes after 24 weeks. The average mass of a rhizome was 675 mg and the majority of plant dry weight was found in the rhizome. When sucrose levels were maintained at nearly constant levels over a 5-week time course, the dry/fresh weight ratio of the rhizome was directly related to the sucrose concentration in the medium. The greatest total plant mass was observed among plants grown with 6% initial sucrose concentration within the 2% to 8% initial sucrose range that was evaluated.

4.2.5 FENUGREEK

Nandal et al. (2007) reported that the treatment combination with sowing on 20th October, spacing of 40 x 5 cm^2 without cutting produced highest seed yield. It was also observed that time of sowing from 20th October to 5th of November with one cutting was the best combination for getting maximum green leaves and seed yield.

4.2.6 FENNEL

A field experiment was conducted by Menaria and Maliwal (2007) at Udaipur (Rajasthan) to study the effect of three plant densities, four fertilizer treatments and two growth regulators on yield maximization of transplanted fennel (*Foeniculum vulgare*). Planting of fennel at a density of 27,770 plants/ha recorded significantly higher yield attributes (umbels/plant, seed yield/plant and test weight), while seed yield/ha was higher with a density of 55, 550 plants/ha (20.89 q/ha). Application of 90 kg N + 40 kg P_2O_5 + 20 kg K_2O + 20 kg S + 5 kg Zn per hectare recorded significantly higher yield attributes and yield of fennel during both the years. Application of naphthalene acetic acid 100 ppm recorded significantly higher yield attributes and seed yield.

4.2.7 CHILI

Chili nursery growing during winter under polyhouse (size 24' x 13' x 6') made of UV-stabilized low-density polythene film of 200 microns (800 gauge) thickness was compared with the other two methods, namely, polycover and no-cover (Kang and Sidhu, 2006). Under each of the three methods, sowing was done on 25 October, 15 November, 30 November and 1 February during 1997–1998, 1998–1999 and 1999–2000 seasons. The results revealed that nursery grown under polyhouse reached transplantable stage in significantly less number of days compared with the other two methods. Polyhouse technique also recorded more number of transplants per unit area and the crop growth from it gave higher early and total fruit yields. A similar trend was observed for all the dates of sowing.

Chili seeds were soaked in solutions of 0.8, 1.6, 2.4 or 3.2 mL monocrotophos/liter or 1, 2, 3 or 4 mL chlorpyrifos/liter for 10 h. All treatments reduced percentage germination compared with the untreated control. Soaking in 2.4 mL monocrotophos/liter had the greatest effect, reducing germination to 45% compared with 78% in the control. Treatments had no effect on plant height, number of leaves per plant or root length at 45 days after sowing (DAS). The lowest concentrations of monocrotophos and chlorpyrifos increased plant DW at 45 DAS (Sujatha and Rao, 1996).

4.3 NUTRIENT MANAGEMENT

4.3.1 BLACK PEPPER

Black pepper is cultivated in a wide range of soils with varying pH and fertility. Ideal condition requires a friable soil rich in humus and essential plant nutrients, with good drainage. Among the nutrients consumed by black pepper, N uptake is the highest followed by K and Ca and the magnitude of the nutrients removed is in the order: N > K > Ca > Mg > P > S > Fe > Mn > Zn. Various sources and levels of manures and fertilizers were evaluated for improving soil nutrient status and uptake by black pepper (Srinivasan et al., 2007).

In an experiment, Hamza et al. (2007) conducted on nutrient diagnosis of black pepper (*Piper nigrum* L.) gardens in Kerala and Karnataka by Diagnosis and Recommendation Integrated System (DRIS) indices at Calicut (Kerala). The nutrient analysis data obtained from extensive (130 samples) surveys of major black pepper growing tracts of Kerala and Karnataka was compared with already worked out soil and leaf nutrient DRIS indices values to find out the deviation of nutrients from the corresponding critical concentrations. The results revealed that the soils in most of the gardens were acidic (pH 4.4 to 6.7). Soil sample analysis showed that 88% gardens had organic carbon (OC) status below the critical values, 74% gardens had Zn and 36% gardens had P and 28% gardens had Ca status below the required levels. Leaf analysis results showed that 46% samples had Mg, 39% samples had Cu and 12% samples had P, K and Zn status below the required critical values. The order of limiting nutrients was: OC > Zn > P > Ca > K > Mg for soil and Mg > Cu > P=K=Zn > Mn for leaf samples. The study revealed the importance of manuring black pepper gardens with organic manures supplemented with secondary and micronutrients.

Mathew and Nybe (2002) carried out a field experiment in Kattappana, Idukki district, Kerala, India, to study the effects of inorganic and organic fertilizers on the performance of black pepper (*Piper nigrum*). The treatments consisted of recommended inorganic fertilizer, 50% N as farmyard manure (FYM) + 50% N as neem cake + 50% P as inorganic fertilizer, 50% N as FYM + 50% N and P as inorganic fertilizers, 50% N as FYM + biofertilizer, 50% N as FYM + 50% P as inorganic fertilizer + Azospirillum, 50% N as FYM + 50% N as inorganic fertilizer + phosphobacterium and VAM (Vesicular arbuscular mycorrhiza) and 50% N and P as inorganic fertilizers + biofertilizer. Yield and spike density were significantly enhanced by the combination of chemical fertilizer and organic fertilizers or biofertilizers over inorganic fertilizers alone.

A five-year study was conducted by Mathew and Nybe (2006) at the College of Horticulture, Kerala Agricultural University, Vellanikkara to study the effect of integration of various nutrient sources on yield, quality and soil and foliar nutrient status in black pepper. Two varieties and 13 nutrient schedules comprising of different combinations of organic, inorganic and biofertilizer sources and a control constituted the treatments. The results indicated significant variation in yield only in the fourth year of treatment imposition indicating the need for continuous application of organic manures and biofertilizers for the expression of treatment effects. The lowest yield during the fourth year was recorded by the no fertilizer control (3.256 kg/ vine/annum), while the yield in the four top ranking treatments, consisting of package of practices and treatments with different levels of integration of the nutrient sources, ranged from 4.254 to 4.721 kg. The study revealed the possibility of satisfying the nitrogen and phosphorous requirement of black pepper through organic means with 50% N as FYM along with application of *Azospirillum*, phosphobacteria and AMF, without yield sacrifice.

4.3.2 SMALL CARDAMOM

Murugan et al. (2007) conducted a field experiment at Cardamom Hill Reserves (CHR), Pampadumpara during 1994–2002 under rain fed situation to study the different levels of nutrients and neem cake under rain fed condition in Southern Western Ghats, India. Experimental results revealed that increasing the levels of nitrogen, phosphorus and potassium had increased the yields of cardamom up to 125:125:200 kg/ha. Application of fertilizer nutrients at the present level of recommendation (75:75:150 kg NPK per ha) in the form of urea, single super phosphate and muriate of potash along with 0.5 kg neem cake per plant had not increased the yield significantly over the control. Application of fertilizer nutrients at the rate of 125:125:200 kg/ha in two splits (just before and after summer monsoon) increased the yield significantly under Pampadumpara rainfall climatology. Among N, P and K, it appears that K is the most important as indicated by a larger absolute value than that of N and P. Therefore, application of fertilizer K is a must to increase the cardamom yield in CHR system. However, rainfall during summer months and number of rainy days had pronounced effect on the production of cardamom. Among rainy seasons, summer rainfall played significant role in increasing the cardamom yield. Higher than the average summer rainfall (366 mm) followed by southwestern monsoon (1162 mm) found to influence the yield significantly. The reason for the higher yield by summer rainfall could be due to increased growing season soil moisture, which is essential for the growth and panicle initiation and subsequent development of flowers and capsule setting. Recently in India, all droughts have very little influence on the rainfall climatology of cardamom hill reserves during the period of southwest monsoon; and therefore, the yield was not affected by droughts but in the summer months' rainfall and its distribution.

Kumar et al. (2009) reported that application of 100% organic manure in the form of FYM enhanced capsule yield by 34.1% over control. As the proportion of inorganic nutrient application increased, the response of yield also increased. Application of 100% inorganic recommended dose of fertilizer (RDF) yielded 188.81 kg/ha followed by 75% inorganic RDF% FYM (144.14 kg/ha). Nutrient content study of two years revealed that potassium content in the leaf was higher followed by nitrogen and phosphorus. The nutrients content in organic treatment was the minimum while that of 100% RDF applied plots was the maximum.

4.3.3   GINGER

Shaikh et al. (2010) reported that recommended dose of fertilizer + 25 t FYM/ha favorably influenced yield and uptake of nutrients by ginger followed by the application of 50% N through recommended dose + 50% N through poultry manure. It is therefore suggested that application of recommended dose of fertilizer + 25 t FYM/ha to ginger planted on flat bed in clay loam soil is best combination.

Under Abia State, Nigeria condition, Nwaogu and Ukpabi (2010) reported that optimum plant height, fresh rhizome yield, and rhizome storability was recorded with, the non-split application of potassium (K) fertilizer @ 50 kg ha^{-1} (with 60 kg ha^{-1} N and @ 15 kg ha^{-1} P) during planting time.

Roy et al. (2008) recorded the maximum plant height, leaf number, clump weight, net returns and benefit cost ratio were observed under 25% N application through poultry manure, and 75% N as urea fertilizer. This was followed by 1/2 N via neem cake + urea 1/2 in terms of yield and net returns, but the next best benefit cost ratio was obtained with neem cake giving 1/4 N + 3/4 urea. Height, leaf number and basal girth were 48.56, 47.18, and 43.50% under intercropping (pre-bearing coconut cv. East Coast Tall garden), which was higher than in the monocrop (41.17, 39.01, and 34.47%, respectively).

4.3.4 TURMERIC

Roy and Hore (2012) recorded a significant difference in the rhizome yield with the application of organic manure – microbial inoculants combination was compared with recommended dose of fertilizers (inorganic). Among different treatment combinations tried, the most effective treatment was vermicompost + Azospirillum + AM (*Glomus fasciculatum*) (28.94 t/ha), followed by compost + Azospirillum + AM (26.93 t/ha), as compared to recommended inorganic NPK (24.11 t/ha). The soil under vermicompost + Azospirillum + AM showed maximum organic carbon content (0.71%) and available potassium (209.09 kg/ha) after harvest; whereas maximum available nitrogen (253.05 kg/ha) and phosphorous (29.44 kg/ha) were noticed in mustard cake + Azospirillum and phosphocompost + Azospirillum + AM, respectively. In general, a higher build up of soil nutrients was observed after the experimentation for organic manure-microbial inoculants combinations compared to inorganic management.

A field experiment was conducted by Pandey et al. (2012) to study the effect of organic and inorganic sources of nutrients on nutrient contents and uptake of turmeric var. PCT-8. Poultry manure @ 5 t/ha recorded highest N,

P and K contents in turmeric rhizomes equally followed by vermicompost and then pig manure each @ 5 t/ha. Vermicompost producing 101.78 q/ha fresh rhizomes took up the maximum nutrients (161.65 kg N, 54.16 kg P, and 110.53 kg K/ha). In case of inorganic sources, 100% RDF (N180P60K120) producing maximum 101.82 q rhizomes/ha resulted in highest percentage of NPK contents as well as their uptake per hectare. The combined input (organic x inorganic sources) further augmented the uptake of these nutrients. In case of rhizome yield, the treatment interactions were non-significant. The findings suggest that due to heavy withdrawal of nutrients by turmeric crop as a result of improved production technology,

Mannikeri et al. (2010) found that application of vermicompost (15.65 t/ha) resulted in the highest composition and uptake of nitrogen (2.1% and 156.7 kg/ha, respectively), phosphorus (0.4% and 31.4 kg/ha, respectively) and potash (3.9% and 283.99 kg/ha, respectively). It also resulted in the highest production of dry matter (7.73 t/ha), fresh rhizome yield (33.6 t/ha), and cured rhizome yield (6.7 t/ha). It was closely followed by the application of pressmud (15 t/ha), poultry manure (6.43 t/ha), and RDF (180:90:90 kg NPK/ha) for all the parameters.

Yamgar et al. (2009) reported that the different sources of K_2O has also found to affect the yield as there was increased yield (8.40 t/ha) with K_2O in the form of sulfate of potash than in the form of muriate of potash (8.00 t/ha). The yields were also found affected and varied according to time of application of K_2O. Significantly higher yield (8.31 t/ha) was obtained with K_2O in split (50% at planting and 50% at the time of earthing up) than that with basal dose (8.08 t/ha).

Nanda et al. (2012) recorded the highest plant height, leaves/plant, tillers/plant, leaf area index, number of fingers/plant and rhizome weight/plant were with the application of 75% NPK + FYM @10 t/ha + Zn @ 5 kg/ha + B @ 3 kg/ha + biofertilizers @ 6 kg/ha. This treatment also resulted in significantly higher fresh and dry rhizome yield, and curcumin content and improvement in soil health parameters.

A field experiment was conducted by Padmapriya et al. (2007) to study the effect of partial shade, inorganic, organic and biofertilizers on biochemical constituents and quality of turmeric. The study was laid out in split plot design, consisting of two main plots viz., open and shade. The sub-plot treatments consisted of different doses of inorganic fertilizers, organic manures, biofertilizers, and growth stimulants constituting of 40 different treatment combinations. The treatment combinations, viz., shade with application of 100%

recommended dose of NPK + 50% FYM (15 t/ha) + coir compost (10 t/ha) + *Azospirillum* (10 kg/ha) + phosphobacteria (10 kg/ha) + 3% panchagavya showed increased total chlorophyll content, total phenol content and registered the highest yield per plot. On the contrary, provision of shade decreased the curing percentage as compared to open condition. Among the quality characters, the highest curcumin (5.57%) and essential oil (5.68%) content were registered in the treatment, shade with application of 50% FYM + coir compost + *Azospirillum* (10 kg/ha) + phosphobacteria (10 kg/ha) + 3% panchagavya.

A field trial was conducted to assess the effect of different organic manures produced in situ in an integrated farming system at Farming Systems Research Station, Sadanandapuram during 2012–13 on the performance of turmeric, variety Suguna, grown as component crop in the system under nutrient recommendation of 30:30:60 kg NPK ha^{-1}. Higher plant growth was recorded in the treatment integrating 75% recommended dose nitrogen as organic manures and remaining 25% as inorganic fertilizers. Rhizome yields were significantly highest (42.71 t ha^{-1}) in the treatment receiving vermicompost along with chemical fertilizers followed by poultry manure substitution (33.08 t ha-1) and sole organics – integration of vermicompost enriched with PGPR Mix I with poultry manure and goat manure (31.61 t ha^{-1}). Net returns per ha ranged between Rs. 1.82 lakh (inorganic fertilizers alone) and Rs. 5.8 lakh (vermicompost + inorganic fertilizers) and benefit cost ratios between 1.76 and 3.19 (Isaac and Varghese, 2016).

4.3.5 CHILI

A pot culture experiment was conducted to assess the effect of vermicomposted vegetable waste, alone and in combination with different organic manures and chemical fertilizer, on the biochemical characters of chili (*Capsicum annuum*). The reducing sugar, free amino acid and phenol contents were higher in the vermicompost treatment on 30 (70.27, 7.98, 14.62 mg/g), 60 (95.51, 17.66, 22.32 mg/g) and 90 DAS (33.67, 3.17, 11.85 mg/g). The protein content was higher in vermicompost treatment on 60 and 90 DAS (113.37 and 79.69 mg/g, respectively), whereas it was higher in vermicompost + farmyard manure (FYM) treatment on 30 (35.73 mg/g) DAS. The carbohydrate content was higher in vermicompost + FYM treatment on 30 and 90 (4.67 and 6.46 mg/g, respectively) DAS, while on 60 DAS; it was higher in the vermicompost treatment

(15.34 m/g). Chlorophyll a (0.23 mg/g), chlorophyll b (0.38 mg/g) and total chlorophyll (0.62 mg/g) were higher in vermicompost + neem cake treatment on 30 DAS. On 60 DAS, higher chlorophyll b (2.61 mg/g) and total chlorophyll (3.62 mg/g) contents were observed in the treatment containing vermicompost alone. On 90 DAS, chlorophyll a (1.01 mg/g) and total chlorophyll (1.92 mg/g) content was higher in vermicompost alone, and chlorophyll b (1.07 mg/g) in the vermicompost + FYM treatment (Yadavand Vijayakumari, 2004).

The effects of adopting a biodynamic calendar for timing the cultural operations and a manurial schedule involving two biodynamic preparations (separately or together) and panchagavyam (a mixture of 5:1 cow dung and ghee in a 5:3:3:5 cow's urine, curd, milk, and water formulation) in conjunction with organic manures as well as "organic manures alone," and the recommended practices of nutrient management (RP) on yield, quality, and economics of chili cultivation were evaluated in Kerala, India. Results show that RP (i.e., application of 20 Mg ha^{-1} farmyard manure + 75:40:25 NPK kg ha^{-1}) significantly improved fruit yield, net returns, and B: C ratio. Although biodynamic calendar and biodynamic preparations had no spectacular effects on the characters studied, application of organic manures generally promoted fruit quality in chili. Indeed, panchagavyam + organic manure demonstrated the maximum shelf life and the 'organic manures alone' (on nutrient equivalent basis) showed the highest ascorbic acid content of chili fruits (Jayasree and George, 2006).

4.3.6   CORIANDER

Under Parbhani, Maharastra condition, Vasmate et al. (2008) reported that application of farm yard manure at the rate of 20 t/ha) recorded maximum height per plant, number of leaves per plant, number of umbels per plant, number of umbels per umbel, number of seeds per umbel, number of seeds per plant, seed yield peer plant, seed yield per plot, seed yield per hectare, test weight, and germination percentage.

Giridhar et al. (2008) found that application of 100% N in combination with *Azospirillum*, PSB and FYM @ 5 t/ha recorded highest seed yield (1004.0 kg/ha) which is significantly superior to 100% N alone (877.7 kg/ha) in coriander. In addition to the increase yield, the treatment recorded an incremental benefit–cost ratio of 1.63. The increase in yield in treatments with *Azospirillum*, PSB and FYM along with 100% RDF may be due to better uptake of nitrogen and phosphorous and enhanced food accumulation.

Choudhary et al. (2008) reported that application of 100% inorganic N + *Azospirillum* at 1.5 kg/ha + 5 t farmyard manure/ha recorded the highest plant height (74.1 cm), number of branches per plant (7.50), number of umbels per plant (25.0), number of umbellets per umbel (6.09), and number of seeds per umbel (46.8) under Jobner, Rajasthan condition. This treatment also recorded a higher seed yield (889 kg/ha) over the other treatments except 75% inorganic N + *Azospirillum* at 1.5 kg/ha + 5 t farmyard manure/ha and 100% inorganic N alone. The highest net return (7732 rupees/ha) was obtained with 100% inorganic N + *Azospirillum* at 1.5 kg/ha + 5 t farmyard manure/ha, while the highest benefit: cost ratio (1.58:1) was obtained with 100% inorganic N alone, followed by 100% inorganic N + *Azospirillum* + 5 t farmyard manure/ha.

Effect of different sources of plant nutrients on productivity of coriander (*Coriandrum sativum*) was studied by Sharma and Jain (2008). Significantly, higher coriander seed yield was recorded with integrated use of 50% NPK (recommended dose of fertilizer NPK @ 40:30:20 kg/ha) through fertilizers + FYM + PSB (16.09 q/ha) followed by 50% NPK + FYM (14.06 q/ha) as compared to inorganic fertilizers alone or in combination with biofertilizers. The highest gross and net returns were also obtained with the application of 50% NPK through fertilizers + FYM + PSB.

In an experiment, Vasmate et al. (2007) found that different spacings of organic manures significantly influenced the growth parameters of coriander. Among the different organic manures, treatment with 20 t FYM/ha recorded the maximum plant height, primary branches, secondary branches, number of leaves, east-west spread and south-north spread. Among the different spacings, 30 x 20 cm was found to be the best treatment in increasing plant height, primary branches, secondary branches, number of leaves, east-west spread and south-north spread. The interaction effect of 20 t FYM/ha + 30x20 cm recorded the maximum east-west spread.

Giridhar et al. (2008) reported that coriander crop is grown as a rainfed rabi crop in vertisols of Andhra Pradesh. The crop generally suffers from periods of moisture stress from the flowering to maturity stage depending on precipitation during November and December months. During these stress periods, the crop shows deficiency symptoms of micronutrients such as copper, manganese, zinc, and iron. The foliar application of $ZnSO_4$ 0.5%, $FeSO_4$ @ 0.5% and combination $ZnSO_4 + FeSO_4 + CuSO_4 + MnSO_4$ all at 0.5%, had significant positive influence on all growth parameters and yield of coriander crop. Among the treatments, $ZnSO_4$ + $FeSO_4$ + $CuSO_4$ + $MnSO_4$ all at 0.5% recorded maximum plant height, number of primary branches and secondary branches, umbels per plant and umbellets per umbel, which are significantly superior to control. Crop maturity differed significantly among the treatments though the difference between the maximum and minimum days was only 3.3 days. The treatments, comprised with $ZnSO_4$ + $FeSO_4$ + $CuSO_4$ + $MnSO_4$ all at the rate of 0.5%, recorded significantly highest yield (940 kg/ha) followed by $FeSO_4$ @ 0.5% (927 kg/ha) and $ZnSO_4$ @ 0.5% (922 kg/ha) which are on par with each other and significantly superior over control (801 kg/ha).

4.3.7 CUMIN

Azizi and Kahrizi (2008) found that seed yield, yield components, biological yield, harvest index, and percentage of seed essential oil were significantly affected by nitrogen fertilizer, plant density and climate of Khoramabad, Iran. The highest those conducted to 2.5 g m^{-2} nitrogen fertilizer, 120 plants m^{-2} plant density and moderate climate. The most principle compounds composing the essential oil were cuminaldehyde (maximum 32.65%) and sum of P-mentha-1, 3-dien-7-al and P-mentha-1, 4-dien-7-al (maximum 55.42%).

Dehaghi and Mollafilabi (2010) reported that in the early stages of growth, due to low use of N by plant, nitrogen has less effect on growth indices and by advancing in time, and development of growth stages of cumin, rate of N use increases in plant and LAI, CGR, LWR, and LAR, and SLA have increased and among rates of N, difference was observed. NAR of photosynthate and RGR during Cumin growth showed a decreasing trend, as well. With respect to obtained results, 100 kg N/ha had lowest effect on 1000 seed weight and yield; and 150 kg N/ha had highest effect on 1000 seed weight and yield. Whereas, highest biomass and number of umbel was obtained in 200 kg N/ha.

4.3.8 FENUGREEK

Application of 80 Kg N/ha and 65 Kg P/ha significantly increased the plant height and number of primary branches per plant of fenugreek under Sonipat,

Haryana condition. The treatment of 80 Kg N/ha with different levels of phosphorus delayed the maturity time as compared to other treatment combination. The results indicated that application of 60 Kg N + 65 Kg P/ha produced highest test weight (10.4 g) as well as seed yield (20.4 q/ha) whereas minimum test weight (8.6 g) and seed yield (15.1 q/ha) was recorded under 20 Kg N/ha and 35 Kg P/ha (Nandal et al., 2007).

Pariari et al. (2009) reported that foliar application of boron @ 0.1% and zinc @ 0.2% twice had been found to be effective in enhancing most of the yield attributes and seed yield of fenugreek. It is also observed that lower concentration is more effective than higher concentration.

Treatment with Zn @ 0.5% + F e @ 0.5% + humic acid @0.05% exhibited the maximum value of all vegetative parameters of fenugreek, i.e., plant height, number of leaves per plant, number of branches per plant and leaf area per plant, as well as green leaf yield per plot and per hectare (Mahorkar, et al., 2008).

The application of farm yard manure showed the significant increase in plant height, total number of nodules per plants, seed index, seed and straw yield, nitrogen, phosphorus, potassium, calcium and sodium uptake by seed and straw (Kumawat and Yadav, 2009). The application of phosphorus increased the yield attributing parameters, yield, root nodules and uptake of N, P, K, Ca and Na. The increasing level of ECIW decreased the plant height, total number of nodules per plant, seed index, seed and straw yield and nutrient uptake by seed and straw.

Ali et al. (2009) studied about the response of fenugreek to fertility levels and biofertilizer inoculations. Significant increase in growth and yield attributes, yield and nutrient uptake and the highest net returns and B:C ratio was recorded with the application of 30 kg N + 26.2 kg P/ha. Amongst, microbial inoculations, dual inoculation with *Rhizobium* + PSB recorded significantly higher growth and yield attributes, yield and nutrient uptake. The highest net returns and B:C ratio was also significantly higher with the combined application of biofertilizers. It may be concluded that significant increase in seed yield of fenugreek can be obtained with the application of 30 kg N + 26.2 kg P with the highest net returns and B:C ratio. Combined inoculation of Rhizobium and PSB was more effective in improving growth, yield, nutrient uptake and net returns and B:C ratio of fenugreek in comparison to their sole application.

The maximum value of N, P content and their uptake, protein content, yield, net returns were recorded by Rathore and Porwal (2008) when fenugreek was sown on 4 November under Udaipur, Rajasthan condition. Microbial inoculation enhanced the N and P contents as well as their uptake by seed and haulm, protein content, net returns and B:C ratio. Higher N and P contents and their uptake by seed and haulm and protein content were registered with 2 manual weeding 20 and 40 DAS but noted at par with pre-emergence application of pendimethalin with B:C ratio of 1.55.

4.3.9 FENNEL

Bhardwaj (2016) studied on the effects of potassium (K) fertilization on transplanted fennel (*Foeniculum vulgare* Mill.) production. Four treatments (K @ 20-, 40-, 60- and 80-kg ha^{-1}) were compared using four replications under CRD. Potassium @ 60 kg ha^{-1} recorded highest value of plant height primary branches plant, secondary branches plant, number of leaves, number of roots plant, tap root length, fresh weight of shoot, fresh weight of root and minimum root: shoot ratio. The maximum number of umbels plant, umbellate umbel, number of seeds umbel, test weight, seed yield, straw yield, and harvest index was also observed in same treatment. All the quality parameters like, volatile and total oil content, soluble sugar, total carbohydrate, crude protein, potash content of seed and soil and overall quality of seeds were higher with application of 80 kg K ha^{-1}, whereas disease incidence was significantly reduced with application of higher dose of K. The maximum gross return, net return and highest benefit: cost ratio was also recorded with application of 60 kg K ha^{-1}.

Phosphatic biofertilizer showed significant effects on essential oil content and anethole and limonene contents in essential oil (except fenchone content) in fennel. The maximum essential oil content in seed and anethole content in essential oil were related to the plots with consumption of 60 kg phosphatic biofertilizer/ha. The lowest limonene content in essential oil was obtained with consumption of 30 kg phosphatic biofertilizer/ha. The highest essential oil content in seed and anethole content in essential oil and minimum fenchone content and limonene content in essential oil were obtained with consumption of 10 tonnes vermicompost/ha (Darzi et al., 2009).

In order to study the effects of biofertilizers on N, P, K concentrations and seed yield in fennel (*Foeniculum Vulgare* Mill.), an experiment was conducted by Darzi et al. (2009) at Damavand, Iran. The highest concentration of N, P and K in seed and seed yield were obtained with mycorrhiza inoculums.

Phosphate biofertilizer also showed significant effects on mentioned traits as the highest N concentration in seed with consumption of 60 kg/ha and maximum concentration of P, K and seed yield with consumption of 30 kg/ha from it were obtained. The highest concentration of N, P and K in seed and seed yield were obtained with application of 10 ton/ha vermicompost. There were positive and synergistic interactions between factors, like interactions between mycorrhizal inoculation and phosphate biofertilizer on N concentration and phosphate biofertilizer and vermicompost on P concentration.

Hans Raj and Thakral (2008) found that application of N at 100 kg/ha, P at 50 kg/ha and K at 50 kg/ha recorded the maximum values of plant height, flowering, umbels per plant and umbellets per umbel of fennel, while seeds per umbel, seed yield (q/ha) and harvest index was found maximum at 75 kg nitrogen, 50 kg phosphorus and 50 kg potash per hectare. However, a dose of 75 kg N, 50 kg P_2O_5 and 25 kg K_2O per hectare was found economical under Hissar, Haryana condition.

4.3.10 BLACK CUMIN

Ozguven et al. (2007) determined the effect of four doses of nitrogen (0, 30, 60 and 90 kg/ha) and three of phosphorus (0, 30 and 60 kg/ha) on yield and quality of black cumin in Turkey. 60 kg/ha nitrogen and 60 kg/ha phosphorus fertilizations gave the highest yield and quality of black cumin. The highest values for plant height, the number of branches, the number of capsules, seed yield, thousand-seed weight, essential oil content and seed fatty oil content were 100.1 cm, 12.73 branches/plant, 22.2 capsules/plant, 1006 kg/ha, 2.35 g, 0.40% and 39.0%, respectively.

Influence of nitrogen (0, 50 and 100 kg/ha), phosphorus (0, 20 and 40 kg/ha) and potassium (0, 30 and 60 kg/ha) on the growth and seed yield of black cumin showed significant differences (Nataraja et al., 2003). Application of nitrogen at 100 kg/ha resulted in the highest values for plant spread (427.75 cm^2) and number of seeds (57.52) per pod. Significant differences were also observed with the interaction of NPK at 50:40:30 kg/ha, producing pods of good size (3.84 cm^2), high 1000-seed weight (2.38 g), and seed yield (17.45 q/ha).

Datta et al. (2001) worked with N-P-K at 15–15–15 (T_1), 30–30–30 (T_2), 45–45–45 (T_3) and 60–60–60 kg/ha (T_4), FYM alone (T_5) and no fertilizer (T6) on black cumin (*N. sativa* cv. Rajendra Kanti). The highest plant heights

(27.25 and 47.50 cm) at 30 and 90 DAS, respectively, were observed under T5 along with the highest number of primary branches, number of fruits per plant, seed weight per fruit, seed weight per plant and seed.

A pot experiment was conducted to study the effects of N as urea (46.5% N) and P as calcium superphosphate (15.5% P_2O_5), singly or in combination with micronutrient mixture (14% Zn EDTA, 13.2% Fe EDTA, and 13.0% Mn EDTA), on the performance of *N. sativa*. The N and P fertilizers were applied at 2.5, 3.3, 4.2 or 5.0 g per pot. Growth parameters (shoot dry weight, number of secondary branches per plant, and plant height), yield and yield components (seed weight, capsule weight, number of seeds per capsule, and number of capsules per plant) were significantly enhanced by N and P fertilizers with or without micronutrients. N at 5 g per pot, and P at 4.2 and 5.0 g per pot in combination with micronutrients recorded the greatest values for the growth parameters, yield and yield components. The micronutrient alone generally enhanced the growth and yield parameters. The application of 4.2 and 5.0 g N and P per pot generally resulted in the highest N, P and K contents of shoots. The highest rates of N and P in combination with micro-nutrients gave the highest seed volatile oil and fixed oil contents (Mohamad et al., 2000).

El-Deen et al. (1997) reported on the response of *N. sativa* to spacing of 20, 30 or 40 cm between hills and P source and rate (calcium superphosphate (15.5% P_2O_5) or triple superphosphate (45.5% P_2O_5) applied at 20, 40 or 60 kg/feddan, and 85% orthophosphoric acid at 0.2, 0.4 or 0.6%. Plant height, number of branches/plant, total carbohydrates content and P content were not significantly influenced by spacing. Stem diameter, volatile oil and fixed oil percentages of seeds increased with increased spacing up to 40 cm. Moderate spacing (30 cm) significantly increased plant dry weight (DW), number of fruits/plant, seed yield/plant, and volatile and fixed oil yield/plant in comparison with the other spacings. Total seed yield, volatile and fixed oil yields/feddan were significantly increased by reduced spacing (increased plant density). All P treatments significantly increased plant height, stem diameter, number of branches, plant DW, number of fruits/plant, volatile and fixed oil percentages, plant P content, and seed yield, volatile and fixed oil yields per plant and per feddan in comparison with control plants. The most effective P source was superphosphate followed by triple superphosphate each at a rate of 60 kg P_2O_5/feddan. The best yields were obtained from a spacing of 20 cm with application of mono or triple superphosphate at a rate of 60 kg P_2O_5/feddan [1 feddan = 0.42 ha].

Das et al. (1991) treated *N. sativa* with 0, 20, 40 or 60 kg N/ha and 0, 20, 30 or 40 kg P_2O_5/ha, alone or in combination. Increasing application of N and P resulted in increased plant height, number of branches/plant, number of capsules/plant, number of seeds/capsule, 1000-seed weight and seed yield, although the difference between 30 and 40 kg P_2O_5/ha was not significant. Highest yield (16.32 q/ha) was obtained with application of 60 kg N + 30 kg P_2O_5/ha. Cost: benefit analysis showed that this combination also gave the best returns.

The effects of B (0.1 or 0.2%) and Zn (0.2 or 0.4%) on the performance of *N. sativa* were studied (Pariari et al., 2003). Both compounds were applied as foliar sprays, singly or in combination, at 3 and 8 weeks after sowing. The number of seeds per capsule and 1000-seed weight was not significantly affected by B and Zn application. Higher number of capsules per plant (16.47) and seed yield (7.60 quintal/ha) were recorded for 0.2% Zn. The combined application of B and Zn at lower concentrations were optimum with regard to plant height (41.60 cm), number of branches (6.7), capsules per plant (16.33), and seed yield (6.03 quintal/ha). B and Zn at higher concentrations resulted in higher 1000-seed weight (2.93 g) than both compounds at lower concentrations and the control (2.90 g).

4.3.11   ONION

The effects of sulfur (0, 15, 30 or 60 kg/ha) applied through gypsum or the slow-release fertilizer sulfur 95 on the composition and yield of onion (cv. Patna Red) were studied in Palampur, India by Jaggi (2004). S application significantly increased bulb and foliage yields, and S content of and uptake by foliage and bulb + foliage. The dry weight of bulb and foliage, and N and S uptake by bulbs and bulb + foliage increased with increasing S rate up to 30 kg/ha. At 30 kg/ha, the bulb yield increased by 105% over no S. The S content of bulbs increased from 0.26 to 0.49%, whereas the N:S ratio decreased from 8.9 to 6.1 following the application of S at 30 kg/ha compared with 15 kg/ha. Bulb yield, bulb N content, and bulb N uptake did not significantly vary with the S source, whereas gypsum had a greater positive effect on S uptake by bulb and foliage, S uptake by bulb + foliage, N uptake by foliage, and N uptake by bulb + foliage than sulfur 95. The S content of bulbs was also higher (0.57%) and the N:S ratio was lower (4.6) when gypsum was applied than sulfur 95 (0.25% and 9.2, respectively).

The effect of N fertilizers on shallot onion (*Allium ascalonicum*) total yield and Mg content in bulbs was investigated in Poland. Urea, ammonium nitrate and calcium nitrate were applied at 100 or 200 kg N ha^{-1}. The fertilizer rates had significant effects on total yield and content of Mg in bulbs. The total shallot yield and Mg content in bulbs increased as the N fertilizer rate increased (Jurgiel-Maecka and Suchorska-Orowska, 2004).

The effects of the application of 100% of the recommended N fertilizer rate (full N) singly or in combination with various vesicular arbuscular mycorrhizas (*Glomus mosseae, Glomus fasciculatum, Glomus epigaes, Glomus deserticola, Glomus macrocarpum, Gigaspora margarita, Gigaspora calospora, Gigaspora gigantea, Endogone duseii,* and *Acaulospora laevis*) on the performance of onion (cv. Pusa Madhavi) were studied in a pot experiment (Jha et al., 2005). Treatments consisting of full N and P, full N and 50% of the recommended P fertilizer rate, or no N and P were used as controls. The application of mycorrhiza with full N significantly increased bulb yield and quality over full N alone. Among the mycorrhizas, E. duseii with full N resulted in the greatest number of leaves per plant (8.2), plant height (62.1 cm), yield per plot (11.1 kg) and yield per hectare (246.4 quintal/ha), and the lowest storage loss after 6 months (43.1%). The yield obtained with full N + E. duseii was comparable to that obtained with full N and P. Almost 50% of the recommended P rate could be substituted by the application of *Gigaspora gigantea, Glomus macrocarpum, Glomus deserticola,* and *Acaulospora laevis*.

The effects of N fertilizer (50, 75 or 100% of the recommended N rate of 100 kg/ha) with or without inoculation of *Azospirillum* were studied in Rajasthan, India by Yadav et al. (2005). Before sowing, seeds were treated with *Azospirillum* at 500 g/ha. Seedlings were dipped for 15 minutes in *Azospirillum* slurry (1 kg *Azospirillum* dissolved in 50 liters of water/ha). Before transplanting, *Azospirillum* (2 kg/ha) was mixed with farmyard manure and incorporated into the soil. Bulb yields were highest with N at 75 (328.4 quintal/ha) and 100 kg/ha (336.5 quintal/ha); and the increase is 11.4 and 14.1%, respectively, over the control. The inoculation of *Azospirillum* resulted in a higher bulb yield (323.7 quintal/ha) over the control (310.9 quintal/ha). The highest net profits were obtained with *Azospirillum* combined with N at 100 (32, 792 rupees/ha) or 75 kg/ha (31, 288 rupees/ha).

4.4 MULCHING

4.4.1 TURMERIC

Kumar et al. (2008) studied on the effect of organic mulches on moisture conservation for rainfed turmeric production in mango orchard. The results indicated that maximum turmeric plant height (104.65 cm), stem girth (8.95 cm), leaf size (35.92 cm x 15.88 cm), dry biomass (2.09 t/ha), dry root weight (5.08 g/plant), number of finger (10.25), finger weight (156.45 g), mother rhizome weight (60.45 g) and fresh yield (7.87 t/ha) were recorded with the application of paddy straw mulch at 1 kg/m². The soil moisture content was higher during rhizome formation, development and maturation stage in plots where paddy straw was applied at 1 kg/m² (1 t/ha).

4.4.2 CHILI

The effects of different mulches (organic or plastic mulch (black LLDPE)) and irrigation levels (0.4, 0.6 or 0.8 irrigation water/cumulative pan evaporation (IW/CPE)) on the yield of chili (*Capsicum annum*) and the economics of plastic mulching were determined in a field experiment conducted in Coimbatore, Tamil Nadu, India during 1995–97. All treatments recorded higher yield and better fruit quality compared to the control, with plastic mulching and irrigation at 0.8 IW/CPE recording the highest yield, number of fruits per plant, and length and circumference of fruits. Plastic mulching recorded the highest net returns (Rs. 74,600/ha) resulting in a Rs. 64,600/ha increase in the net seasonal income.

4.4.3 ONION

A field experiment was conducted in Georgia, USA to determine the effects of irrigation system (drip or sprinkler) and mulch (bare soil, black plastic film or wheat straw) on the bolting, bulb yield and quality of the onion. Individual bulb weight and bulb yields under drip irrigation were similar to those under sprinkler irrigation. There were no consistent differences in the bulb number or yield of plants on plastic film compared to those of plants on wheat straw. Plants on wheat straw had reduced foliar nitrogen content. Total and marketable yields and weight of individual bulbs increased with increasing root zone temperatures up to an optimum of 15.8°C, followed by reductions in yields and individual bulb weight at >15.8°C. Onion bolting increased with decreasing foliage nitrogen content, with plants on wheat straw having the highest bolting incidence. Bolting also increased with decreasing root zone temperatures for the season. The total and marketable yields increased with decreasing mean seasonal soil water potential down to –30 kPa. Irrigation system and mulches had no consistent effect on the soluble solid content or pungency of onion bulbs (Diaz-Perez et al., 2004).

4.5   WATER MANAGEMENT

4.5.1   SMALL CARDAMOM

Ankegowda and Krishnamurthy (2008) evaluated small cardamom accessions for moisture stress. In their study, high biomass and high yielding cardamom genotypes viz., Green gold, Mysore-2, APG 277, Malabar-18, Compound panicle 7 (CP 7), and Hybrid 36 were screened for moisture stress tolerance. Clonally propagated seedlings were planted in cement pots and grown for one and a half year with recommended package of practices under rainout shelter with three replications and two treatments (control and moisture stress). Moisture stress was imposed by withholding irrigation for two months. Data on morphological and physiological parameters related to drought tolerance were recorded at the initiation, middle and end of stress. Plant height and number of leaves per clump did not record significant variation among the accessions at the initiation of stress. Number of dried leaves increased under stress in all genotypes at the middle of stress. Compound panicle 7 recorded higher reduction in biomass at end of stress period compared to all other accessions. Variation in relative water content between the treatments was non significant. Chlorophyll florescence yield reduced significantly under moisture stress treatment compared to control. Results indicate that genotypes Mysore 2, Green gold and Malabar 18 have better adaptability to drought conditions.

4.5.2   GINGER

Rathod et al. (2010) conducted an experiment on effect of levels of fertilizer and irrigation on yield of ginger in vertisols irrigated through micro sprinkler. The results revealed that the highest yield of ginger was recorded in 100% CPE (15.49 t/ha) and 120% of RDF (14.41 t/ha). However, in interaction, the highest yield (17.25 t/ha) was obtained in treatment I1 F2, i.e., 100% CPE with 100% recommended dose of fertilizer over the rest of treatment. The water requirement of ginger crop was 84.24 cm as it produced the highest ginger yield (15.49 t/ha) over rest of the irrigation levels.

4.5.3 CORIANDER

A field experiment was conducted by Singh et al. (2006) in Hisar, Haryana, India using coriander (*Coriandrum sativum*) cv. Hisar Anand. The treatments comprised of four irrigation frequencies (7, 14, 21 and 28 days intervals) and four nitrogen levels (45, 60, 75 and 90 kg/ha). All growth and yield parameters were recorded maximum when irrigation was applied at 21 days interval, which was statistically *at par* with irrigation at 14 days interval. Irrigation at 21 days interval produced 14.9 q/ha coriander seed. Minimum (9.4 q/ha) coriander seed yield was recorded when irrigation was applied at 7 days interval. Maximum seed yield (15.4 q/ha) of coriander was observed at 75 kg N/ha while minimum (11.6 q/ha) was found with 45 kg N/ha.

4.5.4 BLACK CUMIN

The effects of irrigation regimes on the performance of *N. sativa* were studied in Iran. The treatments consisted of no irrigation (control), and irrigation at intervals of 7, 14, and 21 days. Irrigation at 7-days interval resulted in the highest seed yield (1118 kg/ha). The 1000-grain weight and grain/capsule weight ratio did not significantly vary with the irrigation interval. A significant correlation among yield, plant height, number of capsules per plant, and number of seeds per capsule was observed (Akbarinia et al., 2005).

4.5.5 CHILI

In an experiment on chili cv. K2 plants, irrigating once at 7 to 8 days intervals significantly increased fruit size, seed weight, seed yield and seedling root length compared with application at longer intervals of 10–11 or 13–14 days (Vanangamudi et al., 1990). Seed yields were 722, 639, and 610 kg/ha for irrigation at intervals of 7–8, 10–11, and 13–14 days, respectively, and root lengths at these intervals were 8.03, 7.79 and 7.75 cm/seedling, respectively. Increasing N application from 125 to 200 kg/ha significantly increased the following plant attributes: fruit length and circumference, seed number/fruit, seed yield and fruit yield. Neither the frequency of irrigation nor the N application rate affected seed germination. The interaction between irrigation and N was significant for seedling root length only.

The effects of irrigation (40, 60, and 80% drip irrigation and surface irrigation at 0.7 IW/CPE), mulching with black polyethylene sheet and N rate (75, 100 and 125 kg/ha) on the capsaicin and ascorbic acid contents of chili (*Capsicum annuum*) cv. Jwala fruits at the maturity stage (green or red) were studied in Gujarat, India by Panchal et al. (2001). The ascorbic acid content on a dry weight basis was significantly higher with surface irrigation (781.34 mg/100 g), without mulch (752.93 mg/100 g) and 125 kg N/ha (784.11 mg/100 g) at the green maturity stage (858.30 mg/100 g). On the other hand, higher capsaicin content was obtained with surface irrigation (0.86 mg/g), without mulch (0.82 mg/g) and 125 kg N/ha (0.82 mg/g) at the red maturity stage (0.99 mg/g).

4.5.6 ONION

A study conducted in Maharashtra (India) revealed that in the treatment with floppy sprinklers spaced at 3.0 x 3.0 m, the B:C ratio was lowest due to high system and cultivation costs and lower yield. It was seen that the treatment with floppy sprinklers spaced at 6.0 x 6.0 m had the best B:C ratio (3:38). This was due to low system and cultivation costs and higher yield. The B:C ratio was 1.48 in microsprinkler treatment due to high system and cultivation costs. Thus, the adoption of floppy sprinklers at the spacings of 6.0 x 6.0 m was economical as it resulted in maximum benefit cost ratio (3.28) for onion (Gogoi and Firake, 2003).

Two different irrigation treatments were evaluated by Mermoud et al. (2005) at Kamboinse (Burkina Faso), one based on a scientific approach ("optimal" treatment) with different irrigation frequencies (daily and bi-weekly applications) and the second one based on the empirical practices of the farmers of the area ("farmer" treatment). The field data allowed the evaluation and calibration of a deterministic mechanistic model (HYDRUS), which was subsequently used as a prediction tool to study various irrigation management scenarios. The experimental results demonstrate a poor efficiency of the farmers' practices.

The effects of drip irrigation on the performance of processing and fresh-market (cv. Vaquaro; total, colossal, jumbo, medium and repack grades) onions were investigated (Hanson and May, 2004). Both types were grown under subsurface and surface drip irrigation. Irrigation treatments comprised water applications at 60, 75, 90, 105, and 120% of the baseline amount (equivalent to 100% of the potential evapotranspiration). Total yield increased linearly with increasing amount of applied water. The maximum yield for both types of onion when water applied was approximately 889 mm. However, the behavior of the yield-applied water data suggested that yield could continue to increase with applied water amounts exceeding the 120% irrigation treatment. The medium and repack grades of fresh market onions showed little response to irrigation treatments. Surface and sub-surface irrigation treatments did not differ in their effects on yield.

4.6   SHADE MANAGEMENT

4.6.1   BLACK PEPPER

The most popular live stakes used as supports in black pepper (*Piper nigrum* L.) plantations of the humid tropics are *Ailanthus triphysa* (Dennst.) Alston., *Erythrina variegata* L., *Gliricidia sepium* (Jacq.) Steud and *Garuga pinnata* Roxb. Studies on soil properties in the rhizosphere of these tree species are limited. Dinesh et al. (2010) reported that tree rhizospheres also positively affected the activities of enzymes like dehydrogenase, urease, acid phosphatase, aryl sulphatase and β-glucosidase. Principal component analysis (PCA) reflected the strong relationship between microbial activity and the availability of labile and easily mineralizable organic matter, the logical dependence of microbial biomass on soil nutrients and a decrease in substrate use efficiency in soils with low organic substrates. The results imply that among the tree species studied, *G. sepium* and to some extent *G. pinnata* can be used as live supports for the restoration of degraded black pepper plantations and overall improvement in soil quality in the plains of the tropics.

4.6.2 SMALL CARDAMOM

Small Cardamom is a pseophytic cash crop grown on plantation scale under the shade of natural evergreen forests of Western Ghats of South India. The shade canopy provides suitable environment by maintaining humidity and evaporation at suitable level. Trees belonging to 32 families of Angiosperms constitute the major tree flora in the Cardamom hills of South India. It is desirable to maintain a mixed population of medium sized shade trees that facilitate shade regulation and to maintain more or less optimum conditions throughout the year. The main considerations while selecting shade trees are adaptability to climate, rate of growth and ease of establishment. The major shade trees that are suitable for the cardamom tracts of Kerala state of India are *Artocarpus heterophyllus, Toona ciliata, Acrocarpus fraxinifolius, Dysoxylum malabaricum, Palaqium ellipticum, Terminalia tomentosa, Terminalia paniculata, Pterocarpus marsupium, Canarium strictum,* Vitex *altissima, Hopea parviflora,* and *Grewia tiliaefolia* (Radhakrishnan et al., 2010).

4.7 INTERCROPPING

4.7.1 GINGER

Kumar et al. (2010) conducted an experiment to study intercropping of ginger in tamarind plantation compared to sole cropping under irrigated condition. They reported that interception of photosynthetic active radiation (PAR) by ginger crop at 150 days after planting (DAP) as intercrop in tamarind plantation was 25,229 lux compared to 31,643 lux in open area. Significantly higher numbers of rhizomes were recorded under intercropping compared to sole cropping. Ginger grown as intercrop in tamarind plantation recorded higher yield (173.89 g/plant) compared to sole crop in open area (117.17 g/plant).

Sanwal et al. (2006) conducted an experiment on ginger-based intercropping under mid hill agroclimatic conditions of North East Hill Region. They reported that different intercropping systems affected some growth characteristics and yield of ginger except legumes crop as an intercrop. Total yield revealed the highest net monitory return with ginger + cowpea treatment followed by ginger + French bean. The total uptake of N and K differed significantly when ginger intercropped with non-leguminous crop. LER values were always more than one in intercropping systems. Over all, ginger intercropped with cowpea and French bean found most suitable and economically viable system under mid-hill agroclimatic conditions of North East Hill region.

4.7.2   CHILI

In a field trial by Mallanagouda et al. (1995), *C. annuum* was intercropped with garlic, onion or coriander (*Coriandrum sativum*), and plots received the recommended dose of NPK only, the recommended dose of NPK + farmyard manure (FYM), 50% of the recommended dose of NPK + FYM, FYM only, or no NPK or FYM (control). The number of fruits/plant did not differ significantly with different intercrops, but was higher with recommended NPK + FYM than with the other fertilizer treatments (25.47 when averaged between the intercropping systems, compared with 11.60 in controls). Highest DW/fruit was obtained with intercropping with garlic (0.86 g) and with recommended NPK + FYM (0.76 g). Chili dry yield was highest with intercropping with garlic (3.87 q/ha) and recommended NPK + FYM (4.46 q/ha, compared with 1.84 q/ha in controls).

4.8   WEED MANAGEMENT

4.8.1   GINGER

Chatterjee et al. (2011) reported that paraquat @ 10 mL/L followed by one hand weeding at 42 days after planting (DAP) proved to be most effective in reducing the weed density and weed biomass resulting in highest rhizome yield in ginger (34.58 t/ha), followed by two hand weeding treatment (32.31 t/ha), which, however, was *at par* with the treatment comprising with Paraquat @ 10 mL/L followed by one hand weeding at 42 DAP). Considering the benefit cost ratio the treatment with Paraquat 10 mL/L followed by one hand weeding at 42 DAP) was economically more viable than treatment T9 (hand weeding twice at 21 and 42 DAP).

4.8.2   TURMERIC

Ratnam et al. (2012) conducted an experiment was conducted at Lam, Guntur to find out suitable weed management package for turmeric

(*Curcuma longa* L.) in coastal districts of Andhra Pradesh. Pre-emergence application of oxyflourfen @ 0.25 kg/ha followed by post-emergence application of quizalofop ethyl @ 0.05 kg/ha and two hand weeding at 60 and 90 DAS with WCE of 92% significantly recorded the highest fresh rhizome yield (6.6 t/ha) with B:C ratio of 0.61 and was on par with hand weeding at 30, 60 and 90 DAS, which recorded the highest fresh rhizome yield (8.5 t/ha). Uncontrolled weed growth reduced rhizome yield by 80%.

Kaur et al. (2008) reported that losses in rhizome yield of turmeric due to weeds varied from 63.9 to 76.5%. Application of straw mulch @ 9 t/ha significantly reduced weed dry matter and recorded 29.2% higher rhizome yield than 6 t/ha. Averaged over 2 locations, pendimethalin + straw mulch @ 9 t/ha revealed the highest weed control efficiency (84.2%), fresh rhizome yield (29.6 t/ha), herbicide efficiency index (11.2) and benefit:cost ratio (2.30) and was at par with metribuzin and atrazine, both integrated with straw mulch @ 9 t/ha. The fresh rhizome yield with straw mulch @ 9 t/ha + herbicide combination was 48.2, 14.9, 15.3 and 225.5% higher than straw mulch at 6 t, 9 t alone, k6 t/ha + herbicide and unweeded control, respectively. Different weed control measures led to 104.8–289.1, 72.7–220.4, and 90.5–278.1% increase in N, P, and K uptake by the crop, respectively, over the unweeded control. Uncontrolled weeds removed 60.6, 59.9 and 73.6% of total nutrients utilized by both crop and weeds. Integrated use of paddy straw mulch at 9 t/ha with either pendimethalin at 1.0 kg, metribuzin at 0.70 kg and atrazine at 0.75 kg/ha was adjudged very effective for weed control and attaining the highest productivity and profitability in turmeric.

4.8.3 CORIANDER

According to Yadav et al. (2016) Oxadiargyl at 0.06 kg ha^{-1} + hand weeding (HW) at 40 DAS represented the lowest weed density and controlled the weeds to the extent of 94.9%. Two hand weedings done at 20 and 40 DAS and pendimethalin at 1.0 kg ha-1 + HW at 40 DAS were found to be the most superior treatments in reducing density, intensity and dry weight of weeds and increasing weed control efficiency. These treatments controlled the weeds to the extent of 95.1% and 95.4%, respectively at harvest stage than weedy check and showed lower weed infestation of 17.4 and 18.1%, respectively. Two hand weeding treatment gave the highest seed yield (1.37 t ha^{-1}) among all the treatments and was closely followed by pendimethalin at 1.0 kg ha^{-1} + HW at 40 DAS which also increased the seed yield by a margin of 0.84 t ha-1 over weedy check, and registered the lowest weed competition index of 0.7%.

Skotnikov (2008) conducted an experiment on the effect of herbicides on productivity of coriander in Russia. In this experiment, he used herbibides namely, Pulsar (bentazone + MCPB), Basagran M (bentazone), Secator (iodosulfuron-methyl-sodium), Agritox (aqueous solution of alkali and dimethyl-ammonium salts + MCPA), and Furore Super 7.5 (fenoxaprop-P-ethyl) for weed control in coriander. Evidence was obtained that Secator, Agritox, and Furore Super 7.5 gave the best plant density and crop yield resulting in a 2-fold increase in production of the essential oil.

4.8.4 CUMIN

Mehriya et al. (2007) conducted an experiment on the effect of crop-weed competition on seed yield and quality of cumin. The results revealed that weed-free upto 60 DAS gave the lower mean weed density and total weed dry matter at harvest and increased yield attributes viz., mean final plant stand, umbels/plant, seeds/umbel and test weight by 157.9, 147.4, 166.2 and 37.2%, respectively, and mean seed yield by 788.7% over weedy check. This treatment stood at par with complete weed-free and weedy upto 15 DAS with regards to yield attributes and seed yield. The critical period of crop-weed competition was observed between 15 to 60 DAS in cumin. Weed-free environment throughout crop season produced the maximum oil and protein content, and was significantly higher compared to weedy check.

A field experiment was conducted by Mehriya et al. (2008) to determine the effect of different weed management practices on the yield of cumin. *Chenopodium murale, C. album,* and *Rumex dentatus* were the predominant weeds in the experimental field. Oxyfluorfen at 50 or 75 g/ha and oxadiargyl at 50 g/ha, applied at 20 DAS, along with one hand weeding at 35 DAS, being at par among themselves, brought maximum reduction in weed biomass production of *Chenopodium spp.* as well as total weeds, thereby recording (93.9–96.2%) higher weed control efficiency. However, maximum dry weight of R. dentatus was reduced due to the application of paraquat at 0.4 kg/ha + one hand weeding. Integration of oxyfluorfen at 50 or 75 g/ha along with one hand weeding gave comparable seed yield to the weed-free control and consequently provided maximum B:C ratio (2.92). Oxadiargyl +

hand weeding resulted in statistically *at par* seed yield to that obtained with the combined application of oxyfluorfen + hand weeding. Among the sole herbicidal treatments, oxyfluorfen at 50 or 75 g/ha or oxadiargyl at 50 g/ha, applied at 20 DAS, recorded higher WCE (83.3–88.9%) and produced significantly higher growth and yield attributing characters, which led to higher seed yield, net returns, and B:C ratio compared to fluchloralin at 1.0 kg/ha PPI (recommended herbicidal control).

4.8.5 ONION

The effect of pre- or post-planting application of selected herbicides on growth and yield of transplanted onion and weeds was investigated in two field experiments in Jordan. Post-planting application of oxyfluorfen (at 2 l ha^{-1}) at the three-to-four-leaf stage resulted in better onion bulb yield than in the hand-weeded, weed-free control. Pre-planting treatment with oxyfluorfen at 2.5 l ha^{-1} gave higher yields than the weed-free control in irrigated onion, but not under rain-fed conditions. Oxadiazon (at 4 l ha-1) gave higher yields than in weed-free onion as a post-planting treatment under rain-fed conditions, and was second to oxyfluorfen for bulb production. Pre-planting application of pendimethalin (at 4 l ha^{-1}) was also effective, and increased onion yields compared to weed-infested onion plots or other herbicides used under irrigation. Of all herbicides tested under Jordanian conditions, oxyfluorfen and oxadiazon were the best for yield and weed control in onion, followed by pendimethalin, methabenzthiazuron and linuron as pre-planting treatments, and pendimethalin, linuron, and paraquat as post-planting applications (Qasem, 2005).

Nitrate accumulation of vegetables is one of the most important environmental problems recently. There are huge differences among plant species and varieties in nitrate accumulation because of their biological and genetical characteristics. It is known that some herbicides can influence the quantity of nitrate in the plants. A pot experiment was conducted by Nadasy (2003) to study the effect of nitrogen fertilizer levels (0, 15, 30, 60 and 120 mg/kg) and two pre-emergent herbicides: Dual 960 EC (metolachlor) at 2 liters/ha and Stomp 330 (pendimethalin) at 1 liter/ha on nitrate accumulation of green onion (*Allium fistulosum*). The author established that dual and stomp herbicides could influence nitrate accumulation of onion: they decreased the nitrate concentration in the bulbs and contrary increased in the leaves. Green onion accumulated maximum medium quantity of nitrate even if it had excessive nitrogen supply.

4.9 EFFECT OF GROWTH REGULATOR

4.9.1 CORIANDER

Yugandhar et al. (2016) studied the effect of plant growth regulators (PGRs) on growth, seed yield, quality and economics of coriander (*Coriandrum sativum* L.) cv. Sudha. Among different PGRs applied, 75 ppm GA3 resulted in maximum plant height. However, maximum number of primary branches and secondary branches plant-1, number of umbels plant-1, number of umbellets umbel-1, number of seeds umbel-1, seed yield, and B:C ratio was maximum with 250 ppm Cycocel. Minimum number of days to 50% flowering and maturity and maximum carbohydrate content and protein content were noticed with 75 ppm GA3. Similarly, lowest moisture content in seeds was also observed with 75 ppm GA3, while, the essential oil content in seeds was maximum with 50 ppm GA3.

The effect of plant growth regulators and their time of application on growth and yield of coriander was studied by Sarada et al. (2008) in Andhra Pradesh. They reported that Triacontanol @1.0 mL/L recorded maximum plant height, more number of branches and more number of umbellets per umbel. NAA @10 ppm recorded more number of umbels per plant and increased crop duration. Maximum seed yield was recorded with NAA @10 ppm followed by Triacontanol @1.0 mL/L. With regard to number of sprays two sprays at 40 and 60 DAS recorded maximum seed yield, compared to one spray at 40 DAS or three sprays at 40, 60 and 80 DAS.

Meena and Malhotra (2006) reported that under the agro-climatic conditions of Zone IVA (Sub-Humid Southern Plain and Aravali Hills) of Rajasthan, coriander crop should be sown on 15th October with application of 60 kg/ha and foliar spray of NAA 25 @ ppm at 30 DAS, or GA @ 50 ppm at 30 DAS for getting higher leafy yield per plant in coriander grown exclusively for green leaf purpose. Kumar et al. (2006) reported that seed yield per hectare was maximum with a spacing of 30 x 20 cm. Seed yield was improved when plants were sprayed with Cycocel.

4.9.2 BLACK CUMIN

Kumar et al. (2009) reported that ethrel @ 75 ppm proved to be most effective in promoting growth and gave highest seed yield (6.03 q/ha) of black cumin, followed by Ethrel @ 100 ppm (5.59 q/ha) which was statistically *at par*.

Shah et al. (2006) evaluated the effects of GA_3 on black cumin (*Nigella sativa*), sprayed with either deionized water (control) or 10^{-5} M GA_3 at 40 (vegetative stage) or 60 (flowering stage) DAS. It was noted that growth, NPK accumulation and seed yield were maximal when spraying of GA_3 was carried out at 40 DAS. However, spraying at 60 DAS was not much effective in terms of the parameters studied. Moreover, there was a significant difference in spray treatments at various growth stages only when GA_3 was sprayed and not when water was sprayed.

Shah (2007a) reported that surface sterilized seeds of black cumin (Nigella sativa) were soaked in 10^{-6}, 10^{-5} or 10^{-4} M aqueous solutions of gibberellic acid (GA_3) for 5, 10 or 15 h. The plants were then sampled at 50, 70 and 90 DAS for the analysis of shoot length, leaf area, dry mass, carbonic anhydrase (carbonate dehydratase) activity, leaf chlorophyll content, stomatal conductance and net photosynthetic rate. Seed yield was recorded at harvest (130 DAS). All parameters were found to be significantly enhanced by GA_3 treatment, with maximum stimulation being noted following a 10 h soaking treatment with 10^{-5} M GA_3. Moreover, the mentioned parameters were elevated by 70, 68, 65, 39, 44, 37 and 44% compared to the control at the 70 DAS stage. The seed yield was enhanced by 32%.

Shah et al. (2007b) studied on the effect of foliar spray of 10^{-5} M gibberellic acid (GA3) at vegetative stage along with basally applied 0, 40, 60, 80 and 100 kg N/ha on chlorophyll content, net photosynthetic rate, stomatal conductance, leaf N content, leaf area and total dry matter production (monitored at 30 days after spray application) and number of capsules/plant, 1000-seed weight, seed yield/ha, biological yield/ha, harvest index, seed yield merit. Results indicated that, at 0, 40 or 60 kg N/ha, GA_3 did not produce any significant effect, but at basal 80 kg N/ha, GA_3 affected the parameters favorably with the exception of HI. A level of 100 kg N/ha proved supraoptimal. GA_3 sprayed plants exploited nitrogen from the soil more effectively and resulting in enhanced morphophysiological and yield responses.

Shah et al. (2007c) reported that experiments, the test black cumin plants were sprayed with either deionized water (control test) or 10^{-5} M kinetin (KIN) at 40 (vegetative stage) or 60 (flowering) DAS. Spraying with KIN at 40 DAS brought about maximum stimulation of all parameters, whereas spraying at 60 DAS was not much effective. All characteristics recorded at 80 DAS (shoot length, leaf number, leaf area, branch number and dry weight per plant, net photosynthetic rate, stomatal conductance and leaf chlorophyll content), were significantly enhanced by KIN application. Further, at harvest (130 DAS), capsule number per plant, seed yield per plot, and biomass yield per plot, showed a significant increase over the control traits. However, number of seeds per capsule, 1000-seed weight and harvest index remained unaffected.

4.9.3   FENUGREEK

The effect of plant growth regulators and their time of application on growth, yield and economics of fenugreek were studied by Shivran et al. (2016) during winter. Plant growth regulators used were naphthalene acetic acid (NAA) 50 ppm, Triacontanol 1000 ppm and Triacontanol 500 ppm with three times of spray viz., one (40 DAS), two (40 and 60 DAS) and three (40, 60, and 80 DAS) along with absolute control. The highest pods plant^{-1}, test weight, seed yield (1494 kg ha^{-1}), straw yield (3381 kg ha^{-1}), gross returns (Rs. 51,582 ha^{-1}), net returns (Rs. 35586 ha^{-1}) and benefit: cost ratio (2.25) were recorded with NAA 50 ppm which was significantly superior to Triacontanol 1000 ppm and 500 ppm, and water spray. The seed yield and net returns were increased with NAA 50 ppm, Triacontanol 1000 ppm and 500 ppm over spray of water, respectively. Hence, the application of NAA 50 ppm twice at 40 and 60 DAS or thrice at 40, 60 and 80 DAS was found beneficial in terms of increased growth, yield and monetary returns of fenugreek.

A field experiment was conducted at Madhya Pradesh to find out the effect of plant growth regulators on growth and yield of fenugreek (*Trigonella foenum-graecum*). The results indicated that foliar spray of naphthalene acetic acid (NAA) 20 ppm at 25 DAS and 55 DAS resulted in significantly higher growth and seed yield (17.41 q ha^{-1}). The highest benefit: cost ratio (4.20:1) was observed for the treatment, 60 kg phosphorus ha^{-1} + NAA 20 ppm (Gour et al., 2009).

4.10   HARVESTING

4.10.1   CORIANDER

In India, fresh coriander is abundantly available during winters, but has very short shelf life even under refrigerated conditions. This leads to a marked scarcity in availability and a sharp rise in price in the lean period. During peak period, most of the crop is wasted due to improper post harvest processing techniques. The fresh green coriander after proper drying, packaging and storing may be used during lean periods. Different pretreatments and methods were studied and evaluated on the basis of quality and rehydration characteristics. The best pretreatment was found to be dipping for 15 min in solution of 0.1% Magnesium chloride, 0.1% Sodium bicarbonate and 2.0% KMS in water at room temperature and the best method was drying in mini multi rack solar dryer (Kaur et al., 2006).

Alkimim et al. (2016) evaluated the effect of different harvesting times on the physiological quality of *Coriandrum sativum* L. seeds, cv. Palmeira. The experimental was designed in CRD with six treatments (harvest times) and four replications. The first moment to harvest seeds was carried out 15 days after full bloom, when approximately 50% of plants were in flowering phase. Other harvests (22, 29, 36, 43 and, 50 days after bloom) were carried out at 7 days intervals until the seeds reached 16.0% of moisture, which occurred in the 6th harvest. The seeds were analyzed for water content, germination and vigor (first count, seedling emergence and emergence rate index). The harvest season most suitable for *C. sativum* seeds cv. Palmeira occurred between 46 and 50 days after flowering, when the seeds have the highest physiological quality and moisture content between 20 and 16%.

4.10.2 FENNEL

Fruit yield per plant and 1000-fruit weight regularly increased from immature to mature periods, while essential oil content declined with fruit maturity in fennel. The content of trans-anethole, the main component, varied between 81.63% and 87.85%, and the variation was statistically insignificant during maturation stages. Some components, particularly monoterpenes, alpha-pinene, β-myrcene, limonene, and alpha-terpinene, varied significantly ($p<0.05$) during maturation stages (Telci et al., 2009).

4.10.3 ONION

To investigate the effects of physiological maturity of onions (*Allium cepa* L.) at harvest and different topping methods on bulb color, skin retention, and the incidence of storage rots, a field study was carried out by Wright et al. (2001). Onion plants were lifted at five different stages of maturity from 0 to 4 weeks after 50% leaf collapse (top-down). Foliage was removed from the bulbs (topped) either before or after field-curing. Onions that were lifted 3 weeks after 50% top-down and topped before curing had the greatest incidence of rots in store. Increasing harvest maturity increased the mean skin color score of onions, and decreased markedly the mean number of intact outer skins. The timing of foliage removal had no effect on mean skin color score, but onions that were topped before curing had slightly more bulb skins than onions topped after curing. Timing of onion lifting to optimize bulb quality appears to be a trade-off between skin retention and color. These results confirm that traditional method of harvesting onions in New Zealand, where onions are lifted at 60–80% top-down, the bulbs are field-cured, and the foliage is removed after curing, is the simplest method and best compromise to ensure postharvest onion quality and successful storage.

4.10.4   TURMERIC

Optimum stage of harvesting plays an important role in obtaining high quality turmeric in terms of essential oil and curcumin contents. According to Kumar and Gill (2009), harvesting on 12th March produced maximum fresh rhizome yield of 28.94 t ha^{-1} of turmeric which was statistically on par with 20th February (27.61 t ha^{-1}) and 30th January (26.78 t ha^{-1}) harvesting, but was significantly better than all the earlier harvesting dates. A similar trend was observed in processed turmeric yield. The number and weight of rhizomes improved significantly with delay in harvesting. The oil and curcumin content also increased with delay in harvesting.

4.10.5   GINGER

A study was conducted by Kandiannan et al. (2016) under rainfed condition to observe the sprout emergence, tiller production and yield when the ginger was allowed to grow in the second season/year without harvest in first year/season. After harvesting in the second year, fresh rhizome was cleaned and sorted into first year produce and second year produce based on their appearance and texture and weighed separately, their proportion was estimated and multiplication rate from first generation crop to second generation crop was calculated. The mean shoot emergence and tiller production were five and 19, respectively. Average yield in first, second years and total yield were

209, 566 and 775 g plant^{-1}, respectively. The share between first and second year yield were 27.8 and 72.2%, respectively and mean multiplication rate was 3.5 times. Although, yield levels tend to increase in second season, the multiplication rate was much reduced. Hence, the practice of biennial harvest may not have clear yield advantage over regular annual harvest.

4.11   PLANT PROTECTION

4.11.1   BLACK PEPPER

4.11.1.1   Pollu Betle

Birah et al. (20110 reported that integrated module comprised with the application of neem cake application @ 250 g/vine and alternate foliar spray of 0.5% neem seed kernel extract and 0.5% quinalphos reduce the pollu beetle infestation. The pooled data revealed that integrated module (T_2) and organic module (T_1) recorded significantly lower berry damage (3.13 and 8.58%) than farmers' practice (T_3) (14.78%) and untreated control (T_4) (16.19%) during 2008–09. The maximum fresh yield of black pepper, 687.56 and 635.34 kg/ha, was recorded in integrated module followed by 385.50 and 374.67 kg/ha in organic module during 2008–09 and 2009–10, respectively. The yield in untreated control was only 213.33 and 207.59 kg/ha in both the seasons.

4.11.1.2   Yellow Disease

A black pepper (*Piper nigrum*) yellow is a newly recognized disease in Coorg (Kodagu) district of Karnataka, India. Symptoms include yellowing and curling of the leaves. At the advanced stage, vines become yellow and slender, and a generalized decline in yield is evident (Adkar-Purushothama et al., 2009).

Taufik et al. (2011) reported that yellow disease is a complex disease caused by *Fusarium sp, Phytophthora sp.*, and nematodes. Infected plants were quickly killed and were difficult for replanting, causing significant losses for the growers in Indonesia. Various control methods were examined including the use of biconrol agents and cover crop *Arachis pintoi*. Results showed that the treatment of biocontrol and *A. pintoi* promoted vegetative growth of pepper plants, and increased pepper height for up to more five times, and reduced yellow disease incidence to 30%.

4.11.1.3 Phytophthora Foot Root

Two hundred and seventy six plantations in 96 locations in major black pepper (*Piper nigrum* L.) growing areas of Karnataka and Kerala were surveyed by Bhat et al. (2005) for the distribution and incidence of viral disease. The incidence of the disease was highest in Wayanad District (45.4%) followed by Idukki District (29.4%) in Kerala. In Karnataka, Kodagu District (14.9%) had the highest incidence of the disease followed by Hassan District (5.2%). In general, the incidence and severity of the disease was higher in black pepper plantations situated at higher altitudes of Kerala such as Idukki and Wayanad districts. Mosaic, reduction in leaf size and internode length leading to stunting of the vine, and bright yellow mottling along veins were the two foliar symptoms observed on diseased vines. DAC-ELISA of symptomatic black pepper vines with antisera to different viruses confirmed viral infection either by CMV or BSV or by both. All cultivars and improved varieties including hybrids were susceptible to the disease under natural conditions. Vines of all ages raised on all kinds of standards were also found affected by the disease. Among the several weeds found in and around black pepper plantations, a few of them showed typical viral like symptoms, which might act as potential inoculum source. Though 12 species of insects were collected from diseased vines from different locations, no species was specifically associated with diseased vines.

Mammootty et al. (2008) conducted an experiment with fifty genotypes of black pepper including released varieties and cultures were screened against *Phytophthora capsici* foot rot disease in nursery to identify resistant materials. None of the genotypes screened were immune to *P. capsici*. 'Kalluvally II,' 'Panniyur 5,' and 'Kalluvally IV,' however, showed less than 60% leaf infection and were statistically *at par* with 'Balankotta,' 'Cheriyakaniakadan,' and 'Shimoga.' 'Kalluvally II' also suffered relatively lower mortality rates (26.67%), implying some tolerance against *P. capsici*. 'Panniyur 5' was moderately susceptible while the remaining cultivars were clearly susceptible to the disease, as they showed higher leaf infection and mortality.

Lokesh et al. (2008) studied on the efficacy of systemic fungicides and antagonistic organism for the management of Phytophthora foot rot of black pepper in arecanut cropping system. They reported that black pepper vines were less affected when the vines were treated either with metalaxyl gold MZ 64 WP @ 2.5 g/vine or potassium phosphonate (0.5%) as spray (@ 2 L per vine) and drench twice (@ 3 L vine) during before on set of monsoon and second application in the month of August coupled with soil application of antagonistic organism, i.e., Trichoderma harzianum @ 50 g (cfu 10^7) along with 1 kg of neem cake to the root zone of the vine. Application of systemic fungicides alone, i.e., metalaxyl MZ 68 WP @ 2.5 g/vine or potassium phosphonate (0.5%) as spray (@ 2 L per vine) and drench twice (@ 3 L per vine) also showed effectiveness in the combating the disease.

Lokesh et al. (2012) reported that phytophthora foot rot (*Phytophthora capsici* Leonian) of black pepper could also be managed effectively by application to vines with potassium phosphonate (@ 0.3%) as spraying (@ 2 L per vine) and drenching (3 L per vine) and bioagent *Trichoderma harzianum* 50 g with one kg of neem cake as soil application during first week of June and third week of August to the root zone. The protected vines exhibited minimum leaf yellowing, least defoliation, minimum death of vines and highest yield (green berry yield and projected yield). However, bioagents application, i.e., Consortium of bacteria @ 108 cfu/g (for growth, nematode and *Phytophthora* suppression – IISR-6 and IISR 859) as spraying (@ 2 L per vine) and drenching (@ 3 L per vine) and *Trichoderma harzianum* (MTCC 5179) 50 g with one kg of neem cake as soil application around the root zone of the vine twice (June and August) also significantly reduced the disease with respect to less leaf infection, less yellowing, less defoliation and less death of vines.

Vijayaraghavan and Abraham (2007) studied on the mechanism of antagonism of *Trichoderma* spp. on *Phytophtora capsici* causing foot rot of black pepper. On the basis of microscopic examination, *Trichoderma viride, T. longibrachiatum* and *T. harzianum* fungal antagonists were found parasitic on *Phytophthora capsici* as it was evidenced by excessive coiling, penetration and disintegration of the host hyphae.

Experiment on *in vitro* screening of 125 endophytic fungi of black pepper against *Phytophthora capsici* indicated that 23 isolates showed more than 50% inhibition (Sreeja et al., 2016). The nematicidal activity of metabolites from endophytic fungi was also tested on *Radopholus similis* and the isolate BPEF73 (*Daldinia eschscholtzii*) showed highest mortality up to 60%. The sequence analysis of the isolates showed maximum identity with

Annulohypoxylon nitens (BPEF25 and BPEF38), *Daldinia eschscholtzii* (BPEF41 and BPEF73), Fusarium spp. (BPEF72 and BPEF75), *Ceriporia lacerata* (BPEF81), Diaporthe sp. (BPEF11), and *Phomopsis* sp. (BPEF83).

4.11.1.4 Nematode

Aravind et al. (2009) conducted an experiment on isolation and evaluation of endophytic bacteria against plant parasitic nematodes infesting black pepper (*Piper nigrum* L.). In their study, 80 isolates of endophytic bacteria were isolated from different varieties of black pepper (*Piper nigrum* L.) grown at different locations in India. Another 30, isolates were obtained from tissue cultured black pepper plants. These isolates were tentatively grouped into *Bacillus* spp. (32 strains), *Pseudomonads* (26 strains), *Arthrobacter* spp. (20 strains), *Micrococcus* spp. (10 strains), *Curtobacterium* sp. (one strain), *Serratia* (one strain) and twenty unidentified strains based on morphology and biochemical tests. Their nematicidal properties, when tested in an *in vitro* bioassay using *Meloidogyne incognita* juveniles, varied from 0–31.03%. Consortia of these endophytic bacteria were made and evaluated in nurseries for their nematode suppression and growth promotion in black pepper rooted cuttings. All the bacterial consortia were able to suppress nematodes, *M. incognita* and *Radopholus sintilis*, significantly. The maximum number of cuttings (2–3 cuttings/plant) was obtained with phorate treatment followed by treatment with consortia 1 and 4 indicating the potential of these bacteria to be used as nematode biological control agents.

4.11.2 SMALL CARDAMOM

4.11.2.1 Katte Disease

Mosaic or marble or katte disease caused by cardamom mosaic virus (CdMV) is an important production constraint in all cardamom growing regions of the world. Biju et al. (2010) conducted an a survey for detection of CdMV affecting areas of small cardamom in India with 84 cardamom plantations in 44 locations of Karnataka and Kerala were surveyed. The incidence of the disease ranged from 0 to 85%. The incidence was highest in Madikeri (Karnataka) while no incidence was recorded in Peermade (Kerala). In general, incidence and severity of the disease was higher in cardamom plantations of Karnataka. A procedure for total RNA isolation from cardamom and detection of CdMV through reverse transcription-polymerase chain reaction (RT-PCR) using primers targeting the conserved region of coat protein was standardized and subsequently validated by testing more than 50 field cardamom samples originating from Karnataka and Kerala states. The method can be used for indexing the planting material and identifying resistant lines/cultivars before either they are further multiplied in large scale or incorporated in breeding.

The production of small cardamom (*Elettaria cardamomum* Maton) is limited due to number of fungal diseases. *Pythium vexans* de Bary, *Rhizoctonia solani* Kuhn, *Phytophthora meadii* Mc Rae and *Fusarium oxysporum* Schl. are the four major fungal rot pathogens of small cardamom in India causing crop losses ranging up to 50%. Vijayan and Thomas (2006) conducted an experiment on screening of improved selections and hybrids of small cardamom (*Elettaria cardamomum* Maton) for rot tolerance. The screening results showed the hybrid MHC 24 tolerant to all the three rot pathogens viz., *Pythium vexans*, *Phytophthora meadii* and *Fusarium oxysporum*. The hybrid MHC 26 was found to be with least percentage disease index to *P. vexans* followed by MCC 346, MHC 18 and MHC 24. In the case of *R. solani* MCC 85 showed least percent disease index followed by MHC 13, MCC 260, MCC 346 and RRTL-1. The hybrid line MHC 24 showed a least percent disease index to *F. oxysporum* followed by MHC 10, MHC 18 and MCC 200. The RRTL-1 was found susceptible to *P. meadii*. However, it was moderately tolerant to *R. solani* and *P. vexans*. The genotypes MHC 26, MHC 24, MHC 18, MCC 85 and MCC 346 can withstand the infection of all the rot pathogens screened and may be pursued further for exploiting the resistance in small cardamom breeding programme. Among the genotypes screened, none was found resistant to the rot pathogens.

Incidence of root tip rot resulting in severe foliar yellowing of small cardamom (*Elettaria cardamomum* Maton) has become a wide spread disease in recent years. A detailed study was made by Vijayan et al. (2006) on the isolation and identification of the causal organism, its effect on affected plants in expressing disease symptoms and also disease management studies in the field. Isolations of fungus from infected plants revealed association of *Fusarium oxysporum*. The isolated fungus was pathogenic to both the seedlings as well as to mature plants in the field. There was considerable reduction in the contents of chlorophyll 'a' and 'b' in the leaves of infected plants showing prominent yellowing. Disease management studies carried out in the field showed that selected systemic fungicides such as carbendazim, hexaconazole, and thiophanate methyl effectively controlled the disease. The use of consortium of bioagents viz; *Trichoderma harzianum* and *Pseudomonas fluorescens* was also equally effective in reducing the severity of the disease.

Rot diseases such as capsule rot, rhizome rot and root rot and foliar disease like chenthal are the major diseases of small cardamom (*E. cardamomum*). Gopakumar et al. (2006) conducted a field trial for the management of these diseases in disease prone field at Lower Pulney hills, Kodaikanal, TamilnaduResults showed that basal application of *Trichoderma harzianum* + *Pseudomonas fluorescens* + *Bacillus subtilis* and spraying of *P. fluorescens* + *B. subtilis* reduced rhizome rot, root rot and chenthal diseases in cardamom.

4.11.2.2   Cardamom Mosaic Virus

Prasath et al. (2010) conducted an experiment on inheritance of CdMV resistance in cardamom. In their experiment, cardamom lines, NKE 9 and NKE 12, which were resistant to CdMV, were crossed to two susceptible genotypes, viz., CCS 1 and RR 1 to determine the nature of inheritance of resistance. It was revealed from the results that the CdMV resistance in NKE 9 and NKE 12 is genetically governed. The F1 hybrids between resistant and susceptible genotypes were resistant. The segregation pattern for disease reaction in F2 and BC1 generations of the two crosses suggested that CdMV resistance in NKE 9 and NKE 12 could be controlled by two dominant complementary genes. Over all it could be hypothesized that the resistance to CdMV is quantitative, with possibly two major factors, and dependent on gene dosage with completely dominant gene action. This is the first report of CdMV inheritance in cardamom.

4.11.3   GINGER

4.11.3.1   Bacterial Wilt

One hundred and fifty one somaclones of ginger cultivars viz., Maran and Rio-de-Janeiro regenerated through bud culture were screened by Paul et al. (2009) against bacterial wilt disease caused by *Ralstonia solanacearum*. In artificial inoculation of the bacterial wilt pathogen, all the clones took infection, but subsequent germination of rhizomes was observed in four somaclones viz., 970 M, M VI, 364 R, and R XI. Somaclones of Maran were found resistant to the disease as compared to clones of Rio-de-Janeiro. Based on reaction of ginger somaclones to different screening methods and evaluation for yield, three somaclones viz., 970 M, M VI, and 364 R, with high yield showing resistance to bacterial wilt disease could be located which could be used for further field evaluation/production programmes.

4.11.3.2 Rhizome Rot

Sagar et al. (2007) conducted an experiment on management of rhizome rot of ginger by botanicals. In their experiment, efficacy of fourteen plant extracts was evaluated against *Pythium aphanidermatum* and *Fusarium solani* at 5 and 10% concentrations. Among 14 plant-extracts, *Azadirachta indica* showed maximum inhibition of mycelial growth (63.72%) of *P. aphanidermatum*. Among 14 plant extract tested against *F. solani*, maximum inhibition of mycelial growth was noticed in *Ferula asafeotida* powder extract (68.51%) followed by Ocimum leaf extract (60.16%).

Kim et al. (2012) studied on the physico-chemical properties and microbial populations and diversity in soils where rhizome rot disease of ginger frequently occurs. Analysis of soil physical properties revealed that there was a difference between healthy soils and diseased soils. Bulk density and soil moisture were lower in healthy soils. Healthy soils had low pH, ammonium nitrogen, and available phosphoric acid, and diseased soils showed opposite results. The distribution of cellulolytic microorganisms was high in healthy soils. The molecular analysis of soil bacteria showed that healthy soils had a high distribution ratio of Firmicutes and gamma Proteobacteria, and diseased soils had a greater distribution of alpha Proteobacteria.

4.11.4 TURMERIC

4.11.4.1 Root Knot Nematode

Root knot nematode problem in turmeric and ginger crops and it is managed with nematicides, cover crops and organic amendments. Realizing the scope of biological control in these crops, a series of experiments were conducted by Eapen et al. (2008) at Indian Institute of Spices Research, Calicut, Kerala to screen and evaluate various fungal bioagents for control of root knot nematodes (*Meloidogyne incognita*) infesting ginger and turmeric under field conditions. Ten antagonistic fungi were evaluated in different field experiments conducted in root-knot infested ginger and turmeric fields at two locations. The most promising isolates that suppressed root knot nematodes were *Aspergillus nidulans* (Is. 10), *Fusarium oxysporum* (Is. 11), *Paecilomyces lilacinus* (Is. 36), *Trichoderma viride* (Is. 25), *Verticillium lecanii* (Is. 35) and *Pochonia chlamydosporia* (Is. 32). From these, three fungi viz. *F. oxysporum*, *T. viride* and *P. chlamydosporia* were further tested in ginger fields using two delivery systems, soil bed application and seed rhizome dipping generally applicable in dry shed treatment. The final results showed that *P. chlamydosporia* significantly suppressed root knot nematodes in ginger and gave the maximum yield irrespective of the mode of application.

4.11.4.2 Leaf Blight

Hseu et al. (2009) reported that irregular browning lesion of the leaf margin was found on turmeric plants cultivated in Minjeng and Tstun of NanTou County, Taiwan. Water soaked lesions first appeared at leaf hydathodes and then spread along the leaf midrib. In later stage, lesions became darken and usually accompanied with yellowing halo and the color became darker. The same symptoms also appeared in the leaf sheath and finally the plant wilted. A Gram-negative, rod-shaped bacterium was consistently isolated from the diseased tissues except Acidovorax avenae subsp. avenae. The unknown bacterium was identified as *Herbaspirillum huttiensis* (*H. huttiense*) based on its physiological and biochemical characteristics, the Biolog GN MicroPlate Identification System, fatty acid analysis, PCR, 16S rDNA and pathogenicity tests. *In vitro* screening for the efficacy of various agrochemicals to inhibit bacterial growth on NA plates showed that all tested chemicals except Ridomil MZ, including copper bactericides, antibiotics, oxolinic acid, and carbamates, were effective. Among them, tetracycline was the most effective.

4.11.4.3 Turmeric Leaf Roller

Ganguli et al. (2009) conducted an experiment on the biology of turmeric leaf roller skipper butterfly, *Udaspes folus* Cramer in Chhattisgarh. According to them, the total life cycle was completed in 28 days. The mean larval period was 18.5 days with 5 instars. There were a pre-pupal stage of one day and the mean pupal period lasted for 8.5 days. The average adult longevity in the case of male was 4.0 days and female 8.5 days. There were several generations in a year. The butterfly displayed pupal diapause during winter, i.e., from November–December up to March.

4.11.4.4 Rhizome Rot

In an experiment, Khalko and Chowdhury (2008) used three biological control agents, namely *Trichoderma viride*, *Pseudomonas fluorescence* and *Bacillus subtilis*, and farmyard manure (FYM) were used in different combinations both as seed treatment and soil application to develop an effective biological control method against rhizome rot disease of turmeric. The combination of seed treatments with both *Trichoderma viride* and *Pseudomonas fluorescence* along with FYM showed highest disease reduction (63.65% less disease) compared to the control than other treatments. The soil application of *Trichoderma* and *Pseudomonas* along with FYM and the combination of seed treatment and soil application with these two biological control agents also gave very good results in terms of percent reduction in rot over the control. Regarding rhizome yield, combined seed and soil treatment with both *Trichoderma* and *Pseudomonas* along with FYM was the best, recording the highest rhizome yield. *Bacillus subtilis* had a less pronounced effect on disease incidence and rhizome yield.

Bharathi and Sudhakar (2011) reported that soil application of FYM @ 2.5/ha + *Trichoderma viride* @ 5 kg/ha + neem cake @ 0.2 t/ha was the most effective treatment in controlling disease (25 PDI) and was followed by metalaxyl @ 1.5% + neem cake @ 0.2 t/ha (27 PDI). However, the disease incidence was 48% in untreated control. Vermicompost + neem cake was the most effective treatment in recording maximum yields (28.81 t/ha). The low disease incidence could be attributed to mycoparasitization, antibiosis and competition of the biological control agent against the disease causing fungus.

4.11.5 CORIANDER

Bioefficacy of neem products, insecticides and admixtures of neem and methyl demeton against the incidence of coriander aphid, *Hyadaphis coriandari* Das

was studied by Gupta and Pathak (2009). The results indicated that the incidence of coriander aphid was reduced to the lowest extent with a maximum
yield and net profit in the crop treated with neem oil 1% (983 kg/ha and Rs.
12,165 per ha), neem kernel extract (in cow urine) – 3% + methyl demeton –
0.03% (967 kg/ha and Rs. 11,295 per ha), cartap hydrochloride 0.1% (967 kg/
ha and Rs. 9660 per ha) and phenthoate 0.1% (917 kg/ha and Rs. 9000 per ha).

Stemgall of coriander (*Coriandrum sativum* L.) due to *Protomyces mac-
rosporus* causes much damage to the crop. Dabbas et al. (2010) reported
that the seed treatments with *Trichoderma viride* @ 4 g/kg seed + soil treatment with *Trichoderma viride* @ 2 kg/ha gave the lowest disease intensity of
6.12% with maximum grain yield 14.51 q/ha and highest percentage disease
control (51.31) over control treatment.

4.11.6 CUMIN

4.11.6.1 Fusarium Wilt

Sobhanipour et al. (2008) conducted an experiment on biological control
of *Fusarium* wilt of cumin by antagonistic bacteria. In their experiment,
six antagonistic strains of *Pseudomonas* were evaluated for biological control of two *Fusarium oxysporum* f.sp. *cumini* isolates, the causal agent
of wilt cumin (*Cuminum cyminum*), in laboratory and greenhouse conditions. The Pseudomonas strains were identified as *Pseudomonas aeruginosa* (B-28 and H-111), *P. fluorescens* (D-54, Q-229 and K-146), and *P.
putida biovar* B (C- 43) based on their phenotypic features and protein
electrophoretic pattern. Strain B-28 was the most effective antagonist in
dual culture. Autoclaved culture filtrates and volatile compounds of strain
H-111 had the highest growth-inhibiting effect (30.54%) at 25% concentration. All these antagonists can be used with different concentrations of
benomyl fungicide. Cumin seed coating with strain B-28 and mixture of
strains caused 36.25% and 46.25% reduction of number of damping off
and plant infection, respectively. Application of fluorescent pseudomonas
strains can protect cumin wilt, caused by *F. oxysporum* f.sp. *cumini*.

4.11.6.2 Cumin Blight and Wilt

Arora et al. (2008) carried out an experiment on evaluation of cumin varieties
against blight and wilt diseases with time of sowing. From their experiment

it was evident that the period of 10 weeks from sowing was favorable for initiation and further spread of the blight disease. The wilt disease symptoms appeared when the crop was 8 weeks old from the date of sowing. The lowest blight and wilt disease incidence was observed in December-sown crop with 75% relative humidity (RH). The highest blight and wilt disease incidence was observed in October-sown crop with 65% RH. Among the cultivars, none was found totally resistant to either blight or wilt.

An experiment was conducted by Deepak et al. (2008) to assess their possible use as bioagents for several antagonistic fungi on growth of two cumin fungal pathogens under *in vitro* and field conditions. Under *in vitro* conditions maximum inhibition (82.86%) of radial growth of *Fusarium oxysporum* f. sp. *cumini* was observed with the treatment of *Trichoderma harzianum* strain I, whereas maximum inhibition (85.45%) of the mycelial growth of *Alternaria burnsii* was observed in the presence of *Trichoderma harzianum* strain II. The antagonists who showed maximum inhibition of the pathogen in laboratory conditions were applied in field conditions as soil treatment/seed treatment or as foliar spray. The incidence of wilt disease was found to be lowest (PDI 27.40%) when soil was treated with *Trichoderma harzianium* strain I at the rate of 24 g/6 m² (weight of fungus with sorghum seeds). Minimum blight disease incidence was observed when *T. harzianum* strain II was applied to the soil at the rate of 24 g/6 m² (36.15%) or when 10% spore suspension of *T. harzianum* strain II was applied as seed treatment at the time of sowing and as spray at the time of flowering (PDI 35.10%). Thus, treatments of *Trichoderma harzianum* strain I for wilt and *Trichoderma harzianum* strain II for blight diseases of cumin under both the conditions @ 24 g/6 m² or 40 kg/ha seems promising for sustainable management of crop diseases.

Four components of integrated management namely, soil solarization, crop rotation, chemicals and biocontrol agents were tested by Jadeja and Nandoliya (2008) under field condition at Junagadh (Gujarat) for the management of wilt of cumin (*Cuminum cyminum*) caused by *Fusarium oxysporum* f. sp. *cumini*. Growing of sorghum (*Sorghum bicolor*) or maize (*Zea mays*) during kharif season did not reduce wilt incidence during the following rabi season. Soil solarization with 25 micro m LLDPE plastic cover for 15 days in summer proved most effective in reducing wilt incidence to 26.27% as against 44.90% in non-solarization and increasing yield to 396 kg/ha as against 286 kg/ha in non-solarized plots. Application of carbendazim granules @ 10 kg/ha one month after sowing or *Trichoderma viride* in organic carrier @ 62.5 kg ha⁻¹ at sowing time were also effective. Integrating soil solarization followed by growing of sorghum in kharif and application of either carbendazim granules @ 10 kg/ha one month after sowing or application of *T. viride* in organic carrier @ 62. kg/ha was effective for the management of cumin wilt.

4.11.7 FENUGREEK

4.11.7.1 Root Rot

An experiment was conducted by Aiyanathan and Salalrajan (2008) in Tamil Nadu, India, to evaluate the efficacy of organic amendments against root rot disease of fenugreek (*Trigonella foenum-graecum*) caused by *Rhizoctonia solani*. The root rot incidence was reduced to 4.8, 5.7, and 6.0% by the application of neem cake, farmyard manure and poultry manure, respectively (from 25.6% in the untreated control). These amendments reduced root rot incidence by 81.2, 77.7, and 76.5%, respectively, over the control. Goat manure and castor cake did not reduce root rot incidence.

4.11.8 FENNEL

4.11.8.1 Ramularia Blight

Patel and Patel (2008) conducted an experiment to manage Ramularia blight disease (*Ramularia foeniculi*) of fennel (cv. Gujarat Fennel 2) effectively and economically. All fungicidal treatments significantly reduced the disease intensity compared to the control. Among the fungicides, mancozeb was significantly superior which resulted in the minimum disease intensity (21.67%). The next best treatment was Carbendazim + mancozeb (24.68%), tridemorph (26.20%), difenoconazole (32.19%) and propineb (42.21%). The maximum disease control (70.64%) was recorded in the plot sprayed with mancozeb. The next best disease control was with Carbendazim + mancozeb (66.56%), followed by tridemorph (64.50%), difenoconazole (56.38%) and propineb (42.80%). Seed yield was maximum with three sprays of mancozeb (1595 kg/ha), followed by Carbendazim + mancozeb (1341 kg/ha) and were statistically at par with each other. Tridemorph (1103 kg/ha) and difenoconazole (1073 kg/ha) also proved effective in increasing seed yield.

Parashar and Lodha (2008) found that total soluble sugars and reducing sugars were higher in Ramularia blight-infected plant parts, while starch contents were higher in healthy plants. alpha-Amylase and invertase (beta-fructofuranosidase) activities were higher in blight-infected plants than in healthy ones.

4.11.8.2 *Alternaria Petroselini*

According to the Infantino et al. (2009) during 2007, fennel (*Foeniculum vulgare*) plants grown in Metaponto (Matera, Italy) were observed with black depressed lesions on the basal leaves. Isolations were carried out from infected leaves on PDA amended with streptomycin sulfate and ampicillin. An *Alternaria sp.* was consistently isolated, then transferred to potato carrot agar (PCA) under near-UV light for morphological characterization. Based on morphological and molecular analyzes, the pathogen was identified as *Alternaria petroselini*. A pathogenicity test was conducted on adult fennel plants by spraying a conidial suspension of the fungus. One week after inoculation, small dark spots were observed, that soon developed into necrotic areas on the fennel stalks. The same fungus was reisolated from the infected stalks. This is thought to be the first report of *Alternaria petroselini* on fennel in Italy.

4.11.9 CHILI

4.11.9.1 Leaf Curl

The effect of some cultivation practices on leaf curl disease on chili and the efficacy of some insecticides in controlling the vector of the disease (*Bemisia tabaci*) were investigated in West Bengal, India. The insecticide treatments were: T1, profenofos (Curacron); T2, monocrotophos (Nuvacron); T3, carbofuran-3G (Furadon); T3 + T1 (1 or 2 sprays); and T3 + T2 (2 and 3 sprays). The maximum disease percentage (2.10 and 14.81%) was observed during April. T3 + T2 at 2 or 3 sprays was the most effective in controlling the disease and the highest disease reduction was 8.33 and 9.52%, respectively, at 120 days after transplanting. In the winter season, the lowest incidence (2.38%) was also obtained with T3 + T2.

4.11.9.2   Thrips

Field experiments were conducted during kharif seasons in Andhra Pradesh, India to study the efficacy of different seed treatment insecticides against *Scirtothrips dorsalis* in chili cv. LCA-206. Imidacloprid 75% WS at 5 and 8 g/kg seed, acetamiprid 20% SP at 5 and 10 g/kg seed and thiomethoxam (thiamethoxam) 75% WS at 5 and 10 g/kg seed were used as treatments. Thiomethoxam 75% WS at 10 g/kg seed was significantly superior in efficacy against thrips up to 15 DAS, followed by thiomethoxam 75% WS at 8 g/kg seed. At 30 DAS, among the treatments, the efficacy of the insecticides was found to be non-significant, but significantly superior compared to the control. Results showed that none of the treatments had shown any effect on seed germination and phytotoxicity was also found to be nil.

4.11.9.3   Die Back and Fruit Rot

The efficacy of *Trichoderma harzianum*, i.e., partially purified toxin (Th; 1.0%), and sorghum-based (SB; 0.2%) and talc-based (TC; 0.4%) formulations, was evaluated against *Colletotrichum capsici*. On potato dextrose agar, TH resulted in the greatest inhibition of conidial germination (96.2%) and mycelial growth (97.0%) of *C.capsici*. Under greenhouse conditions, *Capsicum annuum* cultivars Pusa Sadabahar and Navjyoti were treated with *T. harzianum* formulations, Neemarin and neem oil during sowing (seed treatment), transplanting (seedling dip), and from the seedling to the transplanting stage (foliar spray). TH was superior in disease and fruit rot control in Pusa Sadabahar (39.6 and 44.4%, respectively) and Navjyoti (48.5 and 53.5%). In a field experiment in Ghaziabad, Delhi, India, during 2002–03, *T. harzianum* formulations, Neemarin and neem oil were used for disease control in both cultivars as seed treatment, seedling dip and foliar sprays. Spraying of *T. harzianum* formulations with Neemarin and neem oil in early morning from September to October inhibited disease development. TH gave the greatest reduction in disease intensity and fruit rot in Pusa Sadabahar (61.1 and 63.7%)

and Navjyoti (56.5 and 46.8%), and the highest fruit yields in both cultivars (27.4 and 29.9 quintal/ha); this treatment also reduced postharvest fruit rot (89.8%) and increased the shelf life of fruits by delaying the appearance of disease symptoms (12.2 days after harvesting). SB and TB were also effective in disease control, in increasing the yield of *Capsicum annuum*, and protection against postharvest fruit rot.

4.11.9.4   Yellow Mite

Fenazaquin (Magister 10 EC; 50 and 100 g a.i./ha), abamectin (Vertimec 1.8 EC; 5 and 9 g a.i./ha), clofentezine (Apollo 50 SC; 250 and 300 g a.i./ha) and dicofol as standard (Colonel-S 18.5 EC; 185 g a.i./ha) were evaluated for their efficacy against the yellow mite (Polyphagotarsonemus latus) in chili (cv. Bullet) at Kalyani, West Bengal, India. Natural infestation by the yellow mite was allowed when the crop (sown during the first week of February) was one month old. The mite population started building up from the second fortnight of March. Acaricidal application was done 3 times at 15-day intervals starting at the time when the population was almost evenly distributed. Abamectin at both dosages recorded the highest mortality (population reduction) of the moving stage as well as different life stages (male, female and larva) of the yellow mite, followed by fenazaquin. Clofentezine was the least effective acaricide.

4.11.9.5   Aflatoxin Contamination

Paterson (2007) studied on prioritization of disease and pest constraints in chili by highlighting aflatoxin concentrations to assist local farmers in control. All samples contained aflatoxin B1 and high levels were obtained from all ground samples. A direct relationship was observed between aflatoxin B1 and aflatoxin B2 concentrations. There was no relation between aflatoxin and *Aspergillus flavus* detection. Chili production in Pakistan may be heavily constrained by aflatoxin contamination. Simply removing *A. flavus* may be insufficient for control. Aflatoxins from chili may be a threat to the health of populations and a constraint on development in Pakistan.

4.11.10 ONION

4.11.10.1 Purple Leaf Blotch and Stemphyllium Blight

Onion leaves are subject to infection by *Alternaria porri*, which causes purple leaf blotch of *Allium* spp. PLB is an important disease of *Allium* spp. worldwide, especially in warm and humid environments causing up to 59% losses in onion bulb yield (Gupta & Pathak, 1988). Stemphylium leaf blight is caused by *Stemphylium vesicarium*. Infection of onion by *Alternaria porri* and *Stemphylium vesicarium* was investigated by Suheri and Price (2000) under a range of controlled temperatures (4–25°C) and leaf wetness periods (0–24 h). Infection of onion leaves occurred after 16 h of leaf wetness at 15°C and 8 h of leaf wetness at 10–25°C, and infection increased with increasing leaf wetness duration to 24 h at all temperatures. Interruption of a single or double leaf wetness period by a dry period of 4–24 h had little effect on lesion numbers. Conidia of *A. porri* and *S. vesicarium* separately or in mixtures caused similar numbers of lesions. *Alternaria porri* and *S. vesicarium* are both potentially important pathogens in winter-grown Allium crops and purple leaf blotch symptoms were considered to be a complex caused by both pathogens.

4.11.10.2 Black Mould Disease

Black mould disease caused by *Aspergillus niger* van Tieghem is a limiting factor in onion (*Allium cepa* L.) production worldwide. Gupta et al. (2012) evaluated some fungicides, plant extracts and bioagents against black mould disease of onion.. Amongst fungicides used as seed treatment, *in vitro* Bavistin (Carbendazim 50% WP, 2.0%) proved to be most effective against *Aspergillus niger* followed by Thiram (80% WP, 2.5%), Captan (50% WP, 2.5%), Indofil M-45 (50% WP, 2.5%) and Topsin M (75% WP, 2.5%) in improving seed germination and vigor index, by reducing pre- and post emergence mortality and number of seedlings showing symptoms. Similarly, among bioagents and plant leaf extract used, *Trichoderma viride* (5.0 mL (108 cfu/mL)/10 g seeds) followed by *Trichoderma pseudokoningii* (5.0 mL (108 cfu/mL)/10 g seeds) and Safeda (5.0 mL/10 g seeds) followed by neem leaf extracts (5.0 mL/10 g seeds) proved to be effective in improving seed germination and vigor index, by reducing pre- and post-emergence mortality and number of seedlings showing symptoms.

4.11.10.3 Bacterial Onion Diseases

Epiphytic microorganisms, isolated from the olive knots, apple fruits and trees, quince, compost and water from different areas were screened for antagonistic activity against *Pseudomonas marginalis*, *Pseudomonas viridiflava*, *Xanthomonas retroflexus*, and *Pantoea ananatis* on onion bulbs. From 77 microorganisms tested for antagonistic properties against bacterial onion diseases, the strain 2066–7 of *Pantoea agglomerans* was selected. Complete control against *Pseudomonas marginalis* and *Pseudomonas viridiflava* at 107 CFU.ml^{-1} concentrations of 2066–7 and an inhibition percent higher than 90% against *Xanthomonas retroflexus* and *Pantoea ananatis* were obtained on wounded onion bulbs inoculated with 105 CFU.ml^{-1} of pathogens under cold conditions. The inhibitions percent were decreased under 25°C and 30°C (Sadik, 2013).

4.11.10.4 Downy Mildew

Efficacy of ten fungicides, viz Aliette, Antracol, Benlate, Cobox, Daconil, Derosal, Dithane, Polyram, Ridomil and Topsin-M, was tested at Peshawar against downy mildew of onion caused by *Peronospora destructor* Berk. Each of the fungicides was sprayed three times at an interval of 10 days following appearance of the disease symptoms. All the fungicides especially Ridomil was the most effective in reducing the disease severity and enhancing yield, followed by Topsin-M, Aliette, and Antracol. Sprays with Ridomil also resulted in the least number of dead plants, greatest plant height, most abundant leaves per plant, and largest number and weight of medium, large and total bulbs. The use of these fungicides is recommended in an integrated control strategy, incorporating other methods such as resistant varieties and prudent cultural practices.

4.11.10.5 Onion Thrips

Thrips are probably one of the most damaging pests of onion worldwide. Ibrahim (2010) studied on the seasonal abundance of onion thrips in Nigeria. Results indicated that November transplant had a peak population of onion thrips in late February (176 thrips/plant); December (416 thrips/plant) and

January (608 thrips/plant) transplants peaked in March, and February (148 thrips/plant) and March (86) transplants had peaks in April. Water traps indicate that the peak population of adult thrips was at the time of harvest in April, similar to November transplant. The early transplant (November) had peak thrips population at maturity and middle transplant recorded the peak population middle of the season and late transplant had their peaks early part of the growing season. Therefore, the findings of this work revealed that onion thrips in Sokoto, Nigeria, breed from January to May with peak in March.

4.11.10.6   Postharvest Rot of Onions

Penicillium aurantiogriseum Dierckx. was found one of the cause of postharvest rot of stored onions (*Allium cepa* L.) in Pakistan (Khokhar and Bajwa, 2015). Infected onion tissues were cultured on Czapek Dox Agar, 2% Malt Extract Agar, Czapek Yeast Autolysate Agar and 25% Glycerol Nitrate Agar at 25°C. *P. aurantiogriseum* was identified as the pathogen on the basis of morphological and molecular characteristics. Pathogenicity tests conducted on healthy onions under laboratory conditions showed typical rot symptoms after seven to fourteen days. This is the first report of postharvest rot of onions caused by *P. aurantiogriseum* in Pakistan.

4.12   POST HARVEST AND PROCESSING

4.12.1   BLACK PEPPER

According to Suresh et al. (2007) significant loss of spice active principles (curcumin, capsaicin and piperine, the active principles of turmeric (*Curcuma longa*), red pepper (*Capsicum annuum*) and black pepper (*Piper nigrum*), respectively) was observed when the spices were subjected to heat processing. Curcumin loss from heat processing of turmeric was 27–53%, with maximum loss in pressure-cooking for 10 minute. Curcumin loss from turmeric was similar even in the presence of red gram. In the presence of tamarind, the loss of curcumin from turmeric was 12–30%. Capsaicin losses from red pepper ranged from 18% to 36%, with maximum loss observed in pressure-cooking of 20 minutes. Presence of either red gram or tamarind or both did not influence the loss of capsaicin. Piperine losses from black pepper ranged from 16% to 34%, with maximum loss observed in pressure-cooking of 10 minutes. The loss was somewhat lower in the presence of red gram. The results of this investigation indicated diminished availability of spice active principles from cooked foods when the food ingredients have been subjected to either boiling or pressure-cooking for few minutes.

Cryogenic grinding of black pepper at different temperatures and feed rates was conducted by Murthy and Bhattacharya (2008) and compared with that of conventional grinding at ambient temperature to ascertain the different quality parameters employing subjective and objective methods. The loss of volatile oil in the case of ambient grinding was about 50% as compared to cryogenic grinding. The loss of monoterpenes was high in ambient grinding, as there was a loss of volatile oil in terms of every monoterpene compounds. Cryogenic grinding technique was superior to ambient grinding in terms of monoterpenes retention in the powder. Sensory assessment of the ground samples indicated that cryogenically ground samples were distinctly high in top notes which represented freshness, and marginally high in basic notes also. A pilot plant model pin mill was employed for cryogenic grinding of black pepper at different feed rates and product temperatures. These two variables had significant effects on dependent variables, viz., volatile oil, and monoterpenes and sesquiterpenes contents. The optimum cryogenic conditions for maximum volatile oil content and a reasonable quantity of monoterpenes were 47 to 57 kg/ha of feed rate, and −20 to −15°C of product temperature.

4.12.2 GINGER

Nath et al. (2013) conducted an experiment to study the effect of slice thickness and blanching time on different quality attributes of instant ginger candy. In their experiment, an attempt was made to optimize the protocol for production of instant ginger candy. The experimental parameters considered were slice thickness (5.0–25.0 mm) and blanching duration (10–30 min) followed by dipping in 40 degrees B and 75 degrees B sugar solutions containing 2.0% citric acid respectively, for 1 and 2 h at 95°C and dried at 60°C for 1 h. The optimum product qualities in terms of hardness (2.08 kg), TSS (73.4%), acidity (1.31%), TSS:acid ratio (56.3), taste score (7.98) and overall acceptability (8.07) were obtained for slice thickness of 10.9 mm and blanching time of 24.9 minute.

Ji and Zhou (2010) conducted an experiment on processing technology of clarified ginger juices. In their experiment, for making clarified ginger juice from Laiwu ginger, the processing conditions, including color protection, starch gelatinization temperature and amylase hydrolysis of the ginger juice were studied. The results showed that the best citrate acid dosage for color protection of the ginger juice were 1.5 g/L and the optimal temperature and time for starch gelatinization were 90–95°C and 30 minutes, respectively. The most suitable conditions of amylase hydrolysis were pH of ginger juice 4.0, temperature 60°C, the amount of amylase 0.2 mL/L, and enzymatic hydrolysis time 30 minutes.

The extraction technology of ginger flavonoid from ginger peel with alcohol soaking was studied by Xu et al. (2012). The results showed that the optimal extraction conditions of ginger flavones were alcohol concentration 75%, temperature 60°C, ratio of materials to solvent 1:45 and extracting times 5.5 hours, under which the yield of ginger flavonoid reached 0.61%.

4.12.3 TURMERIC

The value of turmeric is due to its color and flavor, which is being given by curcumin, volatile oil and oleoresin. The end quality of turmeric is very much dependent on its post-harvest methods. Traditionally, open sun drying is the chief method adopted for processing. Jose and Joy (2009) conducted an experiment on drying of turmeric. In their study, freshly harvested turmeric rhizomes were collected from 30 stations and drying experiments were conducted by adopting three methods: (1) solar tunnel drying; (2) conventional drying; and (3) commercial drying. Various pre-drying and post-drying treatments were conducted. The results proved that conventional processing could maintain the intrinsic quality up to a certain level, but extrinsic quality could not be achieved. Solar tunnel drying method is an effective alternative to traditional open sun drying, where retention of curcumin, volatile oil and oleoresin was high, with less drying time.

4.12.4 CORIANDER

Total phenolic, flavonoid content and antioxidant activity of crude extract of seeds, roots, stem and leaves of coriander plant were determined by Saxena et al. (2016). Maximum phenolic content (62.6 and 50.141 mg gallic acid equivalents (GAE) g-1 extract) was observed in distilled water extract of fresh and dried roots followed by methanol extract (49.53 and 47.32 mg in green and dried stem respectively). Ethyl acetate extract showed more phenolics in dried stem (12.734 mg GAE g-1 extract) and leaves (8.62 mg GAE g-1 extract) as compared to green stem and leaves (1.808 and 5.433 mg, respectively). 1,1-Diphenyl-2-picrylhydrazin scavenging as a measure of antioxidant capacity was more in distilled water extract of green stem (94.49%) followed by methanol crude extract (76.256%) and ethyl acetate extract (59.706%).

Subcritical water extraction (SCWE), hydrodistillation and Soxhlet extraction were compared by Eikani et al. (2007) for the extraction of essential oil from coriander seeds. The results showed that the optimum temperature, mean particle size, and flow rate were 125°C, 0.5 mm, and 2 mL/minute. The subcritical water extraction (SCWE) was compared with both conventional methods in terms of the efficiency and the essential oil composition. Hydrodistillation and soxhlet extraction showed higher extraction efficiencies, but the SCWE resulted to the essential oils more concentrated in valuable oxygenated components.

4.12.5 FENUGREEK

Sawant and Thakor (2008) reported that moisture content decreased with dehydration time. Sun drying of fenugreek required a period of 8 h and mechanical drying varied between 4 to 6 h for temperature ranging from 40 to 60°C. It was found that sample dehydrated at 40°C was most acceptable compared to other samples. It required 6 h for dehydration.

4.12.6 BLACK CUMIN

Ramadan (2002) reported that n-Hexane (H) and chloroform: methanol (CM) were used to extract the oils of Nigella sativa seeds. A combination of column and thin-layer chromatography procedures on silica gel was performed to fractionate the main neutral lipid classes of seed oils. The fatty acid pattern of neutral lipid (NL) fractions, triacylglycerol (TAG) and sterol content were determined. TAGs were the major neutral lipid class (80.8–83.1% of the total NLs), whereas the NL profile was characterized by high levels of free fatty acids (14.3–16.2% of the total NLs). Linoleic acid (C18:2) was the predominant fatty acid followed by oleic (C18:1) and palmitic (C16:0) acids in all examined classes. Six TAG species were determined, but 2 of them, C54:3 (ECN = 48) and C54:6 (ECN = 42), were present to the extent of 74% or above of the total TAG content. The 4-desmethylsterols isolated from the unsaponifiable fractions were β-sitosterol (1135–1182 micro g/g oil) as the main component followed by Delta 5-avenasterol (925–1025 micro g/g oil), and Delta 7-avenasterol (615–809 micro g/g oil). Stigmasterol, campesterol, and lanosterol were detected in small amounts.

4.12.7   CHILI

At moisture content of 9.7, 7.4, 4.6, and 2.9% (db) for green chilies; 8.7, 5.7, 4.5, and 2% (db) for red chilies in open sun drying, indirect solar cabinet dryer, solar tray dryer and hot air oven, respectively. The time required to achieve the above levels of moisture content were 22, 19, 13, and 8 h for green and 21, 18, 13, and 8 h for red chilies in respective dryers. The rate of drying was highest in the solar tray dryer (57 to 0.03) followed by hot air oven (47 to 0.08), indirect solar cabinet dryer (33.8 to 0.24) and open sun drying (31 to 0.01) at similar climatic conditions. The cost of drying chilies was estimated to be Rs. 1.93/h, Rs 2.20/h, Rs. 1.87/h and 4.95 per kg in open sun, indirect solar cabinet, solar tray and hot air oven drying methods, respectively (Wade et al., 2014).

4.12.8   GARLIC

In sun drying, the garlic is spread in a thin layer on the ground and exposed directly to solar radiation, ambient temperature, wind velocity, relative humidity, initial moisture content, absorptive, exposure time and mass of product per unit exposed area (Papu et al., 2014). Effect of storage temperature on allicin content in conventionally-and organically-grown garlic bulbs were studied. Allicin content was higher in organically-grown than conventionally-grown garlic stored at room temperature after six months. At 0°C, allicin content was higher in conventionally-grown than organically-grown garlic. The decrease in allicin content can be explained by the prolonged dormancy period and hindered sprouting that suppressed the metabolic activity (Raslan et al., 2015).

4.12.9 NUTMEG AND MACE

A study was conducted on application of novel drying technology for better color retention of mace. Pulsed microwave assisted hot air drying was investigated at three different power levels 0.5 kW, 1 kW and 1.445 kW with 30 seconds pulsation at a hot air temperature of 45°C and the color values of mace were compared with the market and fresh sample using colorimeter. Further, the major flavor compound, myristicin in mace was analyzed (Meetha et al., 2016).

KEYWORDS

- **agrotechnology**
- **improvement**
- **literature**
- **production**
- **research findings**
- **spice crops**

REFERENCES

Adelberg, J. W., & Cousins, M. M., (2007). Development of Micro and Minirhizomes of Turmeric, *Curcuma longa* L., *in vitro*. *Acta Horticulturae*, *756*, 103–108.

Adkar Purushothama, C. R., Casati, P., & Quaglino, F., Durante, G., & Bianco, P. A., (2009). First report of a 'Candidatus Phytoplasma asteris'-related strain associated with a yellows disease of black pepper (*Piper nigrum*) in India. *Plant Pathology*, *58*(4), 789.

Aiyanathan, K. E. A., & Salalrajan, F., (2008). Organic Amendments for the Management of Root Rot of Fenugreek. *Indian Journal of Plant Protection*, *36*(2), 277–278.

Akbarinia, A., Khosravifard, M., Ashoorabadi, E. S., & Babakhanlou, (2005). Effects of irrigation interval on the yield and agronomic characteristics of black cumin (*Nigella sativa*). *Iranian Journal of Medicinal and Aromatic Plants Research*, *21*(1), 65–73.

Al Mofleh, I. A., Alhaider, A. A., Mossa, J. S., Al Sohaibani, M. O., Al Yahya, M. A., Rafatul-
lah, S., & Shaik, S. A., (2008). Gastro protective effect of an aqueous suspension of
black cumin Nigella sativa on necrotizing agents-induced gastric injury in experimental
animals. *Saudi Journal of Gastroenterology, 14*(3), 128–134.

Ali, A., Sammauria, R., & Yadav, R. S., (2009). Response of fenugreek (*Trigonella foenum-
graecum*) to various fertility levels and biofertilizer inoculations. *Indian Journal of
Agricultural Sciences, 79*(2), 145–147.

Ali, B. H., Blunden, G., Tanira, M. O., & Nemmar, A., (2008). Some phytochemical, pharmacological and toxicological properties of ginger (*Zingiber officinale* Rosc.), a review of
recent research. *Food and Chemical Toxicology, 46*(2), 409–420.

Alkimim, E. R., de S. David, A. M. S., Sousa, T. V., Rodrigues, C. G., & Amaro, H. T. R.,
(2016). Different harvest times and physiological quality of coriander seeds. *R. Bras.
Eng. Agríc. Ambiental, 20*(2), 133–137.

Ankegowda, S. J., & Krishnamurthy, K. S., (2008). Evaluation of small cardamom accessions
for moisture stress. *Journal of Spices and Aromatic Crops, 17*(2), 172–176.

Ankegowda, S. J., (2008). Optimum leaf stage for transplanting small cardamom seedlings
from primary nursery to polybag nursery. *Indian Journal of Horticulture, 65*(2), 252–
254.

Aravind, R., Dinu A., Eapen, S. J., Kumar, A., & Ramana, K. V., (2009). Isolation and evaluation of endophytic bacteria against plant parasitic nematodes infesting black pepper
(*Piper nigrum* L.). *Indian Journal of Nematology, 39*(2), 211–217.

Arora, D., Arora, D. K., Saran, P. L., & Lal, G., (2008). Evaluation of cumin varieties against
blight and wilt diseases with time of sowing. *Annals of Plant Protection Sciences,
16*(2), 441–443.

Attarha, M., Rosbahani, N., & Youssefi, P., (2008). Comparison of the effect of fennel essence
and gripe water syrup in infantile colic. *Scientific Journal of Kurdistan University of
Medical Sciences, 13*(1), 28–35.

Aydin, R., Karaman, M., Cicek, T., & Yardibi, H., (2008). Black cumin (*Nigella sativa* L.)
supplementation into the diet of the laying hen positively influences egg yield parameters, shell quality, and decreases egg cholesterol. *Poultry Science, 87*(12), 2590–2595.

Azizi, K., & Kahrizi, D., (2008). Effect of nitrogen levels, plant density and climate on yield,
quantity and quality in cumin (*Cuminum cyminum* L.) under the conditions of Iran.
Asian Journal of Plant Sciences, 7(8), 710–716.

Bharathi, V., & Sudhakar, R., (2011). Management of rhizome rot of turmeric (*Curcuma
longa* L.) through IDM practices. *Annals of Plant Protection Sciences, 19*(2), 400–402.

Bhardwaj, R. L., (2016). Response of transplanted fennel (*Foeniculum vulgare* Mill.) to
potassium fertilization. *Journal of Spices and Aromatic Crops, 25*(2), 149–158.

Bhat, A. I., Devasahayam, S., Venugopal, M. N., & Bhai, R. S., (2005). Distribution and
incidence of viral disease of black pepper (*Piper nigrum* L.) in Karnataka and Kerala,
India. *Journal of Plantation Crops, 33*(1), 59–64.

Biju, C. N., Siljo, A., & Bhat, A. I., (2010). Survey and RT-PCR based detection of Cardamom mosaic virus affecting small cardamom in India. *Indian Journal of Virology,
21*(2), 148–150.

Birah, A., Someshwar Bhagat, Tripathi, A. K., & Srivastava, R. C., (2011). Impact of pest
management modules against pollu beetle, *Longitarsus nigripennis* Motschulsky in
black pepper in bay islands. *Indian Journal of Entomology, 73*(2), 110–112.

Chatterjee, R., Kumara, J. N. U., Chattopadhyay, P. K., & Bhattacharya, S. P., (2011). Chemical and cultural method of weed control in ginger. *Indian Agriculturist, 55*(1–2), 39–42.

Choudhary, G. R., Jain, N. K., & Jat, N. L., (2008). Response of coriander (*Coriandrum sativum*) to inorganic nitrogen, farmyard manure and biofertilizer. *Indian Journal of Agricultural Sciences, 78*(9), 761–763.

Darzi, M. T., Ghalavand, A., & Rejali, F., (2009). The effects of biofertilizers application on N, P, K assimilation and seed yield in fennel (*Foeniculum vulgare* Mill.). *Iranian Journal of Medicinal and Aromatic Plants, 25*(1), 1–19

Darzi, M. T., Ghalavand, A., Sefidkon, F., & Rejali, F., (2009). The effects of mycorrhiza, vermicompost and phosphatic biofertilizer application on quantity and quality of essential oil in Fennel (*Foeniculum vulgare* Mill.). *Iranian Journal of Medicinal and Aromatic Plants, 24*(4), 396–413.

Das, A. K., Sadhu, M. K., & Som, M. G., (1991). Effect of N and P levels on growth and yield of black cumin (*Nigella sativa* Linn.). *Horticultural Journal, 4*(1), 41–47, 134.

Datta, S., Mini Poduval, Basak, J., & Chatterjee, R., (2001). Fertilizer trial on cumin black (*Nigella sativa* L.) on alluvial zone of West Bengal. *Environment and Ecology, 19*(4), 920–922.

Deepak, P., Saran, L., & Lal, G., (2008). Control of wilt and blight diseases of cumin through antagonistic fungi under *in vitro* and field conditions. *Notulae Botanicae, Horti-Agrobotanici Cluj Napoca, 36*(2), 91–96.

Dehaghi, M. A., & Mollafilabi, A., (2010). Effects of different rates of N fertilizer on physiological indices of growth and yield components of cumin. *Acta Horticulturae, 853*, 69–76.

Devasena, T., (2009). Antitumorigenic action of fenugreek seeds. *Asian Journal of Bio-Science, 4*(1), 83–87.

Diaz-Perez, J. C., Randle, W. M., Boyhan, G., Walcott, R. W; Giddings, D., Bertrand, D., Sanders, H. F., & Gitaitis, R. D., (2004). Effects of mulch and irrigation system on sweet onion: I. Bolting, plant growth, and bulb yield and quality. *Journal of the American Society for Horticultural Science, 129*(2), 218–224.

Dinesh, R., Srinivasan, V., Hamza, S., Parthasarathy, V. A., & Aipe, K. C., (2010). Physico-chemical, biochemical and microbial properties of the rhizospheric soils of tree species used as supports for black pepper cultivation in the humid tropics. *Geerma, 158*(3–4), 252–258.

Dutta, D., Bandyopadhyay, P., & Maiti, D., (2009). Effect of P fertilization and growth regulators on yield, nutrient uptake and economics of fenugreek (*Trigonella foenum-graecum* L.). *Research on Crops, 9*(3), 599–601.

Eapen, S. J., Beena, B., & Ramana, K. V., (2008). Evaluation of fungal bioagents for management of root-knot nematodes in ginger and turmeric fields. *Journal of Spices and Aromatic Crops, 17*(2), 122–127.

Eikani, M. H., Golmohammad, F., & Rowshanzamir, S., (2007). Subcritical water extraction of essential oils from coriander seeds (*Coriandrum sativum* L.). *Journal of Food Engineering, 80*(2), 735–740.

El-Deen, E., & Ahmed, T., (1997). Influence of plant distance and some phosphorus fertilization sources on black cumin (*Nigella sativa* L.) plants. *Assiut Journal of Agricultural Sciences, 28*(2), 39–56.

Ganguli, J., Soman, D., & Ganguli, R. N., (2009). Biology of turmeric leaf roller skipper butterfly, *Udaspes folus* Cramer in Chhattisgarh. *Journal of Applied Zoological Researches, 19*(1), 9–10.

Giridhar, K., Sarada, C., Reddy, T. Y., (2008). Influence of micronutrients on growth and yield of coriander (*Coriandrum sativum*) in rainfed vertisols. *Journal of Spices and Aromatic Crops, 17*(2), 187–189.

Giridhar, K., Sarada, C., & Reddy, T. Y., (2008). Efficacy of biofertilizers on the performance of rainfed coriander (*Coriandrum sativum*) in vertisols. *Journal of Spices and Aromatic-Crops, 17*(2), 98–102.

Gogoi, M., & Firake, N. N., (2003). Economics of floppy sprinkler irrigation method for onion. *Journal of Maharashtra Agricultural Universities, 28*(2), 200–201.

Gopakumar, B., Dhanapal, K., & Thomas, J., (2006). Potential of a consortium of biocontrol agents for disease management in small cardamom (*Elettaria cardamomum* Maton.). *Journal of Plantation Crops, 34*(3), 476–479.

Gour, R., Naruka, I. S., Singh, P. P., Rathore, S. S., & Shaktawat, R. P. S., (2009). Effect of phosphorus and plant growth regulators on growth and yield of fenugreek (*Trigonella foenum-graecum* L.). *Journal of Spices and Aromatic Crops, 18*(1), 33–36.

Gupta, R. B. L., & Pathak, V. N., (1988). Yield losses in onions due to purple leaf blotch disease caused by *Alternaria porri*. *Phytophylactica, 20,* 21–23.

Gupta, M. P., & Pathak, R. K., (2009). Comparative efficacy of neem products and insecticides against the incidence of coriander aphid, *Hyadaphis coriandari* Das. *Agricultural Science Digest, 29*(1), 69–71.

Gupta, R., Khokhar, M. K., & Lal, R., (2012). Management of the Black Mould Disease of Onion. *J Plant Pathol Microb, 3*(5), 1–3.

Hamza, S., Srinivasan, V., & Dinesh, R., (2007). Nutrient diagnosis of black pepper (*Piper nigrum* L.) gardens in Kerala and Karnataka. *Journal of Spices and Aromatic Crops, 16*(2), 77–81.

Hans, R., & Thakral, K. K., (2008). Effect of chemical fertilizers on growth, yield and quality of fennel (*Foeniculum vulgare* Miller). *Journal of Spices and Aromatic Crops, 17*(2), 134–139.

Hanson, B. R., & May, D. M., (2004). Response of processing and fresh-market onions to drip irrigation. *Acta Horticulturae, 664,* 399–405.

Hseu, S. H., Sung, C. J., & Lin, C. Y., (2009). Occurrence of leaf blight of turmeric caused by *Herbaspirillum huttiensis* in Taiwan. *Plant Pathology Bulletin, 18*(1), 13–22.

Ibrahim, N. D., (2010). Seasonal abundance of onion thrips, *Thrips tabaci* Lindeman. in Sokoto, Nigeria. *Journal of Agricultural Science, 2*(1), 107–114.

Infantino, A., Giambattista, G. di. Pucci, N., Pallottini, L., Poletti, F., & Boccongelli, C., (2009). First report of *Alternaria petroselini* on fennel in Italy. *Plant Pathology, 58*(6), 1175.

Isaac, S. R., & Varghese, J., (2016). Nutrient management in turmeric (*Curcuma longa* L.) in an integrated farming system in southern Kerala. *Journal of Spices and Aromatic Crops, 25*(2), 206–209.

Jadeja, K. B., & Nandoliya, D. M., (2008). Integrated management of wilt of cumin (*Cuminum cyminum* L.). *Journal of Spices and Aromatic Crops, 17*(3), 223–229.

Jagadev, P. N., Panda, K. N., & Beura, S., (2008). *Zingiber officinale* A fast protocol for *in vitro* propagation of ginger of a tribal district of India. *Acta Horticulturae, 765,* 101–107.

Jaggi, R.C., (2004). Effect of sulphur levels and sources on composition and yield of onion (*Allium cepa*). *Indian Journal of Agricultural Sciences, 74*(4), 219–220.

Jayasree, P., & George, A., (2006). Do biodynamic practices influence yield, quality, and economics of cultivation of chili (*Capsicum annuum* L.)? *Journal of Tropical Agriculture, 44*(1–2), 68–70.

Jha, A. K., Netra Pal., & Saxena, A. K., (2005). Screening of vesicular arbuscular mycorrhizas for onion. *Indian Journal of Horticulture, 62*(4), 411–412.

Ji QingZhu, & Zhou, T., (2010). Processing technology of the clarified ginger juice. *Modern Food Science and Technology, 26*(8), 850–854.

Jose, K. P., & Joy, C. M., (2009). Solar tunnel drying of turmeric (*Curcuma longa* Linn. syn. *C. domestica* Val.) for quality improvement. *Journal of Food Processing and Preservation, 33*(S1), 121–135.

Jurgiel-Maecka, G., & Suchorska-Orowska, J., (2004). The effect of nitrogen fertilization on magnesium content in shallot bulbs and yield. *Journal of Elementology, 9*(3), 329–335.

Kandiannan, K., Prasath, D., & Sasikumar, B., (2016). Biennial harvest reduces rhizome multiplication rate and provides no yield advantage in ginger (*Zingiber officinale* Roscoe.). *Journal of Spices and Aromatic Crops, 25*(1), 79–83.

Kang, B. S., & Sidhu, B.S., (2006). Studies on growing off-season chili nursery under polyhouse. *Annals of Biology, 22*(1), 39–41.

Kapoor, I. P. S., Singh, B., Singh, G., Heluani, C. S., de. Lampasona, M. P., & de Catalan, C. A. N., (2009). Chemistry and *in vitro* antioxidant activity of volatile oil and oleoresins of black pepper (*Piper nigrum*). *Journal of Agricultural and Food Chemistry, 57*(12), 5358–5364.

Kassaian, N., Azadbakht, L., Forghani, B., & Amini, M., (2009). Effect of fenugreek seeds on blood glucose and lipid profiles in type 2 diabetic patients. *International Journal for Vitamin and Nutrition Research, 79*(1), 34–39.

Kaur, K., Bhullar, M. S., Kaur, J., & Walia, U.S., (2008). Weed management in turmeric (*Curcuma longa*) through integrated approaches. *Indian Journal of Agronomy, 53*(3), 229–234.

Kaur, P., Kumar, A., Arora, S., & Ghuman, B. S., (2006). Quality of dried coriander leaves as affected by pretreatments and method of drying. *European Food Research and Technology, 223*(2), 189–194.

Khalko, S., & Chowdhury, A. K., (2008). Biological control of rhizome rot disease of turmeric. *Journal of Mycopathological Research, 46*(1), 127–128.

Khokhar, I., & Bajwa, R., (2015). Penicillium aurantiogriseum Dierckx: First Report on Causing a Postharvest Rot of Onions (*Allium cepa* L.) in Pakistan. *Int. J. Curr. Res. Biosci. Plant Biol., 2*(7), 88–91.

Kim, G., Lee, E. M., & Ahn T. Y., (2012). Physico chemical properties and microbial populations land diversity in soils where rhizome rot disease of ginger frequently occurs. *Acta Horticulturae, 938*, 85–90.

Krishnamurthy, K. S., & Chempakam, B., (2009). Investigation on the influence of seedling's physiological attributes on productivity in black pepper. *Indian Journal of Horticulture, 66*(1), 95–100.

Kumar, B., & Gill, B. S., (2009). Effect of method of planting and harvesting time on growth, yield and quality of turmeric (*Curcuma longa* L.). *Journal of Spices and Aromatic Crops, 18*(1), 22–27.

Kumar, D., Pandey, V., & Nath, V., (2008). Effect of organic mulches on moisture conservation for rainfed turmeric production in mango orchard. *Indian Journal of Soil Conservation, 36*(3), 188–191.

Kumar, J. N. U., Chatterjee, R., & Sharangi, A. B., (2009). Effect of ethrel on growth and yield of black cumin (*Nigella sativa* L.) under Gangetic alluvial soil of West Bengal. *Journal of Crop and Weed, 4*(2), 31–32.

Kumar, K., Singh, G. P., Singh, N., & Nehra, B. K., (2006). Effect of row spacing and cycocel on growth and seed yield of coriander cv. Hisar Anand. *Haryana Journal of Horticultural Sciences, 35*(3–4), 350.

Kumar, M. D.,Devaraju, K. M., Madaiah, D., & Shivakumar, K. V., (2009). Effect of integrated nutrient management on yield and nutrient content by cardamom (*Elettaria cardamomum* L. Maton.). *Karnataka Journal of Agricultural Sciences*, *22*(5), 1016–1019.

Kumar, R. D., Sreenivasulu, G. B., Prashanth, S. J., Jayaprakashnarayan, R. P., Nataraj, S. K., & Hegde, N. K., (2010). Performance of ginger in tamarind plantation (as intercrop) compared to sole cropping (ginger). *International Journal of Agricultural Sciences*, *6*(1), 193–195.

Kumawat, R. M., & Yadav, K. K., (2009). Effect of FYM and phosphorus on the performance of fenugreek (*Trigonella foenum-graecum* L.) irrigated with saline water. *Environment and Ecology*, *27*(2), 611–616.

Lokesh, M. S., Patil, S. V., Gurumurthy, S. B., Naik, N., & Palakshappa, M. G., (2012). Phytophthora foot root (*Phytophthora capsici* Leonian.) of black pepper management through fungi toxicant and consortium in western Ghats of Karnataka. *Internationa Journal of Plant Protection*, *5*(1), 157–159.

Lokesh, M. S., Hegde, H. G., Naik, N., (2008). Efficacy of systemic fungicides and antagonistic organism for the management of Phytophthora foot rot of black pepper in arecanut cropping system. *Journal of Spices and Aromatic Crops*, *17*(2), 114–121.

Mahorkar, V. K., Meena, H. R., Jadhao, B. J., Panchbhai, D. M., Dod, V. N., & Peshattiwar, P. D., (2008). Effect of micronutrients and humic acid on growth and leaf yield of fenugreek. *Plant Archives*, *8*(1), 303–306.

Majdalawieh, A. F., & Carr, R. I., (2010). *In vitro* investigation of the potential immunomodulatory and anticancer activities of black pepper (Piper nigrum) and cardamom (*Elettaria cardamomum*). *Journal of Medicinal Food*, *13*(2), 371–381.

Mallanagouda, B., Sulikeri, G. S., Murthy, B. G., & Prathibha, N. C., (1995). Performance of chili (*Capsicum annuum*) under different intercropping systems and fertility levels. *Indian Journal of Agronomy*, *40*(2), 277–279.

Mammootty, K. P., Abraham, K., & Vijayaraghavan, R., (2008). Screening black pepper (*Piper nigrum* L.) varieties/cultivars against Phytophthora disease in the nursery. *Journal of Tropical Agriculture*, *46*(1–2), 70–72.

Mannikeri, I. M., Dharmatti, P. R., & Shashidhar, T. R., (2010). Influence of different organic manures on nutrient uptake, dry matter production and yield of turmeric (*Curcuma longa*). *Journal of Ecobiology, 26*(3–4), 355–359.

Mathew, J., & Nybe, E. V., (2002). Integrated nutrient management for sustainable production in black pepper. *Proceedings of the 15ᵗʰ Plantation Crops Symposium Placrosym XV*, (Eds. Sreedharan, K., l., Vinod Kumar, Jayarama; Chulaki, B. M.) Mysore, India, from 10–13 December, 381–382.

Mathew, J., & Nybe, E. V., (2006). Integration of nutrient sources on nutrient status, yield and quality of black pepper. *Journal of Plantation Crops*, *34*(3), 340–343.

Meena, S. S., & Malhotra, S. K., (2006). Effect of sowing time, nitrogen and plant growth regulators on green leaf yield of coriander. *Haryana Journal of Horticultural Sciences*, *35*(3–4), 310–311.

Meetha, J. N., Muhammadali, P., Joy, M. I., Mahendran, R., & Santhakumaran, A., (2016). Pulsed microwave assisted hot air drying of nutmeg mace for better color retention. *Journal of Spices and Aromatic Crops*, *25*(1), 84–87.

Mehriya, M. L., Yadav, R. S., Jangir, R. P., & Poonia, B. L., (2007). Effect of crop-weed competition on seed yield and quality of cumin (*Cuminum cyminum* L.). *Indian Journal of Weed Science*, *39*(1–2), 104–108.

Mehriya, M. L.,Yadav, R. S., Jangir, R. P., & Poonia, B. L., (2008). Effect of different weed management practices on weeds and yield of cumin. *Annals of Arid Zone, 47*(2), 139–144.

Menaria, B. L., & Maliwal, P. L., (2007). Maximization of seed yield in transplanted fennel (*Foeniculum vulgare* Mill.). *Journal of Spices and Aromatic Crops, 16*(1), 46–49.

Mermoud, A., Tamini, T. D., & Yacouba, H., (2005). Impacts of different irrigation schedules on the water balance components of an onion crop in a semi-arid zone. *Agricultural Water Management, 77*(1–3), 282–295.

Mohamed, Medani, R. A., & Khafaga, E. R., (2000). Effect of nitrogen and phosphorus applications with or without micronutrients on black cumin (*Nigella sativa* L.) plants. *Annals of Agricultural Science Cairo, 3*(Special), 1323–1338.

Muhammad, G., Sajid, M., Nasir, M., Nyla, J., Kausar, R., Jabeen, K., & Gulshan, A., (2008). Composition and antimicrobial properties of essential oil of *Foeniculum vulgare. African Journal of Biotechnology, 7*(24), 4364–4368.

Murthy, C. T., & Bhattacharya, S., (2008). Cryogenic grinding of black pepper. *Journal of Food Engineering, 85*(1), 18–28.

Murugan, M., Backiyarani, S., Josephrajkumar, A., Hiremath, M. B., & Shetty, P. K., (2007). Yield of small cardamom (*Elettaria cardamomum* M) variety PV1 as influenced by levels of nutrients and neem cake under rain fed condition in southern western Ghats, India. *Caspian Journal of Environmental Sciences, 5*(1), 19–25.

Nadasy, E., (2003). Connection between herbicide treatments and nitrate accumulation of green onion. Bulletin OILB/SROP, *26*(3), 271–276.

Nagalakshmi, S., Palanisamy, D., Eswaran, S., & Sreenarayanan, V.V., (2002). Influence of plastic mulching on chili yield and economics. *South Indian Horticulture, 50*(1/3), 262–265.

Nampoothiri, S. V., Praseetha, E. K., Venugopalan, V. V., & Menon, A. N., (2012). Process development for the enrichment of curcuminoids in turmeric spent oleoresin and its inhibitory potential against LDL oxidation and angiotensin-converting enzyme. *International Journal of Food Sciences and Nutrition, 63*(6), 696–702.

Nanda, S. S., Mohapatra, S., & Mukhi, S. K., (2012). Integrated effect of organic and inorganic sources of nutrients on turmeric (*Curcuma longa*). *Indian Journal of Agronomy, 57*(2), 191–194.

Nandal, J. K., Tehlan, S. K., Malik, T. P., & Mehra, R., (2007). Effect of nitrogen and phosphorus levels on growth and seed yield of fenugreek. *Haryana Journal of Horticultural Sciences, 36*(3–4), 413–414.

Nandal, J. K., Dahiya, M. S., Gupta, V., & Singh, D., (2007). Response of sowing time, spacing and cutting of leaves on growth and seed yield of fenugreek. *Haryana Journal of Horticultural Sciences, 36*(3–4), 374–376.

Nath, A., Deka, B. C., Jha, A. K., Paul, D., & Misra, L. K. (2013). Effect of slice thickness and blanching time on different quality attributes of instant ginger candy. *Journal of Food Science and Technology Mysore. 50*(1), 197–202.

Nwaogu, E. N., & Ukpabi, U. J., (2010). Potassium fertilization effects on the field performances and post-harvest characteristics of imported Indian ginger cultivars in Abia State, Nigeria. *Agricultural Journal, 5*(1), 31–36.

Ozguven, M., Kirpik, M., Koller, W. D., Kerschbaum, S., Range, P., & Schweiger, P., (2001). Yield and quality characteristics of black cumin (*Nigella sativa* L.) in the Cukurova region of South Turkey. *Zeitschrift fur Arznei and Gewurzpflanzen, 6*(1), 20–24.

Padmapriya, S., Chezhiyan, N., & Sathiyamurthy, V. A., (2007). Effect of shade and integrated nutrient management on biochemical constituents of turmeric (*Curcuma longa* L.). *Journal of Horticultural Sciences, 2*(2), 123–129.

Panchal, S. C., Bhatnagar, R., Momin, R. A., & Chauhan, N. P., (2001). Capsaicin and ascorbic acid content of chili as influenced by cultural practices. *Capsicum and Eggplant Newsletter, 20*, 19–22.

Pandey, G., Pandey, R., Ahirwar, K., & Namdeo, K. N., (2012). Effect of organic and inorganic sources of nutrients on nutrient contents and uptake of turmeric (*Curcuma longa* L.). *Crop Research, 44*(1–2), 243–245.

Papu S., Singh, A., Jaivir, S., Sweta, S., Arya, A. M., & Singh, B. R., (2014). Effect of Drying Characteristics of Garlic—A Review. *J Food Process Technol, 5*, 318. doi: 10.4172/2157-7110.1000.318.

Parashar, A., & Lodha, P., (2008). Quantification of total carbohydrates and related enzymes in Ramularia blight infected fennel plants. *Annals of Plant Protection Sciences, 16*(2), 438–440.

Pariari, A., Khan, S., & Imam, M. N., (2009). Influence of boron and zinc on increasing productivity of fenugreek seed (*Trigonella foenum graecum* L.). *Journal of Crop and Weed, 5*(2), 57–58.

Pariari, A., Sharangi, A. B., Chatterjee, R., & Das, D. K., (2003). Response of black cumin (Nigella sativa L.) to the application of boron and zinc. *Indian Agriculturist, 47*(1–2), 107–111.

Patel, D. S., & Patel, S. I., (2008). Management of Ramularia blight of fennel caused by *Ramularia foeniculi* Sibilla. *Indian Phytopathology, 61*(3), 355–356.

Paterson, R. R. M., (2007). *Aflatoxins Contamination in Chili Samples from Pakistan.* Food Control, *18*(7), 817–820.

Paul, R., Shylaja, M. R., & Abraham, K., (2009). Screening somaclones of ginger (*Zingiber officinale*) for bacterial wilt disease. *Indian Phytopathology, 62*(4), 424–428.

Prasath, D., Venugopal, M. N., & Parthasarathy, V. A., (2010). Inheritance of cardamom mosaic virus (CdMV) resistance in cardamom. *Scientia Horticulturae, 125*(3), 539–541.

Qasem, J. R., (2005). Chemical control of weeds in onion (*Allium cepa* L.). *Journal of Horticultural Science and Biotechnology, 80*(6), 721–726.

Radhakrishnan, V. V., Madhusoodanan, K. J., & Mohanan, K. V., (2010). Shade trees and shade management in Cardamom. *Journal of Non Timber Forest Products, 17*(4), 433–435.

Rahman, I., Nuruzzaman, M., Habiba, S. U., & Uddin, A. F. M. J., (2010). Evaluation of micro propagated ginger plantlets in different soil composition of pot culture. *International Journal of Sustainable Agricultural Technology, 6*(5), 18–21.

Ramadan, M. F., & Morsel, J. T., (2002). Neutral lipid classes of black cumin (*Nigella sativa* L.) seed oils. *European Food Research and Technology, 214* (3), 202–206.

Raslan, M., AbouZid, S., Abdallah, M., & Hifnawy, M., (2015). Studies on garlic production in Egypt using conventional and organic agricultural conditions. *African Journal of Agricultural Research, 10*(13), 1631–1635.

Rathod, S. D., Kamble, B. M., Phalke, D. H., & Pawar, V. P., (2010). Effect of levels of fertilizer and irrigation on yield of ginger in vertisols irrigated through micro sprinkler. *Advances in Plant Sciences, 23*(1), 173–175.

Rathore, H. S., & Porwal, M. K., (2008). Nutrient uptake, protein content and relative economics of fenugreek *(Trigonella foenum graecum)* as influenced by sowing dates, microbial inoculation and weed management. *Indian Journal of Agricultural Sciences, 78*(6), 560–562.

Ratnam, M., Rao, A. S., & Reddy, T. Y., (2012). Integrated weed management in turmeric (*Curcuma longa*). *Indian Journal of Agronomy, 57*(1), 82–84.

Raziq, F., Alam, I., Naz, I., & Khan, H., (2008). Evaluation of fungicides for controlling downy mildew of onion under field conditions. *Sarhad J. Agric., 24*(1), 85–92.

Reddy, A. V., Sreehari, G., Kumar, A. K., & Reddy, K. M., (2006). Testing of certain new insecticides for seed treatment purpose in chili against thrips (*Scirtothrips dorsalis*). *Research on Crops, 7*(2), 529–531.

Roy, S. S., Hore, J. K., Bandopadhyay, A., & Ghosh, D. K., (2008). Effect of different organic manures with varying levels of nitrogen on the growth and yield of ginger grown as intercrop in pre-bearing coconut garden. *Indian Coconut Journal, 38*(9), 18–22.

Roy, S. S., & Hore, J. K., (2012). Effect of organic manures and microbial inoculants on soil nutrient availability and yield of turmeric intercropped in arecanut gardens. *Journal of Crop and Weed, 8*(1), 90–94.

Sadik, S., Mazouz, H., Bouaichi, A., Benbouazza, A., & Achbani, E. H., (2015). Biological Control of Bacterial Onion Diseases using a Bacterium, *Pantoea agglomerans* 2066-7. International *Journal of Science and Research, 4*(1), 103–111.

Sagar, S. D., Kulkarni, S., & Hegde, Y. R., (2007). Management of rhizome rot of ginger by botanicals. *International Journal of Plant Sciences, 2*(2), 155–158.

Saleem, F., Khan, M. T. J., Mumtaz, A. M., Khan, K. I., Bashir, S., & Jamshaid, M., (2008). Antimicrobial activity of the extracts of seeds of *Trigonella foenum-graecum*. *Pakistan Journal of Zoology, 40*(5), 385–387.

Sangeeth, K. P., Bhai, R. S., & Srinivasan, V., (2008). Evaluation of indigenous *Azospirillum* isolates for growth promotion in black pepper (*Piper nigrum* L) rooted cuttings. *Journal of Spices and Aromatic Crops, 17*(2), 128–133.

Sanwal, S. K., Yadav, R. K., Yadav, D. S., Rai, N., & Singh, P. K., (2006). Ginger-based intercropping: highly profitable and sustainable in mid hill agroclimatic conditions of North East Hill Region. *Vegetable Science, 33*(2), 160–163.

Sarada, C., Giridhar, K., & Reddy, T. Y., (2008). Effect of bio-regulators and their time of application on growth and yield of coriander (*Coriandrum sativum*). *Journal of Spices and Aromatic Crops, 17*(2), 183–186.

Sarkar, P. K., Sarkar, H., Sarkar, M. A., & Somachoudhury, A. K., (2005). Yellow mite, Polyphagotarsonemus latus (Banks): a menace in chili cultivation and its management options using biorational acaricides. *Indian Journal of Plant Protection, 33*(2), 294–296.

Sawant, A., & Thakor, N. J., (2008). Studies on dehydration of fenugreek (Methi). *Agriculture Update, 3*(3–4), 342–345.

Saxena, S. N., Rathore, S. S., Maheshwari, G., Saxena, R., Sharma, L. K., & Ranjan, J. K., (2016). Analysis of medicinally important compounds and antioxidant activity in solvent extracts of coriander (*Coriandrum sativum* L.) plant parts. *Journal of Spices and Aromatic Crops, 25*(1), 65–69.

Shah, S. H., (2007a). Photosynthetic and yield responses of *Nigella sativa* L. to pre-sowing seed treatment with GA_3. *Turkish Journal of Biology, 31*(2), 103–107.

Shah, S. H., (2007b). Physiological effects of pre-sowing seed treatment with gibberellic acid on *Nigella sativa* L. *Acta Botanica Croatica, 66*(1), 67–73.

Shah, S. H., Ahmad, I., & Samiullah, (2006). Effect of gibberellic acid spray on growth, nutrient uptake and yield attributes during various growth stages of black cumin (*Nigella sativa* L.). *Asian Journal of Plant Sciences, 5*(5), 881–884.

Shah, S. H., (2007c). Effect of kinetin spray on growth and productivity of black cumin plants. *Russian Journal of Plant Physiology*, *54*(5), 702–70.

Shaikh, A. A., Desai, M. M., Shinde, S. B., & Tambe, A. D., (2010). Yield and nutrient uptake of ginger (*Zingiber officinale* Rosc.) as affected by organic manures and fertilizers. *International Journal of Agricultural Sciences*, *6*(1), 28–30.

Sharangi, A. B., & Kumar, R., (2011). Performance of rooted cuttings of black pepper (Piper nigrum L.) with organic substitution of nitrogen. *International Journal of Agricultural-Research*, *6*(9), 673–681

Sharangi, A. B., Kumar, R., & Sahu, P. K., (2010). Survivability of black pepper (*Piper nigrum* L.) cuttings from different portions of vine and growing media. *Journal of Crop and Weed, 6*(1), 52–54.

Sharma, P., Kadu, L. N., & Sain, S. K., (2005). Biological management of dieback and fruit rot of chili caused by *Colletotrichum capsici* (Syd.) Butler and Bisby. *Indian Journal of Plant Protection., 33*(2), 226–230.

Sharma, R. K., & Jain, N., (2008). Effect of different sources of plant nutrients on productivity of coriander (*Coriandrum sativum*). *Haryana Journal of Agronomy*, *24*(1–2), 4–6.

Shivran, A. C., Jat, N. L., Singh, D., Sastry, E. V. D., & Rajput, S. S., (2016). Response of fenugreek (*Trigonella foenum graecum* L.) to plant growth regulators and their time of application, *Journal of Spices and Aromatic Crops, 25*(2), 169–174.

Singh, R., & Chowdhury, A. K., (2004). Impact of some cultivation practices on chili (*Capsicum annuum* L.) leaf curl virus disease and performance of different insecticides for its management. *Journal of Interacademicia, 8*(4), 523–527.

Singh, A.,Yadav, A. C., Mehla, C. P., Singh, J., & Singh, V. P., (2006). Response of coriander to irrigation and nitrogen levels. *Haryana Journal of Horticultural Sciences, 35*(3–4), 312.

Singh, B., Kaur, R., & Singh, K., (2008). Characterization of Rhizobium strain isolated from the roots of *Trigonella foenum-graecum* (fenugreek). *African Journal of Biotechnology, 7*(20), 3671–3676).

Skotnikov, P. V., (2008). The effect of herbicides on productivity of coriander. *Zashchita i Karantin Rastenii, 8*, 23.

Sobhanipour, A., Etebarian, H. R., Roostaee, M. A., Khodakaramian, G., & Aminian, H., (2008). Biological control of Fusarium wilt of cumin by antagonistic bacteria. *Journal of Agricultural Sciences Guilan, 1*(10), 41–50.

Sreeja, K., Anandaraj, M., & Bhai, R. S., (2016). *In vitro* evaluation of fungal endophytes of black pepper against *Phytophthora capsici* and *Radopholus similis*. *Journal of Spices and Aromatic Crops, 25*(2), 113–122.

Srinivasan, K., (2007). Black pepper and its pungent principle-piperine: a review of diverse physiological effects. *Critical Reviews in Food Science and Nutrition, 47*(8), 735–748.

Srinivasan, V., Dinesh, R., Hamza, S., & Parthasarathy, V. A., (2007). Nutrient management in black pepper (*Piper nigrum* L.). *Perspectives in Agriculture, Veterinary Science, Nutrition and Natural Resources, 2*(62), 14.

Suheri, H., & Price, T. V., (2000). Infection of onion leaves by *Alternaria porri* and *Stemphylium vesicarium* and disease development in controlled environments. *Plant Pathology, 49*, 375–382.

Sujatha, T., & Rao, G.V.H., (1996). Seed treatment with insecticides in chili. *Seed Research, 23*(1), 53–54.

Suresh, D., Manjunatha, H., & Srinivasan, K., (2007). Effect of heat processing of spices on the concentrations of their bioactive principles: turmeric *(Curcuma longa),* red pepper

(Capsicum annuum) and black pepper *(Piper nigrum)*. *Journal of Food Composition and Analysis, 20*(3–4), 346–351.

Taufik, M., Khaeruni, A., Wahab, A., & Amiruddin., (2011). Biocontrol agents and Arachis pintoi promote the growth of black pepper *(Piper nigrum)* and reduce the incidence of yellow disease. *Menara Perkebulnan, 79*(2), 42–48.

Telci, I., Demirtas, I., & Sahin, A., (2009). Variation in plant properties and essential oil composition of sweet fennel *(Foeniculum vulgare* Mill.) fruits during stages of maturity. *Industrial Crops and Products, 30*(1), 126–130.

Thangaselvabai, T., Joshua, J. P., Justin, C. G. L., & Jayasekar, M., (2010). Effect of different spacings on yield and yield attributes of black pepper var. Panniyur-I. *Advances in Plant Sciences, 23*(1), 111–112.

Thankamani, C. K., Dinesh, R., Eapen, S. J., Kumar, A., Kandiannan, K., & Mathew, P. A., (2008). Effect of solarized potting mixture on growth of black pepper *(Piper nigrum* L.) rooted cuttings in the nursery. *Journal of Spices and Aromatic Crops, 17*(2), 103–108.

Thankamani, C. K., Srinivasan, V., Hamza, S., Kandiannan, K., & Mathew, P. A., (2007). Evaluation of nursery mixture for planting material production in black pepper *(Piper nigrum* L.). *Journal of Spices and Aromatic Crops, 16*(2), 111–114.

Tiwari, B. K., & Agrawal, G. P., (2012). Turmeric: a wonder drug in Ayurvedic era. *World Journal of Pharmacy and Pharmaceutical Sciences, 1*(1), 161–181.

Vanangamudi, K., Subramanian, K. S., & Baskaran, M., (1990). Influence of irrigation and nitrogen on the yield and quality of chili fruit and seed. *Seed Research, 18*(2), 114–116.

Vasmate, S. D., Kalalbandi, B. M., Patil, R. F., Digrase, S. S., & Manolikar, R. R., (2007). Effect of spacings and organic manures on growth of coriander. *Asian Journal of Horticulture, 2*(2), 266–268.

Vasmate, S. D., Patil, R. F., Manolikar, R. R., Kalabandi, B. M., & Digrase, S. S., (2008). Effect of spacing and organic manures on seed of coriander *(Corianderium sativum* L.). *Asian Journal of Horticulture, 3*(1), 127–129.

Vijayan, A. K., & Thomas, J., (2006). Screening of improved selections and hybrids of small cardamom *(Elettaria cardamomum* Maton) for rot tolerance. *Journal of Plantation Crops, 34*(3), 508–511.

Vijayan, A. K., Abraham, S. M., & Thomas, J., (2006). Studies on root tip rot disease of small cardamom and its management. *Journal of Plantation Crops, 34*(3), 486–488.

Vijayaraghavan, R., & Abraham, K., (2007). Potential of *Trichoderma* spp. on the management of phytophthora rot in black pepper nursery. *Journal of Plant Disease Sciences, 2*(1), 1–4.

Vijayaraghavan, R., & Abraham, K., (2007). Mechanism of antagonism of *Trichoderma* spp. on *Phytophthora capsici* causing foot rot of black pepper. *Journal of Plant Disease Sciences, 2*(2), 124–125.

Wade, N. C., Wane, S. S., & Kshirsagar, S. M., (2014). Comparative study of drying characteristics in chilies. *Ind. J. Sci. Res. and Tech., 2*(3), 105–111.

Wadikar, D. D., & Premavalli, K. S., (2012). Optimization of ginger based ready to drink appetizer by response surface methodology and its shelf stability. *Journal of Food Processing and Preservatioln, 36*(6), 489–496.

Wadikar, D. D., Nanjappa, C., Premavalli, K. S., & Bawa, A. S., (2010). Development of ginger based ready-to-eat appetizers by response surface methodology. *Appetite, 55*(1), 76–83.

Wright, P. J., Grant, D. G., & Triggs, C. M., (2001). Effects of onion *(Allium cepa)* plant maturity at harvest and method of topping on bulb quality and incidence of rots in storage. *New Zealand Journal of Crop and Horticultural Science, 29*(2), 85–91.

Xu Qing, L., Zhou Yong, Q., Zhan, Y., & Zeng Qing, Z., (2012). Research on extraction technology of flavonoid from ginger peel. *Modern Food Science and Technology, 28*(8), 998–(1001).

Yadav, B. D., Khandelwal, R. B., & Sharma, Y. K., (2005). Use of biofertilizer (Azospirillum) in onion. *Indian Journal of Horticulture, 62*(2), 168–170.

Yadav, R. H., & Vijayakumari, B., (2004). Impact of vermicompost on biochemical characters of chili (*Capsicum annuum*). *Journal of Ecotoxicology and Environmental Monitoring, 14*(1), 51–56.

Yadav, S. S., Choudhary, I., Yadav, L. R., & Sharma, O. P., (2016). Weed management in coriander (*Coriandrum sativum* L.) at varying levels of nitrogen. *Journal of Spices and Aromatic Crops, 25*(1), 18–25.

Yamgar, V. T., Shirke, M. S., Kamble, B. M., Salunkhe, S. M., & Tambe, B. N., (2009). Studies on effect of potash levels, sources and time of application on yield of turmeric (*Curcuma longa* L.) cv. Salem. *Advances in Plant Sciences, 22*(1), 123–125.

Yugandhar, V., Reddy, P. S. S., Sivaram, G. T., & Reddy, D. S., (2016). Influence of plant growth regulators on growth, seed yield, quality and economics of coriander (*Coriandrum sativum* L.) cv. Sudha. *Journal of Spices and Aromatic Crops, 25*(1), 13–17.

Zarai, Z., Boujelbene, E., Ben Sale N., Gargouri, Y., & Sayari, A., (2013). Antioxidant and antimicrobial activities of various solvent extracts, piperine and piperic acid, could be used as natural antioxidant and antibacterial agents in both food preservation and human health. *LWT Food Science and Technology, 50*(2), 634–641.

CHAPTER 5

PROMISES OF ORGANIC SPICES

CONTENTS

India is the largest producer, consumer, and exporter of spices. Indian spice is itself a brand throughout the globe. Due to the intervention of modern technology for target oriented production system in Indian agriculture huge amount of synthetic fertilizers and pesticides Indian spices is also sometimes reported stopped for export towards the major importing countries for the presence of residues beyond the tolerable limit. The conscious consumers of our country are also demand for residue free spices and spice products. In most cases consumers are ready to pay more for good quality residue free spices. In this situation a thrust is being seen among the spices growers to produce spices by means of organic agriculture since last decade. According to FAO/WHO Codex Alimentarius Commission organic agriculture is a holistic production management system, which promotes and enhances agro-eco system health, including biodiversity, biological cycles and soil biological activity. It emphasizes the use of management practices in preference to the use of off-farm inputs. This is accomplished by using, where possible, agronomic, biological, and mechanical methods, as opposed to using synthetic materials, to fulfill any specific function within the system.

Many techniques like inter-cropping, mulching, and integration of crops and livestock – are practiced under various agricultural systems in organic farming. Properly managed, organic farming is one of the sustainable farming systems, which reduces or eliminates soil and water pollution and helps conserve water and soil on agricultural lands.

Many of our ancient literatures like Rigveda, Ramayana, Mahabharata, Arthasashthra has mentioned the organic way of cultivating crops. Lampkin et al. (1999) mentioned that, organic agriculture is a production system, which avoids or largely excludes the use of synthetic fertilizers, pesticides, growth regulators, and livestock feed additives. It relies on crop rotation, crop residues, animal manure, legumes, green manure, off-farm organic waste and aspects of biological pest control (Bhattacharyya, 2004). Today, organic agriculture is gaining due importance in the agriculture sectors. In Austria and Switzerland, organic agriculture has come to represent as much as 10% of the food production system, while USA, France, Japan and Singapore are fixing growth rates that exceed 20% annually for organic food production. The yield level of most of all the crops especially spices is very low after stopping the use of artificial. Pest suppression and fertility problems are common before restoration of full biological activity like as growth in beneficial insect populations and nitrogen fixation from legumes etc. In some cases it requires years to restore the ecosystem of the production system to the point where organic production is possible. In such cases integrated approaches that allow judicious use of synthetic inputs may be more suitable as start-up options. Strategically entire farming systems are converted into organic production unit step by step, so that the entire operation is not put at risk of low yield. In India spices growers produce over 60 different types of spices because of varied agro-climatic conditions and soil types. About 60 lakhs MT of spices are produced in India, of which, about 6.9 lakhs MT (11%) is exported to more than 150 countries. The demand for organic spices and spice products is growing rapidly in developed countries like Europe, USA, Japan and Australia. 1 to 1.5% of the total food consumed by the population of such countries are organic food especially spices and herbs and this demand are steadily increasing day-by-day.

To promote the production of organic spices, the Spices Board of India has documented the technology of production of organic spices featuring the organic concepts, principles, basic standards, production guidelines, documentation, inspection and certification which was approved by the National Standards Committee constituted by the members of IFOAM in India. Research programmes on organic production of most of the demanded spices are being conducted by different government agencies, state agricultural universities and other private organizations. The Spices Board encourages non-governmental organizations, farmers' groups, Self Help Groups to adopt organic farming techniques in spices and to promote organic spices business. Due to the adoption of such technologies to some extent, the infrastructure development for production of organic inputs is on stream in prominent growing areas.

5.1 WHY ORGANIC SPICES?

Organic spices are gaining importance day-by-day due to prominence bad effects of inorganic and synthetic based conventional farming on the human health. Organic spice trade is booming all over the world for increasing health consciousness among the people. Organic spice production has the following specific advantages:

- Organically produced spices are safe, nutritious, and good for human health with eco-friendly technology.
- Organic spices, being qualitatively superior are highly remunerative because of higher demand in the international market.
- The fertility status of the soil is maintained for a longer period of time.
- Organically produced spices are found more resistant to pest and diseases compared to their inorganically produced counterpart.

Many researchers concluded that organic way of cultivation is beneficial for human beings, for crops in general and spices in particular. The adverse effects of continuous use of high dose of chemical fertilizers on soil health and environment were realized (Kamal and Yousuf, 2012). Organic manures and biofertilizers offer an alternative to chemical fertilizers and increasingly used in spice crop production including turmeric (Srinivasan et al., 2000).

Organic source of nutrients are recommended for retaining productivity of soil, reducing usage of chemical fertilizers, improving soil health and minimize environmental pollution (Hossain and Ishimine, 2007). Application of organic manures also quickly increases soil microbial biomass and their activity (Dinesh et al., 2010). Meena et al. (2007) worked on residual effect of phosphorus from organic manures. Soil microorganisms and their activities play important roles in transformation of plant nutrients from unavailable to available forms and also helpful for improvement of soil fertility (Yamawaki et al., 2013). Application of biofertilizers like *Azospirillium* is helpful for fixation of substantial amount of atmospheric nitrogen and supplies to the crop and increases soil fertility. Application of PSB increases the uptake of phosphorus, which readily fixed in the soil (Wang and Qui, 2006). The spices from organic farming can also be considered as value added products. They fetch premium price in the international market. The price may be higher by 20–50% or more and in certain cases even 100% than the spices grown from conventional farming (Arya, 2000). India has the tremendous scope for production of organic spices. At present Spice Board provides the guidelines for production and export of organic spices. During the recent times organic spices share 1–1.5% of the global spice markets.

The major importers of the organic spices are United States, Europe and Japan. Organic certification is a growing world demand for organic food and it is an important task for export of organic spices.

5.2 GUIDELINES FOR PRODUCTION OF ORGANIC SPICES

1. A minimum area of 0.4 ha is the eligibility requirement for organic spice growers having a valid organic certificate issued by an international agency.
2. The conversion periods to organic spice production from conventional cultivation is at least two years for annual spice (e.g., Seed spices, ginger, turmeric, etc.) and three years for perennial spice crop (e.g., black pepper, cardamom, tree spices, etc.) production.
3. Organic spices should be cultivated on basis of community approach in large areas or zone.
4. Legume crop should be incorporated in the cropping system of organic farming to restore and improve fertility level. In crop rotation, crops belonging to the same family should be avoided.

5. Mixed farming system integrating crop husbandry and live stock farming is the most ideal where the livestock is also maintained following organic standard.
6. Crop residues and farm wastes should be recycled through composting so that fertility is restored and maintained at a high level.
7. Planting material for raising organic spice cultivation should be collected from organic sources. In the, absence, the initial planting material can be collected from conventional sources.
8. An isolation distance of at least 25 m width is to be maintained along boundary/periphery of the conventional farm.
9. Pests can be controlled by parasites, predators, botanicals and bio-control agents.
10. Disease can be controlled by crop rotation, soil solarization, using botanicals and using biocontrol agents.

5.3 ORGANIC BLACK PEPPER CULTIVATION

Black pepper is well known as 'king of spices' and popular throughout the globe for its pungent taste with typical flavor. Since the time of Portuguese sailors who used to do business with Indian spices black pepper was popular as component Indian spice flavor. After the introduction of modern agricultural system the production of spices also included the uses of inorganic fertilizers, which resulted in the less interest in traditional Indian spices among the western users due to the presence of inorganic input residues. Again during the last decade interest among the growers has been raised for organic production of black pepper.

Tropical hot and humid climatic condition is best suited for black pepper and about 20°C well-distributed rainfall is required. A temperature range of 10 to 40°C is ideal and the crop can be grown upto 1400 m from MSL. Dry spell for at least one month is required during flowering for better berry set. It can be cultivated in red loam, sandy loam or lateritic loam soil. However, well-drained virgin soil rich in humus having the pH range of 5.0 to 6.5 is ideal for higher yield and quality. For new planting, varieties that are resistant or at least tolerant to diseases, pests and nematode infection should be selected. In that case different Panniyur varieties (Panniyur-1 to Panniyur-5) including the higher yielding varieties like Sreekara, Subhakara, Panchami, Pournami, etc. can be selected for organic cultivation. Before planting of cuttings livestock should be reared according to organic standards. All crop residues and farm wastes available on the farm should be recycled to restore the soil fertility a high level. Composting, including vermi-composting of organic residues, should be carried out as much as possible. At least 10 kg FYM with 10 kg cow dung manure or poultry manure should be applied per plant with 500 g ash and rock phosphate each in every year. VAM and biofertilizers based on bacteria and algae may also be incorporated. Green ground cover of leguminous plant should be maintained as far as possible particularly in pure crop cultivation. Prophylactic measures should be taken to reduce the incidence of diseases and pests. An isolation belt of at least 25 m wide must be maintained around an organic holding. Some common biofertilizers that may be used in pepper include *Rhizobium, Azospirillum, Azotobacter,* etc. To take advantage of Rhizobium, leguminous cover crops should be grown in the pepper holdings and as green manure crops on the boundaries. Azosiprillum can be applied as a co-inoculant for leguminous cover crops or green manure crops. The recommended dose for pepper is 15–20 g of inoculant per plant. Azotobacter produces growth-promoting substances like vitamins of the B group, indole acetic acid, and gibberellic acid. The recommended dose for pepper is 15–20 g per plant.

For control of pests and diseases biological control measures are recommended which includes the use of natural enemies to manage pests and causal organisms of diseases. Natural enemies or antagonists and may be referred to as microbial insecticides or biopesticides. Some of the commonly used microbes are bacteria such as *Bacillus thuringiensis* acts against caterpillars, beetles; viruses such as nuclear polyhedrosis virus (NPV) acts against certain caterpillars, and fungi such as *Trichoderma viridae* against *Phytophthora capcici* (the fungus that causes foot-rot). Botanical pesticides for pepper production includes Pyrethrins (*Chrysanthemum sp.*), Rotenone (*Derris sp.*), Nicotine sulphate (*Tobacco Nicotinia sp.*) and Azadiractin (*Azadiracta indica*) Chili pepper, Mexican marigold, garlic and black pepper etc., are also used as botanicals. Black pepper berries become ready to harvest 7 to 8 months after planting in the main field when one or two berries in a spike start to ripe by turning color. Usually harvesting starts from November and continues upto February–March. Yield normally varies from 2–3 kg dry berries per plant a plant per year.

5.4   ORGANIC CARDAMOM CULTIVATION

Cardamom is among the most popular aromatic spice used today for its fragrant seeds directly used in quality sweets, many popular commercial beverages and other products. More traditionally, this spice has been enjoyed in its native India for thousands of years in a wide range of dishes, deserts and teas. Cardamom is native to tropical regions and can be grown where minimum temperatures approximately 35°C or higher. It can also be grown with care indoors or in a greenhouse or overwintered carefully if grown in a large, deep container that can be moved indoors as needed. It prefers a rich, loamy, slightly acidic soil with a pH approximately 6.1–6.6. To sow, the smaller seeds should be sown in a light but rich starting medium buried shallowly beneath the surface of the soil (approximately 1/8"). Cardamom requires a steady supply of moisture and cannot tolerate drought. If growing in a greenhouse, it should be kept humid and maintained carefully. Cardamom is not tolerant of cold, but should be kept in a location with many hours of partially occluded sunlight.

Land is prepared by two or three thorough ploughing to bring the soil to fine tilth. In case of direct sowing, three to four ploughing are undertaken and sowing is done along with the last ploughing. The soil can be treated with *Azatobacter* or *Azospirillum* @ 2.5–3 kg per hectare mixed with 150 kg of FYM and broadcasted in the field. Farmyard manure @ 10–15 tonnes and 3–4 tonnes per hectare of vermin-compost are added during planting. Organic manure such as farmyard manure collected from own farm is applied @ 10–15 tonnes per hectare. Use of biofertilizers can also be resorted to in combination with organic inputs. Cardamom requires approximately 3 years of growth to produce capsules containing seeds. After the flowers mature, they will gradually dry out as capsule develops. These can be collected when capsules begin to turn green, and later dried on screens over the course of 6–7 days. Turn frequently.

The seeds can be collected once pods are dry and easy to break open. Place pods into a bowl, and carefully thresh by applying light pressure to break up dried seedpod. Separate seed from chaff by winnowing with a small fan, or by placing into a medium screen and gently shaking back and forth while lightly pressing extraneous matter through. Seeds can be collected in same manner as with harvesting. Seeds are stored in a sealed container in a dry, cool location out of direct sunlight for optimum life.

5.5   ORGANIC CULTIVATION OF GINGER

Well-drained friable loamy soils rich in humus with good drainage are ideal for ginger cultivation. As ginger is one of the exhaustive crops it is essential to convert the whole farm as organic with ginger as one of the crops in rotation rather to grow ginger in the same field year after year. It is generally recommended to keep 25 m isolation distance between organic ginger field and conventional farm. Ginger can be cultivated organically as an inter or mixed crop provided all the other crops are grown following organic methods including at least a leguminous crop in rotation. In that case ginger-banana-legume or ginger-vegetable-legume can be adopted.

Carefully preserved seed rhizomes collected from organic farms and free from pests and diseases can be used for planting. Seed materials from high yielding local varieties may also be used in the absence of organically produced seeds at the beginning. Seed rhizomes may be treated with any microbial inoculation to prevent rhizome rot or any other bacterial or fungal disease instead of use of any chemicals. After preparing the lands beds of 15 cm height, 1 m width and of convenient length may be prepared giving at least 50 cm spacing between beds. Soil solarization of the beds by polythene sheets is beneficial to check pests and disease infestation. Seed rhizomes are planted in shallow pits at a spacing of 20–25 cm in both direction with 25 g powdered neem cake and well rotten cattle manure or compost mixed with *Trichoderma*.

Mulching with green or dried leaves (@ 10–12 t/ha at the time of planting, @ 5 t/ha at 40th and 90th day after planting) is an essential operation to enhance germination of seed rhizomes and also to prevent washing off soil due to heavy rain, addition of organic matter to the soil and conserve moisture during the later part of the cropping season. Cow dung slurry or liquid manure may be poured on the bed after each mulching to enhance microbial activity and nutrient availability. Weeding is done depending on the intensity of weed growth. Application of well rotten cow dung manure or FYM or compost @ 5–6 t/ha are recommended as basal dose during planting of rhizomes in the pits. Compost enriched with additional application of natural phosphorus and potassium is highly useful. Application of neem cake @ 2 t/ha is also desirable.

Soft rot or rhizome rot caused by *Pythium aphanidermatum* is a major disease of ginger. As water stagnation pre-disposes the plants to infection, good drainage should be confirmed to avoid such disease. Seed rhizomes should be free from any infection. Soil solarization is also another good practice, which can reduce the fungus inoculum. *Trichoderma* may be applied at the time of planting and subsequently if necessary. Affected clumps are to be removed carefully along with the soil surrounding the rhizome to reduce the spread of the disease. The bacterial wilt caused by *Pseudomonas solanacearum* can be managed by treating the seed rhizomes with streptocycline (200 ppm) for 30 minutes and shade drying before planting. Soil drenching of Bordeaux mixture (1%) may be helpful in infected fields. The shoot borer (*Conogethes punctiferalis*) is the most important pest of ginger, which appears during July–October can be controlled by killing caterpillars and spraying neem oil (0.5%) at fortnightly intervals.

The crop is harvested in about eight to ten months when fully mature leaves turn yellow and start drying up gradually. Clumps are lifted carefully with a spade or digging fork and rhizomes are separated from dried leaves, roots and adhering soil. The average yield of fresh ginger per hectare varies with varieties ranging from 15 to 25 tonnes. The rhizomes to be used as seed material big and healthy rhizomes from disease-free plants are selected immediately after harvest should be preserved carefully by spreading layers of leaves of *Glycosmis pentaphylla* and to get good germination, the seed rhizomes are to be stored properly in pits in layers of sand and saw dust (i.e., put one layer of seed rhizomes, then put 2 cm thick layer of sand or saw dust) under shade. Walls of the pits may be coated with cow dung paste.

5.6 ORGANIC PRODUCTION OF TURMERIC

Turmeric requires a warm and humid tropical climate within 1500 mm above MSL, 20–30°C temperature range and 1500 mm well distributed rainfall. Rich loamy, friable soil with good content of organic matter and good drainage are the best. It cannot stand water stagnation or alkalinity. It is generally cultivated organically as an intercrop along with other crops in organic production system. In some areas, turmeric is grown as an intercrop with mango, jack and litchi or with coconut and arecanut. Turmeric is rotated with sugarcane, chili, onion, garlic, other vegetables, pulses, wheat, ragi, etc. 20–25 m buffer zone is maintained between organic turmeric field and conventional non-organic field.

Carefully preserved seed rhizomes of turmeric collected from organic farms and free from pests and diseases can be used for planting similar to ginger. Seed materials from high yielding local varieties may also be used in the absence of organically produced seeds at the beginning. The fingers and mother rhizomes are cut into 4–5 cm long pieces with at least one sound bud. After preparing the lands beds of 15 cm height, 1 m width and of convenient length may be prepared giving at least 50 cm spacing between beds. Soil solarization of the beds by polythene sheets is beneficial to check pests and disease infestation. Seed rhizomes are planted during April–July in shallow pits or on the top of ridges (45–60 cm spaced) at a spacing of 15–20 cm between the plants with 25 g powdered neem cake per planting point. About 2500 kg rhizomes are required for planting one hectare of land Seed rhizomes may be put in shallow pits and covered with well rotten cattle manure or compost mixed with *Trichoderma*.

Mulching the beds with green leaves (@ 4–5 tonnes per acre during planting and @ 2 tonnes/acre at 50th day after planting) is an important practice in organic turmeric production for better germination of seed rhizomes. Leaves of leguminous crops rich in nitrogen, phosphorus rich Acalypha weed and potassium rich Calotropis can be used as mulching materials. Cow dung slurry may be spreaded on the bed after each mulching to enhance microbial activity and nutrient availability.

Turmeric is a heavy feeder crop. Application of well rotten cow dung or compost @ 10 tonne per hectare may be given as basal dose with application of neem cake @ 2.5 tonnes per hectare in different splits. If shoot borer incidence can be controlled by cutting infested shoots and picking out larvae and killing. Neem oil @ 0.5% may also be sprayed at fortnightly intervals. Leaf spot and leaf blotch can be controlled by restricted use of Bordeaux mixture @ 1%. Seed treatment with *Trichoderma* at the time of planting can check the incidence of rhizome rot. The crop has to be harvested in about 7 to 9 months after planting. Usually the land is ploughed and the rhizomes are gathered by hand picking or the clumps are carefully lifted with a spade. Harvested rhizomes are cleaned to remove soil and other extraneous matters. The average yield of fresh rhizome is 20–25 tonnes per hectare. The fresh rhizomes are processed into dried one after boiling and drying under sun. The recovery of dry product varies from 20–25% depending upon the variety and the location where the crop is grown. Dried turmeric has a poor appearance and rough dull color outside the surface with scales and root bits. Smoothening and polishing the outer surface by manual or mechanical rubbing improves the appearance. The color of the turmeric always attracts the buyers. In order to impart attractive yellow color, turmeric suspension in water is added to the polishing drum and uniformly suspension coated rhizomes are again dried in the sun.

5.7   ORGANIC CHILI CULTIVATION

Chili requires a warm and humid climate for its growth and dry weather during the maturation of fruits for best quality. A temperature ranging from 20–25°C is ideal for chili. Heavy rainfall results poor fruit set and high humidity leads to rotting of fruits. High temperature and low relative humidity increases the transpiration during flowering resulting in shedding of buds, flowers and small fruits. Chili can be cultivated organically as an inter crop or mixed crop under organic methods. It is desirable to include a leguminous crop in rotation with chili. Chili is propagated by seeds by raising seedlings. Seeds of high yielding varieties with tolerance to pests collected from certified organic farms are to be used. Seeds should not be treated with any chemical fungicides or pesticides.

Land is prepared by two or three thorough ploughing to bring the soil to fine tilth. In case of direct sowing, three to four ploughing are undertaken and sowing is done along with the last ploughing. The soil can be treated with *Azatobacter* or *Azospirillum* @ 2.5–3 kg per hectare mixed with 150 kg of FYM and broadcasted in the field. Farmyard manure @ 10–15 tonnes and 3–4 tonnes per hectare of vermin-compost are added during planting. Suspension of 200 gm N biofertilizer and 200 gm Phosphotika in 300–400 mL of water are mixed thoroughly. This mixture is properly added with 10 kg seeds and then dried in shade. Seeds are then sown immediately. For the treatment of seedlings 1 kg each of two biofertilizers is mixed in sufficient quantity of water roots of seedlings are dipped in this suspension for 30–40 min before transplanting. The seeds may also be treated with *Trichoderma* and *Psuedomonas* sp. @ 10 g per kg of seed to prevent incidence of seedling rot in the nursery. 400 g of seeds would be sufficient for raising nursery for transplantation in an area of acre. Biological seed treatment with antagonistic *Pseudomonas fluorescens* reduces the bacterial wilt incidence under field conditions.

Fresh seeds are sown in well-prepared nursery beds. Although chili can be grown by direct broadcasting the seeds in the main field, transplanting method is preferred for better quality and survival. Seeds treated with Trichoderma are sown in raised nursery beds and covered with thin layer of sand. The nursery bed is usually raised from ground level and is prepared by thorough mixing with compost and sand. The seeds germinate in 5–7 days and become ready to transplant in the main field within 40–45 days after germination. Seedlings are transplanted in shallow trenches/pits or on ridges/ level lands at a spacing of 60 x 45 cm or 45 x 30. Organic manure such as farmyard manure collected from own farm is applied @ 10–15 tonnes per hectare. Use of biofertilizers can also be resorted to in combination with organic inputs.

Application of neem seed kernel extract (NSKE) should be done to protect the plants from thrips, aphids and mites. 10 kg of neem seed kernels may be boiled in 15 l of water. 200 mL of this extract may be mixed in 15 L of water and four- to five-sprays may be given to control sucking pests. Seed extracts of *Melia azadirachta* along with *Urtica dioica* can also be used for control of pests. Release of larvae of *Chrysoperla cornea*, a biocontrol agent, once in 15 days is also helpful in controlling thrips and mites. Fruit borer can be managed to a certain extent by adoption of biocontrol measures like installation of pheromone traps in the field @ 15 nos. per hectare helps to monitor the adult moths. 4–5 spraying with nuclear polyhedrosis virus (NPV) @ 200 LE (larval equivalent)/acre is beneficial to control the early larval stage of the pod borers. The egg masses of *Spodoptera* borer can be mechanically collected and destroyed. *Trichogramma*, an egg parasite, may be released two days after appearance of moths. Restricted use of *Bacillus thuringiensis* @ 0.4 kg/acre is also beneficial. Fruit rot & Die back caused by *Colletotrichum capsici* and bacterial wilts are the two major can be checked by careful seed selection and adoption of phytosanitary measures, seed treatment with *Trichoderma*. For effective disease control, 10 g of *Trichoderma* or *Pseduomonas* sp. per liter of water should be used for spraying.

Mature dark green fruits should be harvested for fresh use or for preparing chili pickle. For dry chili and for making chili powder, picking should be done when the fruit is fully red ripe. Ripe fruits are harvested at frequent intervals. About 5–6 pickings are necessary for dry chili and 8–10 pickings for green chili. The yield of fresh chili varies from 7.5–10 tonnes per hectare and 25–35 kg of dried fruits may be obtained from 100 kg fresh fruits. Recycled and reusable materials like clean jute bags or other biodegradable materials shall be used for packaging of dried chili. Care should be taken to stack the bags at 50–60 cm away from the wall. Storing chili for longer periods may lead to deterioration of color and quality. The product may be stored for 8–10 months in cold storages.

5.8   ORGANIC SEED SPICES PRODUCTION

Seed spices are the indispensable components of most of all culinary items throughout the globe. Seed spices are not only used as whole but also used as different processed products like powder, oleoresin etc. Essential oils are also extracted from the seed spices for flavoring and confectionaries. However, different seed spices and their products are less demanded as the foreign export consignment having residual component of inorganic toxic inputs over the tolerable limit. Thus, the demand gradually targeted for organically grown seed spices in the export market. Indian domestic market is also has considerable demand of organic seed spices. Although the production tune of organic seed spices is limited only in few pockets of seed spices growing belt of Rajasthan and Gujarat. This is mainly due to lack of information about the demand and price of organic seed spices and lack of awareness of technology of organic seed spices production. In the following subsections, the brief technologies of organic production of different major seed spices are discussed.

5.8.1   CORIANDER

Organic green leaf can be produced throughout the year, but dry cold weather free from frost and rain during the winter is most suitable for seed production. It can be cultivated best in loamy to sandy loam soil rich in organic matter. Soil should be free from weed and previous history of several pests and diseases. Varieties like Gujarat Coriander-2, Rajendra Swati, Sadhana, etc. are suitable for organic cultivation. Land is ploughed 3–4 times with fine tilth during the first half of October. 10–15 kg seeds are required to plant one-hectare area. Seeds are soaked in water or in dilute fresh cow urine for overnight. Seeds are sown at a depth of 3–4 cm giving 30–40 cm spacing in rows and thinned out to maintain 10–15 cm within rows. About 12 tonnes of farmyard manure can be applied during land preparation. Application of bone meal and ash also improves the plant growth and seed yield. Green manuring before cultivation is beneficial for green yield as well as seed yield. Application of 10 to 15 tonnes vermicompost significantly increased the concentration of trace elements (Fe, Mn) in leaves and Zn content even with 5 tonnes vermicompost (Reddy et al. 1993). Good herbage yield of 6067 kg ha^{-1} with

15 t ha^{-1} of vermicompost and highest seed yield of 1, 314 kg ha^{-1} was obtained with 20 t ha^{-1} of vermicompost (Vadiraj et al. 1998). Diluted well rotten neem cake can also be applied to reduce the disease infestation in standing crop. The eco-friendly technique of using 5% onion leaf extract as foliar spray three times protected the plants from powdery mildew. The *Fusarium* wilt incidence can be better managed by seed pelleting with *Trichoderma viride* (with 106 CFU @4 g/kg of seed) plus neem cake application (150 kg/ha). Crop becomes ready within 90–110 days and harvested by pulling the whole plant. The seeds are threshed, winnowed, cleaned, dried and packed in gunny bag. Seed yield of organic coriander is about 300–450 kg/ha.

5.8.2 FENUGREEK

It can be cultivated in all places in the tropics as well as temperate region during the winter. Though fenugreek can be cultivated in all types of soil but clayey loam soil, rich in organic matter and pH ranging from 6.0–7.0 is best suited. After 3–4 ploughing and proper harrowing field is divided into uniform beds. Varieties like Rajendra Kranti, RMt-1, etc. can be grown under organic cultivation. Seeds are sown during October–November in lines at a spacing of 20–30 cm and 20–25 seeds are required to sow one-hectare area. Seeds can be soaked in diluted cow urine to reduce the chance of damping off. Seedlings are thinned out to maintain 10–15 cm distance within rows. Basal application of 15 tonnes/ha of farmyard manure along with rock phosphate and ash is done. Vermi-compost @ 2 tonnes/ha can be applied 15–20 days after germination for better plant growth. Diluted well rotten mustard cake or neem cake is applied for higher seed yield. The combined application of *Azotobacter* with *Rhizobium* and *Pseudomonas striata* improved the growth, yield and nodulation (Parakhia et al., 2000). To control damping off, prophylactic application of 1% Bordeaux mixture was found effective. To control aphids, application of fish oil, rosin soap or neem seed kernel extract (3%), neem oil or tobacco decoction (0.05%) was found effective. Plants are pulled out when pods become dried. Plants are dried in sun, threshed; seeds are cleaned, dried and packed in gunny bags. 600–700 kg seeds can be obtained from one-hectare land under organic cultivation.

5.8.3   CUMIN

Cumin prefers cool dry climate and can be grown up to 3000 m from MSL. Deep friable well drained loamy to sandy loam soil rich in organic matter. 8 kg seeds are sown in well-prepared clod free one-hectare land during November. Varieties like Gujarat Cumin-1, Rajasthan Zeera-19, etc. can be selected for organic cultivation. Use of *Azospirillum* or *Azotobacter* in combination with 5 t/ha sheep manure per ha is found effective. 12 tonnes/ha of farmyard manure along with ash and rock phosphate or dolomite are recommended as basal application. Vermi-compost @ 2 tonnes/ha can be top dressed at 15–20 days after germination followed by application of well rotten mustard cake or neem cake. Wilt disease can be controlled by treating seeds with *Trichoderma* @4 g/kg seed, summer fallow of cultivated land and soil solarization during summer before sowing. Crop rotation with cluster bean- cumin, cluster bean-wheat and cluster bean-mustard also proved to be better for management of cumin wilt. Late sowing minimizes incidence of blight. As *Fusarium* wilt resistant variety, GC-3 may be used. Crop becomes ready to harvest within 100–110 days. Seeds are packed in ploythene lined gunny bags after threshing and dying. 300–400 kg seeds can be obtained from one-hectare area under organic cultivation.

5.8.4   FENNEL

It prefers mild winter and can be grown up to 2500 m from MSL. Well-drained loamy soil rich in organic matter or black soil is suitable for cultivation of fennel. Varieties like Co-1, Gujarat Fennel-1, S-7-9, etc. can be grown under organic cultivation. 9–12 kg seeds/ha are required for broadcasting and 3–4 kg/ha for transplanting of seedlings of 5–6 weeks age. In broadcasting average 30–45 cm distance is maintained after thinning. 60 x 30 cm spacing is maintained in transplanted seedlings. Basal application of 15 tonnes/ha of farmyard manure along with rock phosphate and ash is done. Application of 10 t/ha sheep manure or 4 t vermicompost along with seed inoculation by *Azotobacter* recorded higher growth and yield attributes over recommended doses of fertilizer. Diluted well rotten mustard cake or neem cake is applied for higher seed yield. Sugary disease can be controlled by reducing the number of irrigation and the use of resistant variety *viz.*, Gujarat Fennel-1. *Ramularia* blight can be controlled with incorporation of *Trichoderma viride* (4 g/kg of seed). Damping off in nursery can be controlled by prophylactic spray of 1% Bordeaux mixture. Crop matures in 7–8 months and harvesting is done at full mature but unripe stage. Umbells are cut with stems, dried, threshed, cleaned and packed in polythene lined gunny bags. Average yield of fennel under organic cultivation is 200–250 kg/ha.

5.9   CERTIFICATION OF ORGANIC SPICES

Certification for organic spice farms is an essential pre-requisite for marketing the produce as organic especially in the international markets. Realizing the potential for export of organic products, Ministry of Commerce, Govt. of India has set up National Programme for Organic Production (NPOP) in April 2000. A National Steering Committee (NSC) was set up by the Ministry of Commerce under NPOP. Spices Board is active member in NPOP. Maintaining strict quality control standards, India's National Programme for Organic Production (NPOP), ensures the export of only those products, which qualify the National Standards for Organic Production. National Accreditation Body [NAB] set up under NPOP is responsible for accreditation of organic certification bodies in the country. As per the National Accreditation Policy, all the certifying agencies operating in India are to obtain accreditation from National Accreditation Body (NAB). At present, there are 24 certification bodies accredited under NPOP.

5.10   FUTURE DIRECTIONS

Efficient use of inputs (time, method and quantity) is the major challenge not only to the growers of different crops but also to the scientists and policy makers. Adoption of system-based production rather than crop-based one is the prime necessity today. Globally the emphasis should be consecrated to quality production rather than quantity. So far due to lack of availability of suitable biopesticides, it is not possible to practice organic cultivation in several spices. Organic productions of these spices will not only boost the economy of this region but also sustain the productivity of natural resources. More organized research coupled with coordinated and holistic approaches are necessary for carrying forward the organic mission.

KEYWORDS

- certification
- future directions
- importance
- organic spices
- production
- scenario

REFERENCES

Arya, P. S., (2000). Value added spice products. In: *Spice Crops of India*. Kalyani Publishers, 71–77.

Bhattacharyya, P., (2004). *Organic Food Production in India—Status, Strategy and Scope*. Agribios India, 1–182.

Dinesh, R., Srinivasan, V., Hamza, S., & Manjusha, A., (2010). Short-term incorporation of organic manures and biofetilizers influences biochemical and microbial characteristics of soil under an annual crop (*Curcuma longa* L.). *Bioresource Tech.*, *101*, 4697–4702.

Hossain, A., & Ishimine, Y., (2007). Effect of farm yard manure on growth and yield of turmeric (*Curcuma longa* L.) cultivated in dark red soil, red soil and gray soil in Okinawa, Japan. *Plant Prod. Sci.*, *10*(1), 146–150.

Kamal, M. Z. U., & Yousuf, M. N., (2012). Effect of organic manures on growth, rhizome yield and quality attributes of turmeric (*Curcuma longa* L.). *The Agriculturist*, *10*(1), 16–22.

Kumar, N., Khader, J. M. A., Rangaswami, P., & Irulappan, I., (2010). Part I. Spices. In: *Introduction to Spices, Plantation Crops, Medicinal and Aromatic Plants*, Publ. Oxford & IBH Publishing Company Pvt. Ltd., New Delhi, India (ISBN: 81-204-1137-4), pp. 101–703.

Lampkin, N., Foster, C., Padel, S., & Midmore, P., (1999). The policy and regulatory environment for organic farming in Europe. In: *Organic Farming in Europe: Economics and Policy Dabbert, Haring*, Zano Heds, London, Volume 1.

Meena, S., Senthilvalavan, P., Malarkodi, M., & Kaleeswari, R. K., (2007). Residual Effect of Phosphorus from Organic Manures in Sunflower. *Research Journal of Agriculture and Biological Sciences*, *3*(5), 377–379.

Parakhia, A. M., Akbari, L. F., & Andharia, J. H., (2000). Seed bacterization for better quality and more yield of fenugreek. *Gujarat Agric. Univ. Res. J.*, *25*, 34–38.

Peter, K. V., (2006). *Handbook of Herbs and Spices*, Publ. Woodhead Publishing, New Delhi, India (ISBN: 9781845690175), Vol. 3, pp. 520+xxviii.

Pruthi, J. S., (2003). Major Spices of India, *Crop Management and Postharvest Technology*, Publication Division, Indian Council of Agricultural Research, New Delhi, pp. 514+xxii.

Reddy, B. S., Rao, C. V., Rivenson, A., & Kelloff, G., (1993). Chemoprevention of colon carcinogenesis by organosulfur compounds. *Cancer Res.*, *53*, 3493– 3498.

Shangmugavelu, K. G., Kumar, N., & Peter, K. V., (2010). *Production Technology of Spices and Plantation Crops,* Vedam Books Pvt. Ltd., New Delhi, India (ISBN: 8177541617), pp. 546.

Srinivasan, V., Sadanadan, A. K., & Hamza, S., (2000). An INM approach in spices with special references on coir compost. In.: *International Conference on Managing Natural Resources for Sustainable Agricultural Production in 21ˢᵗ Century.* Resource Management, New Delhi, Vol. 3, pp. 1363–1365.

Vadiraj, B. A., Siddagangaiah, D., & Potty, S. N., (1998). Response of coriander (*Coriandrum sativum* L.) cultivars to graded levels of vermicompost. *Journal of Spices and Aromatic Crops*, *7*(2), 141–143.

Wanj, B., & Qui, Y. L., (2006). Phylogenetic distribution and evaluation of mycorrhizas in land plants. *Mycorrhiza*, *16*, 299–363.

Yamawaki, K., Matsumura, A., Hattori, R., Tarui, A., Hossain, M. A., Ohashi, Y., & Daimon, H., (2013). Effect of inoculation with arbuscular mycorrhizal fungi on growth, nutrient uptake and curcumin production of turmeric (*Curcuma longa* L.). *Agril. Sci.,* *4*(2), 6–71.

CHAPTER 6

FUTURE THRUST AREAS

CONTENTS

6.1 INTRODUCTION

India is endowed with rich diversity of spice crops, which makes the country a cynosure for the explorers and traders from foreign countries and contributing to the prosperity and wealth of our country since time memorial. The same dominance, well supported by the ample diversity of agroclimatic conditions is still there with renewed interests, consequences, threats as well as promises.

The transformation of spices from a mere seasoning or flavoring item to an inseparable entity to day-to-day's cooking throughout the globe is phenomenal. India is leading the global spice market from the front row and the Indian legacy with the spice crops is gradually going global. Apart from the production of spices both in quantitative and qualitative terms, the scientists are working tirelessly to evolve newer and smart technologies almost every day to keep up the momentum. Some of the noticeable achievements in this line are improved soil-less method for production of healthy planting materials in many spice crops, accreditation of nurseries, novel delivery method of biocontrol agents through encapsulation and biofortification, refining plant architecture (small bushy plant types) for making perennial spices suitable for urban horticulture, site-specific nutrient management plans and micro-nutrient formulations for targeted yield in major spices, micro-irrigation/

fertigation for precision and optimal farming, bringing automation and IT into precision farming, vertical column method for quality production of black pepper vines for planting, whole genome sequencing and annotation of Phytophthora infection, DNA barcoding technique to detect adulterant in spice products, use of solar energy for majority of farms as well as processing, business planning and development unit through the identification of industry oriented production hub, etc.

Amid the sensational progress and ambitious vision, there are challenges that threaten spice industry in general and marketing in particular. Shifting of interests of growers to more profitable and or less risky crops due to escalating cost of cultivation, cyclic market fluctuations at international and national levels, insufficient inflow of information among the different stakeholders in the industry, emergence and epidemics of pests and diseases, pesticide residues and mycotoxin contaminants in the products and lack of MRL and ADI standards in some of the pesticides used in spices, new stricter legislations and regulations, severe adulteration and incidents of contaminants in spices (aflatoxin, pesticide, illegal dyes, microbials, etc.) and last but not the least, climate change resulting in drought or excess moisture, high or low temperature during critical periods, etc. has affected spice industry and trade. This is further complicated by uncompetitive production costs in comparison to other countries such as Vietnam, Indonesia and China.

Crop-wise priorities are considered as the major points of interest since long which still needs further refinements. In black pepper, the "King of Spices," some priority areas have already been identified viz., development of Phytophthora, nematode, pollu resistant and low input efficient varieties with high quality parameters, Popularizing bush pepper technology in urban horticulture and home gardens to meet part of domestic demand and availability of safe and fresh pepper, popularization and use of piperine as bio-enhancer for higher returns, etc. In cardamom, the "Queen of Spices," those are development of thrips resistant genotypes, development and popularization of biocontrol agents for control of shoot and capsule borer, identification and deployment of varieties with synchronous flowering to reduce number of harvests, biannual replanting for increased production on short-term basis, etc. In rhizomatous spices like ginger and turmeric the factors to be considered are development of varieties resistant to rhizome rot/bacterial wilt (*Pythium* and *Ralstonia*) and varieties suitable for vegetable types and high quality in case of ginger and development of dry rot (*Pythium*) and leaf disease (*Taphrina*) resistant turmeric varieties with high curcumin preferably with golden yellow color, isolation and popularization of high value products and phytochemicals, protected soil-less cultivation, popularization of mechanization in farm operations and processing, popularization and use of turmeric as a major ingredient in formulations/drugs for cancer and Alzheimer's treatment. Tree spices are also very much important contributing significantly. The issues needing scientific intervention are development of bisexual and off-season types for increased yield and reduced aflatoxin, canopy architecture and development of early flowering, dwarf types for ultra high density planting and urban horticulture, developing low myristicin types for food industry, isolating and utilization of myristicin and elemicin derivatives for cancer treatment, in nutmeg. In clove, these are multiplication and deployment of bold 'King' and 'Madagascar' types for increased market preference, popularization of dwarf types in homestead and urban gardens, etc. In cinnamon, development of low coumarin types with high regeneration ability, high production and harvesting technologies to harvest large number of cinnamon pencils (quills) and extra time earning opportunities for women at home and development and popularization of cinnamon products for diabetes control are vital. Identification and deployment of nutritionally rich and high HCA types garcinia and establishing commercial plantations along with development and popularization of garcinia juice as health drinks for obesity control are two important considerations. In tamarind, high density planting of dwarf high yielding types in waste lands for increased production and dry land agriculture and developing sweet types for confectionary as well as spice purpose are required. The seed spices are always considered a promising area. Deployment of disease resistant, dual purpose, multi-cut coriander varieties and protected cultivation of leafy coriander in urban roof tops and households for quick returns (one crop in 30 days), development of wilt and blight resistant cumin varieties and a holistic effort to grow cumin in non-traditional areas, development of fennel varieties with synchronous flowering (whole umbel) and maturity of seeds/grains harvesting of fennel at different stages for value addition and different uses (e.g., early harvesting for sweet fennel for mouth fresheners), developing dual purpose, determinate high value fenugreek types for use as functional foods, development of high yielding varieties of ajwain and black cumin are to be done properly. For other spices like mint some more high menthol yielding varieties are to be developed and in saffron, one of the costliest spices in the world, protected cultivation throughout the year using controlled temperature regimes and aeroponics are to be properly taken into consideration (Krishnamurthy et al., 2015).

Spices and herbs in ancient civilizations and biblical times were considered the superfood with mythical medicinal power. We are at the cusp of a major revolution with consumer interest in spices and herbs attaining a new high. Spices could become the next multi-vitamins. They could even gain superfood status in consumer perception of healthy foods similar to fruits and vegetables today. From the first through fourth centuries, Arabians developed techniques to distill essential oils from aromatic plants. Around the ninth century, Arab physicians used spices and herbs to formulate syrups and flavoring extracts. Growing global demand for "natural," "clean," and "safe" foods is also beginning to drive greater interest in spices and herbs.

Health and wellness is beginning to emerge as another key driver that could greatly expand the growth of spice and herb consumption. The aging world population is increasingly focused on nutrition, and is using spices and herbs as natural remedies and for prevention. New York Times (June 30, 2008) has identified eleven foods as the best ones namely beets cabbage, Swiss chard, cinnamon, pomegranate, juice, dried plums, pumpkin seeds, sardines, turmeric, frozen blueberries and canned pumpkin. Surprisingly cinnamon tops the list including another important spice, turmeric.

A futuristic view of the future is that ethnic foods especially spicy ones (i.e., Latino, Asian, Indian) will be more popular, as the global "natural" and "clean" movement continues to gain steam, spices will be used increasingly in packaged foods and displayed on labels worldwide as a source of "good" ingredient, the internet will bring Ayurvedic and Chinese medicine to the people in their own home world wide, the "green and sustainable" mega trend will favor usage of natural spices and herbs, media will pay greater attention to the wellness news of spices, consumers will continue to want more antioxidants in their diet and polyphenols in the "multi spices and herbs" will be recognized as the new "multivitamins." Scientists are working to understand the major role inflammation plays in development of most diseases viz., neurological diseases, pulmonary diseases, cancer, cardiovascular diseases, Alzheimer's diseases, diabetes II, arthritis, autoimmune diseases. Studies are also underway to determine the role spices and herbs (viz., sage, ginger, fenugreek, rosemary, turmeric, black cumin, cinnamon, red pepper, garlic, and other spices) could have as antiinflammatories. Other scientists are studying whether spices and herbs could have a role in reducing cancer. Researchers have predicted that common spices such as garlic, curry, ginger and chili play a role in reducing cancer incidence. Studies in China, Europe and the US have consistently found lower cancer (especially colon/stomach) rates with garlic consumption (Challier et al., 1998; Fleischauer et al., 2001; Hsing et al., 2002). Sage and turmeric may be taken to improve memory. Studies in Britain concluded sage could potentially help those suffering from age- or disease-related declines in cognitive function (memory, attention and mood). Study of 1000 elderly Asians showed those who ate curry (which contained turmeric) had better cognitive performance and it improved as curry consumption increased (Cao et al., 1998). Cinnamon and other spices may be used to help regulate blood glucose. Blood glucose regulation can be tied to weight management. Several human studies seem to indicate cinnamon can help regulate blood glucose (though results aren't consistent). Other spices (such as turmeric, sage, cinnamon, rosemary, marjoram, and tarragon) have also regulated blood glucose in test tube and animal studies. Antioxidants are extraordinary workhorses in our bodies. Studies suggest they provide a range of benefits, including giving our immune system a boost. They also appear to reduce inflammation, which is increasingly recognized as a first step in heart disease, cancer, diabetes and other chronic disease.

McCormick has identified seven "super spices"—cinnamon, ginger, oregano, red pepper, rosemary, thyme, and yellow curry powder each contributing a concentrated source of antioxidants. Preliminary studies indicate that spices and herbs have antiinflammatory properties that may hold tremendous potential in promoting good health. Other studies suggest spices and herbs may help curb our hunger and boost our metabolism, which might make it easier for us to manage our weight.

The future strategies should include diversification of spice industry, adding new spices and spice products to the export basket, value addition of spices and their derivatives, establishing international spice brand and building acceptable benchmark standards in quality and safety parameters. A closer linkages and coordination is to be established with agencies like International Pepper Community (IPC), International Spice Trade Association (ASTA), European Spice Association (ESA), International Spice Group (ISG), Food and Agricultural Organization (FAO), etc. The spice industries should always conform the norms of AGMARK, ASTA, BIS, ESA, WHO, USFDA, and cleanliness specifications by EU and other major importing countries. In line with the global standards, quality certification system conforming to HACCP norms should also to be formulated.

Mainly driven by the concern for food and environmental safety, organic products in all sectors are gaining ground across the world. The apex body of any country dealing with spices should bestow committed support and encourage organic farming by working out clear-cut guidelines. Many spices are now being utilized in the form of alternative medicine to alleviate issues like diabetes, obesity, cardiovascular problems, cognitive disorder, etc. The native wisdom of many cultures attributing the medicinal and nutraceutical properties of useful spices should be consolidated and scientifically validated by means of latest research taking into account the fact that the global nutraceutical market is growing with time. Geographical and varietal influences on the concentration of bioactive molecules of spices should be given due importance. Advanced tools and the popular *omics* technique in the extraction and purification may be adopted.

The demand for spices and spice extracts is booming globally. The demand for a variety of traditional cuisines of different cultures and ethnic groups is driving the demand for various spices and spice extracts to meet the rising demand. All these parameters are likely to boost the Indian spices exports in the coming years. Spice crops based economy has a vibrant growth in ensuring livelihood security particularly to the small holder producers. The estimated growth rate for spices demand in the world is around 3.2%, which is just above the population growth rate. As spice crops is labor intensive, the industry is expected to create surplus employment potential in production alone. Further, there is substantial scope for creating new jobs through value addition in spices. A coordinated involvement of farmers, scientists, researchers, policy makers, funding agencies as well as corporate sectors is required in providing production, improvement, protection and post harvest value addition of this promising sector to exploit the emerging world markets. If we can utilize the prevailing opportunities, the spice industry and marketing will see through exponential growth ensuring wealth and prosperity both to the country as well as to the stakeholders associated with it.

KEYWORDS

- **omics technique**
- **phytophthora infection**
- **spice crops**
- **super spices**

REFERENCES

Cao, Y., D'Olhaberriague, L., Vikingstad, E. M., Levine, S. R., & Welch, K. M., (1998). Pilot study of functional MRI to assess cerebral activation of motor function after poststroke hemiparesis. *Stroke, 29*, 112–122, Doi: 10.1161/01.STR.29.1.112.

Challier, B., Perarnau, J. M., & Viel, J. F., (1998). Garlic, onion and cereal fibers as protective factors for breast cancer: a French case-control study. *Eur J Epidemiol., 14*(8), 737–747.

Fleischauer, A. T., & Arab, L., (2001). Garlic and cancer: a critical review of the epidemiologic literature. *J Nutr., 131*(3S), 1032–1040.

Hsing, A.W., Chokkalingam, A. P., Gao, Y. T., Wu, G., Wang, X., Deng, J., Cheng, J., Sesterhenn, I. A., Mostofi, F. K., Chiang, T., Chen, Y. L., Stanczyk, F. Z., & Chang C., (2002). Polymorphic CAG/CAA repeat length in the AIB1/SRC-3 gene and prostate cancer risk: a population-based case-control study. *Cancer Epidemiol. Biomarkers Prev., 11*, 337–341.

Krishnamurthy, K. S., Biju, C. N., Jayashree, E., Prasath, D., Dinesh, R., Suresh, J., & Nirmal Babu, K. (Eds.), (2015). Souvenir and Abstracts, National Symposium on Spices and Aromatic Crops (SYMSAC VIII), Towards 2050, *Strategies for Sustainable Spices Production,* Indian Society for Spices, Kozhikode, Kerala, India, pp. 263.

INDEX

Printed and bound by CPI Group (UK) Ltd, Croydon, CR0 4YY

23/10/2024

01777704-0010